Synthesis Lectures on Mathematics & Statistics

Series Editor

Steven G. Krantz, Department of Mathematics, Washington University, Saint Louis, MO, USA

This series includes titles in applied mathematics and statistics for cross-disciplinary STEM professionals, educators, researchers, and students. The series focuses on new and traditional techniques to develop mathematical knowledge and skills, an understanding of core mathematical reasoning, and the ability to utilize data in specific applications.

Hidefumi Katsuura

Introduction to Analysis

Theorems and Examples

 Springer

Hidefumi Katsuura
San Jose State University
San Jose, CA, USA

ISSN 1938-1743   ISSN 1938-1751 (electronic)
Synthesis Lectures on Mathematics & Statistics
ISBN 978-3-031-67953-7   ISBN 978-3-031-67954-4 (eBook)
https://doi.org/10.1007/978-3-031-67954-4

This Springer imprint is published by the registered company Springer Nature Switzerland AG
The registered company address is: Gewerbestrasse 11, 6330 Cham, Switzerland

If disposing of this product, please recycle the paper.

Preface

This textbook is composed of lectures I gave in the Introduction to Analysis class at San Jose State University. I am writing this textbook for readers who have had an introductory calculus sequence in colleges or universities. Its subject is the details of theoretical aspects of calculus rather than applications of it. Many theorems in this book are familiar to the readers. However, because we want to establish detailed explanations of theorems that are familiar, please pretend not to know any theorems until they are examined and proven in this textbook.

About the use of "I" and "we" in this textbook: When I write research papers, I only use "we" in them to maintain neutrality. On the other hand, this textbook reflects essentially my style and opinion. When expressing these opinions I use "I". So there is no "I" in proofs of theorems. But in this preface and other places in this textbook where I share my point of view and my own thinking, "I" shows up regularly.

Our main objective is to discuss limits of sequences, series, and their convergence. But, doing so requires the knowledge of real numbers. What is a real number? To begin, we take natural numbers as given, and use the completeness axiom to establish real numbers in Chap. 2. For readers, this is probably the most difficult part of the whole book. And then in Chap. 3, we study sequences and series to better understand real numbers. To be more precise, convergent sequences and series are the only way to express most real numbers. This is why sequences and series are critical topics.

There are many textbooks for studying introductory analysis. But their coverage on real numbers is rather vague. For example, often the existence of $\sqrt{2}$, a notation of the positive number whose square is 2, is justified by saying "since it is the diagonal length of a unit square by the Pythagorean theorem". At the time of Pythagoras, all they knew was that $\sqrt{2}$ was not a rational number. But how do you know such a number exists? The Pythagorean school could not answer that. So when some mathematicians justify it by saying "since it is the diagonal length of a unit square", that does not seem right to me. What axiom of real numbers is needed to show the existence of $\sqrt{2}$? We say that it is the completeness axiom. But most analysis textbooks do not prove this using the completeness axiom. Our goal in Chap. 2 is to prove the existence of $\sqrt{2}$ using the completeness axiom, and it is one of the most difficult proofs that we treat.

In Chap. 3, Cauchy sequences are also introduced. Cauchy sequences are often introduced without any applications, and therefore, it is not easy to see their importance in analysis. We generalize the proof of the existence of $\sqrt{2}$ to the power of numbers of the type r^p, where $r > 0$ and p a rational number. Then we explain r^p when p is an irrational number (as in $2^{\sqrt{2}}$) using Cauchy sequences. Perhaps, this may explain and help to see the importance of Cauchy sequences. And this explanation of r^p when p is an irrational number is crucial to understand why the exponential function r^x is continuous, for example, in Chap. 4. Many introductory textbooks on analysis neglect to prove the continuity of functions of the types r^x and x^r. But I think they are just as important as proving the extreme value theorem and the intermediate value theorem in Chap. 4.

We explain the number e defined as $e = \sum_{n=0}^{\infty} \frac{1}{n!}$, and prove that $e = \lim_{n\to\infty} \left(1 + \frac{1}{n}\right)^n$ in Chap. 3. Then we prove that $e^x = \sum_{n=0}^{\infty} \frac{x^n}{n!}$ in Chap. 4. This gives us a **plausible** reason why we can deduce $\frac{d}{dx} e^x = e^x$ in Chap. 5. And we use this result beyond that point. However, the actual proof of $\frac{d}{dx} e^x = e^x$ is in Chap. 7. I am using this as a motivation in order to engage readers as well as a main goal of this textbook.

Most calculus textbooks take a convenient shortcut to establish $\frac{d}{dx} e^x = e^x$ by defining e to be the number such that $\lim_{h\to 0} \frac{e^h - 1}{h} = 1$. This makes the *derivation* of $\frac{d}{dx} e^x = e^x$ quick and easy. But first things first, how do we know the existence of the limit $\lim_{h\to 0} \frac{e^h - 1}{h}$ let alone the existence of a number e satisfying the limit condition $\lim_{h\to 0} \frac{e^h - 1}{h} = 1$? How can we even approximate e from this definition? This definition may be fine for an introductory calculus to simply tell the inexperienced readers that e is approximately 2.7. But I cannot even think about how to start answering these questions without using $\frac{d}{dx} e^x = e^x$. Approximating e using this definition alone seems impossible. On the other hand, by replacing ∞ by 3, $e = \sum_{n=0}^{\infty} \frac{1}{n!}$ gives $\sum_{n=0}^{3} \frac{1}{n!} = 1 + 1 + \frac{1}{2} + \frac{1}{6} = 2 + \frac{2}{3} \approx 2.67$, we arrive at 2.67 to be a good approximation of e. We will also prove the existence of the limit. The aim of this textbook is to establish a theoretical background in calculus. Hence, I consider the take-it-for-granted definition of e to be questionable. By taking $e = \sum_{n=0}^{\infty} \frac{1}{n!}$ as the definition of e, the proving of $\frac{d}{dx} e^x = e^x$ becomes one of the most important motivations. For this, we introduce the uniform convergence of sequences and series of functions in Chap. 7.

In Chap. 5, I construct a continuous, but nowhere differentiable function f using a sequence of functions $\langle f_n \rangle_{n=1}^{\infty}$. We do not prove these in Chap. 5. However, we prove that this function f is continuous in Chap. 7. It is not easy to come up with a non-trivial example of sequence of functions that requires the concept of uniformly Cauchy sequence in Chap. 7. I thought this sequence $\langle f_n \rangle_{n=1}^{\infty}$ would be a good non-trivial example. And this will lead us to the continuity of the limit function f. I hope this will help in understanding the importance of uniformly Cauchy sequence of functions. We do not prove that this function f is nowhere differentiable. It is beyond the scope of this textbook.

Chapter 6 is on integrations. It is the continuation of the introduction of anti-derivatives introduced in Chap. 5. We will introduce and use Darboux integration over Riemann integration. The reason for this is that it is easier to use Darboux integration than to use Riemann integration in order to prove theorems in this chapter. While Darboux and Riemann integrations are equivalent definitions, proving this equivalence in a general setting is rather difficult. We only prove the equivalence of Darboux and Riemann integrations of a continuous function.

This textbook is also an introduction to proofwriting. So I tried to be extremely careful with my words. I want the students to have a precise writing style. I have tried hard to avoid words like "obviously", "easily", "trivial", or "clearly" because these words are not only condescending but also lead "easily" to elementary mistakes. When I have made mistakes in mathematics, it has almost always been when I thought something was obviously clear and easy.

I take care to quantify all letters introduced in proofs by "for every", "for all", "for some", "there exists", and "let". These are important words because they introduce **subjects** of sentences and paragraphs. Without clear definitions of letters used in proofs, readers cannot follow their arguments. Knowing the **subject** of a sentence, of a paragraph, or of a story is very important. If you are trying to prove a theorem, you have to state the subject clearly. We will study the continuity of a function in this textbook. But since we have some notion of continuity from a previous calculus class, let me use continuity to explain a typical mistake. I often hear even from some other math teachers that "a function f is continuous at a if the limit is $f(a)$ as x approaches a". **What is the problem in this sentence?** It does not have a well-formatted **subject** after "if". It should be written properly as "a function f is continuous at a if the limit **of $f(x)$** is $f(a)$ as x approaches a". Without the words "of $f(x)$", the quoted sentence is ambiguous.

In order to set sail in this direction, we start this textbook with a discussion of the basics of set theory. Chapter 1 includes sets, counting arguments, functions, and mathematical induction. Yes, let me emphasize mathematical induction. Many, if not most mathematicians, give **incorrect** proofs when using mathematical induction by giving incorrect **subjects** in proofs with mathematical induction. And we discuss this in detail in Sect. 1.4.

In most studies other than mathematics, people have little problem identifying **subjects**. In a story, a person is introduced with his or her name. Without names, and just using 'he' or 'she', it would be very hard to understand the story. But, for some reason, in mathematics, the goal of obtaining "numerical solutions" to problems is emphasized so much that people seem to forget the storytelling side of mathematics. And when a variable "x" is introduced, often it is not explained what it means. Conclusions are important, but the storytelling task of mathematics to show **how to get to the conclusions** is even more important.

Besides subjects, please pay attention to words like "and" and "or". The symbols like "=", ">", "∈" and "⊂"are important because they signal relationships—one of the keys to understanding and engaging in any story. However, unlike stories found in literature where ambiguities often enrich the story, an equation is a **sentence** that must be written without ambiguity. The equation $2x - 6 = 0$ has the subject "$2x - 6$", "=" signals the relationship, and "0" is the conclusion, and it reads "$2x - 6$ is equal to zero". The object of this equation is usually to find x that satisfies the equation. The solution to the equation $2x - 6 = 0$ is $x = 3$. "$3 = x$" is not an acceptable solution since "x is 3", while "3" could be other things besides "x". The solutions to the equation $x^2 - 3 = 0$ are "$\sqrt{3}$ and $-\sqrt{3}$". However, "$x = \sqrt{3}$ **or** $x = -\sqrt{3}$", and **not** "$x = \sqrt{3}$ **and** $x = -\sqrt{3}$" since x cannot be $\sqrt{3}$ and $-\sqrt{3}$ simultaneously.

I refuse to use symbols like ∀, ∃, or ⇒ since the use of "for every", "there exists", or "implies" are key phrases in theorems and in proofs. They deserve to be written explicitly using words not mathematical symbols.

Exercise problems are an integral part of learning mathematics. Also, which problems and when students are working on them is crucial in the logical progression of this textbook. Creation of theorems is often motivated by examples. So knowing examples is a very important part of learning mathematics. Many examples are given here. Some examples are also given in the form of exercise problems. Instead of gathering all the exercises at the end of each section, I have incorporated them as parts of the narrative. Problems are intended not only to check your understanding of concepts you have just learned but also to prepare you to prove subsequent theorems, or to explain why some hypotheses in a theorem are necessary. So please at least read the statements of problems.

In some problems, I ask not to "prove by contradiction". This is not because I do not approve of proofs "by contradiction". Some theorems cannot be proven without "contradictions". However, I know the particular problems I have selected can be proven without "contradiction". And I believe that if a statement can be proven without "contradiction", then a proof is usually better if it is written without "contradiction". Very often, people prove theorems by contradiction because the authors do not want to spend the time to edit their writing and thoughts. It is important to edit essays for an English class. Similarly, it is important to edit your proofs in mathematical proofs. Proofs are essays. Please edit your proof. This will make you a better mathematician.

My hobby is woodworking using Japanese hand tools, which influences my approach to mathematics. I admire a Japanese master carpenter named Tsunekazu Nishioka. He was the head carpenter who oversaw the maintenance of the 1300-year-old Horyuji Temple in Nara, Japan, in the twentieth century. Reflecting on the skills of a woodworker he had employed over the years, Nishioka wrote that the worse the woodworker, the more hand tools he had. This stuck with me. By practicing how to use tools well, not by buying additional tools for special purposes, you become a better woodworker.

I think this resonates with mathematical learning. I have given many examples and kept the theorems needed in this textbook close to a bare minimum so the readers can learn and apply them well. And I refrained from including theorems just because they are in many of the standard introductory analysis textbooks. Readers can learn about these theorems later from other textbooks, or find them on their own for the joy of their own journey to discovery.

San Jose, USA   Hidefumi Katsuura
June 2024

Acknowledgements

I would like to express my appreciation for the encouragement and help from Michael Dalby, Samih Obaid, and Myrna Schlegel. I also sincerely thank the staff at Springer Nature for working on this book.

Contents

Set Theory

1.1 Sets

Sets play an important role in the development of mathematics. To us, sets are used to express mathematical ideas rigorously. So, we start this textbook by introducing some set theory. And by doing so, we hope to lead you into real numbers and real analysis.

Set theory existed before Georg Cantor, but he made set theory into a branch of mathematical study around 1890–1900. We will introduce some of his ideas in this textbook.

What is a set? This is a naïve question but a fundamental and difficult one to answer. It is philosophically deep. This may sound strange but we **do not define what a set is. We assume a set to be known intuitively**, as in the set of natural numbers $\mathbb{N} = \{1, 2, 3, \ldots\}$, the set of integers $\mathbb{Z} = \{0, \pm 1, \pm 2, \pm 3, \ldots\}$, the set of rational numbers $\mathbb{Q} = \left\{\frac{p}{q} : p, q \in Z \text{ and } q \neq 0\right\}$, and the set of real numbers, denoted by $\mathbb{R}$. These sets of numbers will be studied in detail in the next chapter, but for now, we assume that we know them.

Definition 1.1.1 We often denote a set by an upper-case alphabet. The set without any element is said to be the empty set and denoted by $\emptyset$. . (We do not write $\emptyset$ for the numerical zero, 0.) Let X and Y be nonempty sets.

(1) If x is an element of X, we write $x \in X$, and we read "x is an element of X".
(2) If x is not an element of X, we write $x \notin X$, we read $x \notin X$ as "x is not an element of X".
(3) The set $X \cup Y$ consists of all the elements in the sets X or Y. Hence, $x \in (X \cup Y)$ if $x \in X$ **or** $x \in Y$. We read the set "$X \cup Y$" as "X union Y".

H. Katsuura, *Introduction to Analysis*, Synthesis Lectures on Mathematics & Statistics,
https://doi.org/10.1007/978-3-031-67954-4_1

(4) The set $X \cap Y$ consists of all elements in both the sets X and Y. So, $x \in (X \cap Y)$ if $x \in X$ **and** $x \in Y$. We read the set "$X \cap Y$" as "X <u>intersection</u> Y".

(5) If $x \in X$ **but** $x \notin Y$, we write $x \in (X - Y)$, and we read that the set $X - Y$ as "X <u>minus</u> Y" or "<u>complement</u> of Y relative to X". The statement "$x \in X$ **but** $x \notin Y$" is the same as saying "$x \in X$ **and** $x \notin Y$". The use of "but" here is to emphasize the complement as exclusive of Y.

(6) If $x \in Y$ for every $x \in X$, then we write $X \subset Y$ or $Y \supset X$. If we want to emphasize X, we write $X \subset Y$, and read it as "X is <u>contained</u> in Y" or "X is a **subset** of Y". If we want to emphasize Y, then we write $Y \supset X$, and read it as "Y <u>contains</u> X".

(7) We write $X = Y$ if $X \subset Y$ and $Y \subset X$. We read $X = Y$ as "X is <u>equal</u> to Y" or "X is Y".

Remark 1.1.1 It is unfortunate that, in many math courses, the grades are given based only on numerical conclusions. And especially in entry-level mathematical education, math courses rarely focus on the derivations of conclusions. Therefore, many people mistakenly think of mathematics as symbolic manipulations that people have to learn how to solve problems, and the reasoning behind obtaining conclusions is often neglected. This is very unfortunate. And because of this, the transition from a basic calculus course to a proof-oriented course like this one becomes difficult for many.

It may sound silly or unnecessary to tell you how to read symbols in the above definitions. But it is important to approach writing mathematics like writing with words. Writing proofs or solutions is the same as writing English essays, except that you are using mathematical notations/symbols rather than words for shorthand to convey your thoughts. Solutions should be essays, and these essays to convey your thought are **proofs**. Once we begin to think of solutions as essays, and think about how these essays to convey our thought are **proofs**, we realize we have been writing proofs all along when we write solutions to almost all mathematics problems. It was just that we did not emphasize the solutions' status as proofs.

How do you write proofs? All you have to do is to write essays. The only difference between an English essay and a proof is that mathematical symbols are used in proofs for shorthand. Keeping in mind that mathematical symbols are shorthand, make a practice of **reading** (using words) these symbols as you write. An equation "$1 + 2 = 3$" is a sentence "one plus two is three". As I indicated in the preface of this book, please keep in mind what the **subject** is. Many people lose sight of the subjects of their proofs because they are not used to reading their own mathematical writings. So, we usually do not write "$3 = 1 + 2$" since we usually are not interested in conveying "three **is** one plus two". Very often, difficulty in proving theorems stems from not knowing or losing sight of the subjects. On the other hand, if you identify the subjects, proofs can often be completed smoothly. Please put words to what you are writing. If you cannot read what you are writing, please do not expect anyone else to be able to read what you write. If it is not an essay, then it is not a proof. Even scratch work should be written in a form that can be read. Neatness is important and pays off in the end.

Axiom If X is a set, then $\emptyset \subset X$.

(This axiom says that the empty set $\emptyset$ is a subset of any set. Every set has the empty set $\emptyset$ as a subset. Also $X \subset X$ for any set X. Hence, $\emptyset \subset \emptyset$. The empty set plays a role in set theory like the number zero in arithmetic as in the next problem.)

Problem 1.1.1 Let X be a set. Prove that the following are true:

(1) $X \cup \emptyset = X$, (2) $X \cap \emptyset = \emptyset$,

(3) $X - \emptyset = X$, (4) $\emptyset - X = \emptyset$, (5) $X - X = \emptyset$.

(These seem too "simple" to require proofs. However, some of these are rather difficult to prove. The words like "simple", "easy", "trivial", "obvious" are used often to disguise the weakness in the arguments by many mathematicians. So, I avoid these words in this textbook unless I want to convey "not easy".)

Solution to (1): This is a set equality. So, we have to use Definition 1.1.1 (7). That is, we have to show that $X \cup \emptyset \subset X$ and $X \subset X \cup \emptyset$ by Definition 1.1.1 (6).

Let us start by showing $X \cup \emptyset \subset X$. Let $x \in (X \cup \emptyset)$. Then $x \in X$ or $x \in \emptyset$. But $x \notin \emptyset$ since $\emptyset$ is a set without any elements. Hence, $x \in X$. Therefore, $X \cup \emptyset \subset X$.

Conversely, we will show that $X \subset X \cup \emptyset$. Let $x \in X$. Then $x \in X$ or $x \in \emptyset$. Hence, $x \in (X \cup \emptyset)$. This shows that $X \subset X \cup \emptyset$.

Therefore, we have shown that $X \cup \emptyset = X$.

Definition 1.1.2 Let $a \leq b$ be real numbers. (Keep in mind that the notation"∞" is a symbol for infinity. At this moment, it has no meaning. It is not a number.) We frequently simplify set notations using interval notations as follow:

$$(a, \infty) = \{x \in \mathbb{R} : x > a\}.$$

(This means that (a, ∞) denotes the set of all numbers larger than a.)

$$[a, \infty) = \{x \in \mathbb{R} : x \geq a\}.$$

(This means that $[a, \infty)$ denotes the set of all numbers larger than **or equal to** a.)

$$(-\infty, a) = \{x \in \mathbb{R} : x < a\}, \quad (-\infty, a] = \{x \in \mathbb{R} : x \leq a\},$$

$$(a, b) = \{x \in \mathbb{R} : a < x < b\}, \quad [a, b) = \{x \in \mathbb{R} : a \leq x < b\},$$

$$(a, b] = \{x \in \mathbb{R} : a < x \le b\}, \text{ and } [a, b] = \{x \in \mathbb{R} : a \le x \le b\}.$$

(Think of $(-\infty, \infty) = \mathbb{R}$ as the number line x-axis.)

Definition 1.1.3 Let $n \ge 1$ be an integer. Let X_k be a set **for each** $k = 1, 2, 3, \ldots$.

(1) The set $\overset{n}{\underset{k=1}{\cup}} X_k$ (<u>union</u> of X_k from $k = 1$ to n) consists of all elements in X_k for some
$k = 1, 2, 3, \ldots, n$. Hence, $x \in \overset{n}{\underset{k=1}{\cup}} X_k$ if $x \in X_k$ for **some** $k = 1, 2, 3, \ldots, n$. We often
write $X_1 \cup X_2 \cup X_3 \cup \cdots \cup X_n$ in place of $\overset{n}{\underset{k=1}{\cup}} X_k$. We write $x \in \overset{\infty}{\underset{k=1}{\cup}} X_k$ if $x \in X_k$ for
some $k = 1, 2, 3, \ldots$.
(2) The set $\overset{n}{\underset{k=1}{\cap}} X_k$ (<u>intersection</u> of X_k from $k = 1$ to n) consists of elements that belong to X_k
for all $k = 1, 2, 3, \ldots, n$. Hence, $x \in \overset{n}{\underset{k=1}{\cap}} X_k$ if $x \in X_k$ for **every (all)** $k = 1, 2, 3, \ldots, n$.
We often write $X_1 \cap X_2 \cap X_3 \cap \cdots \cap X_n$ in place of $\overset{n}{\underset{k=1}{\cap}} X_k$. We write $x \in \overset{\infty}{\underset{k=1}{\cap}} X_k$ if $x \in X_k$
for **every (each, all)** $k = 1, 2, 3, \ldots$.

Remark 1.1.2 In the above definition, please notice the difference between "$x \in X_n$ for **some** $n = 1, 2, 3, \ldots, k$" in part (1) and "$x \in X_n$ for **every** $n = 1, 2, 3, \ldots, k$" in part (2). The statements "let x be ...", "for some ...", "for every ...", "there is ...", and "for all ..." are called **quantifiers**. In mathematics, the use of these quantifiers is essential since quantifiers define **subjects** in sentences. It is like identifying someone in a story. If a letter, say x, is introduced, you have to explain what it is. No letter should be written without explaining what it is. These letters are similar to introducing people in a novel. Also, as in Definition 1.1.1, parts (3), (4) and (5), you must pay extra attention to the use of "**and**" and "**or**".

Example1.1.1 (1) The solutions to equation $x^2 - 1 = 0$ are not "$x = 1$ **and** $x = -1$" since x cannot be 1 and -1 simultaneously. The solutions to equation $x^2 - 1 = 0$ are "$x = 1$ **or** $x = -1$". We also say that "$x^2 - 1 = 0$ holds for $x = 1$ **or** $x = -1$". And the statement "there exists an x such that $x^2 - 1 = 0$" is equivalent to "$x^2 - 1 = 0$ for **some** x".

   The statement "$x^2 - 1 > 0$ is true **for** x such that $x > 1$ **or** $x < -1$" can be improved. Simply saying "for x" does not make clear if it means "for all x" or "for some x". Hence, it should be "$x^2 - 1 > 0$ is true **for all** x such that $x > 1$ **or** $x < -1$", or "$x^2 - 1 > 0$ is true **for all** x such that $x \in (-\infty, -1) \cup (1, \infty)$".
   And $x^2 - 1 < 0$ holds for all x such that $(x < 1$ **and** $x > -1)$. In short, $x^2 - 1 < 0$ holds for every $-1 < x < 1$, or $x^2 - 1 < 0$ is true **for all** x such that $x \in (-1, 1)$.

(2) Note that $\{1, 1\} = \{1\}$, and the set $\{1, 1\}$ is a set with **one** element. The set $\{\emptyset\}$ has **one** element and it is not the empty set. So, $\{\emptyset, \{\emptyset\}\}$ has two elements, each element being a set.

Problem 1.1.2 Let $a \leq b \leq c$. Prove the following:

(1) $(a, b) = (a, \infty) \cap (-\infty, b)$.
(2) $(a, b] \cup [b, c) = (a, c)$.
(3) $[a, \infty) = \bigcup_{n=0}^{\infty} [a + n, a + n + 1)$.

Definition 1.1.4 If x is a real number, we define the $\underline{\text{absolute value}}$ of x, denoted by $|x|$, to be x if $x \geq 0$, and $-x$ if $x < 0$. In symbols, we write $|x| = \begin{cases} x \text{ if } x \geq 0, \text{ and} \\ -x \text{ if } x < 0 \end{cases}$. We read "$|x|$" as "the absolute value of x".

Problem 1.1.3 (1) If $x \in \mathbb{R}$, then prove that $|x| \geq 0$, and $|x| \geq x$.

(2) Use Definition 1.1.4 to prove that $\{x \in \mathbb{R} : |x| < 1\} = (-1, 1)$.
    (We usually drop "$\in \mathbb{R}$", and say "prove that $\{x : |x| < 1\} = (-1, 1)$".)

(3) Use Definition 1.1.4 to prove $\{x : |x| > 1\} = (-\infty, -1) \cup (1, \infty)$.

(4) Prove that $\{x : x^2 - 1 < 0\} = (-1, 1)$.

(5) Prove that $\{x : x^2 - 1 > 0\} = (-\infty, -1) \cup (1, \infty)$.

(6) Sketch the set $\{(x, y) \in \mathbb{R}^2 : |x| + y = 1\}$. This is usually stated as "graph the equation of $|x| + y = 1$". (I suppose you know the answer to this question. However, the point of this exercise is to use Definition 1.4 correctly to see the reason why the set graphs the way it does rather than just stating the answers.)

(7) Sketch the graph of $|x| + |y| = 1$ in the plane. (Hint: it looks like a diamond.)

The next two problems are for the challenge.

(8) Sketch the set $\{(x, y, z) \in \mathbb{R}^3 : |x| + |y| = 1\}$. This is usually stated as sketch the surface of $|x| + |y| = 1$ in three-dimensional space. (Hint: it looks like a diamond cylinder.)

(9) Sketch the surfaces of $|x| + |y| + |z| = 1$ in three-dimensional space. (Hint: It's a regular octahedron.)

Solution to (2): By Definition 1.1.1 (7). We have to prove $\{x \in \mathbb{R} : |x| < 1\} \subset (-1, 1)$ and $(-1, 1) \subset \{x \in \mathbb{R} : |x| < 1\}$.
    Let $x \in \{x : |x| < 1\}$. Since $x \geq 0$ or $x < 0$, we consider two cases.

(Case 1): Suppose $x \geq 0$. Then $|x| = x$. Since $|x| < 1$, we have $x < 1$. But $x \geq 0$. Hence, $0 \leq x < 1$ so that $-1 < x < 1$. Thus, $x \in (-1, 1)$.

(Case 2): Suppose $x < 0$. Then $|x| = -x$. Since $|x| < 1$, we have $-x < 1$. But then $-x < 1$ implies that $x > -1$. Since $x < 0$, we have $-1 < x < 0$ so that $-1 < x < 1$. Thus, $x \in (-1, 1)$.

Therefore, from Cases 1 and 2, we have shown that $\{x : |x| < 1\} \subset (-1, 1)$.

Conversely, let $x \in (-1, 1)$. Then $-1 < x < 1$. Again, we consider two cases, $x \geq 0$ or $x < 0$.

(Case 3): Suppose $x \geq 0$. Then $|x| = x$. Since $x < 1$, we have $|x| < 1$ so that $x \in \{x : |x| < 1\}$.

(Case 4): Suppose $x < 0$. Then $|x| = -x$ or $x = -|x|$. Since $-1 < x$, we have $-1 < -|x|$. Hence, $|x| < 1$ so that $x \in \{x : |x| < 1\}$.

Cases 3 and 4 show that $(-1, 1) \subset \{x : |x| < 1\}$.

Therefore, we have shown $\{x \in \mathbb{R} : |x| < 1\} = (-1, 1)$.

Remark 1.1.3 You should be able to solve inequalities
$x^2 - x \leq 0$, $\frac{x^2-4}{x^3+1} \geq 0$, $x^2 + 4x \geq 10$, etc., quickly.

Next, we would like to talk about the triangle inequality. It has several forms. The basic triangle inequality is the following.

Let $\triangle ABC$ be a triangle. Then

(1) $|AC| \leq |AB| + |BC|$.

Using vectors, we have the equation $\overrightarrow{AB} + \overrightarrow{BC} = \overrightarrow{AC}$. The notation $\left|\overrightarrow{AB}\right|$ is to mean the length $|AB|$ as in (1). By replacing $|AC|$ by $\left|\overrightarrow{AB} + \overrightarrow{BC}\right|$, we can rewrite (1) as the next triangle inequality of vectors

(2) $\left|\overrightarrow{AB} + \overrightarrow{BC}\right| \leq \left|\overrightarrow{AB}\right| + \left|\overrightarrow{BC}\right|$.

Because of the resemblance to the above inequality (2), the next theorem is also called the triangle inequality.

Theorem 1.1.1 *(The Triangle Inequality)* Let $a, b, c \in \mathbb{R}$. Then

(1) $|a + b| \leq |a| + |b|$, and

(2) $|a - b| \geq ||a| - |b||$. (This means $|a - b| \geq |a| - |b|$, and $|a - b| \geq |b| - |a|$.)

(3) $|a - b| \leq |a - c| + |c - b|$.

Proof of (1): Since $|a| \geq a$ and $|b| \geq b$, we have $|a| + |b| \geq a + b$. Similarly, since $|a| \geq -a$ and $|b| \geq -b$, we have $|a| + |b| \geq -a - b = -(a + b)$. Since $|a + b| \geq a + b$ or $|a + b| \geq -(a + b)$, we have $|a| + |b| \geq |a + b|$.

Proof of (2): We have $|a| = |(a - b) + b| \leq |a - b| + |b|$. Solving for $|a - b|$, we have $|a - b| \geq |a| - |b|$. Similarly, we have $|a - b| \geq |b| - |a|$. Therefore, $|a - b| \geq ||a| - |b||$.

Proof of (3): This is left to the readers.

Theorem 1.1.1 is fundamental that it is our first theorem in this textbook. It will be used frequently in the subsequent chapters.

Problem 1.1.4 Negating definitions often helps us to understand definitions.
(1) If X is not a subset of Y, we write $X \not\subset Y$. By negating Definition 1.1.1 (6), state the definition of $X \not\subset Y$.
(Answer: $X \not\subset Y$ if there is an element $x \in X$ such that $x \notin Y$.)
(2) If X is not equal to Y, we write $X \neq Y$. By negating Definition 1.1.1 (7), state the definition of $X \neq Y$. As we have said earlier, note that $\{\emptyset\} \neq \emptyset$.

Problem 1.1.5 Let X, Y, Z be sets. Let X_n be a set for each $n = 1, 2, 3, \ldots$. Prove the following statements:

(1) $X \cup (Y \cap Z) = (X \cup Y) \cap (X \cup Z)$.
(2) $X \cap (Y \cup Z) = (X \cap Y) \cup (X \cap Z)$.
(3) $X - (Y \cap Z) = (X - Y) \cup (X - Z)$.
(4) $X - (Y \cup Z) = (X - Y) \cap (X - Z)$.
(5) $X \cup \left(\bigcap_{n=1}^{\infty} X_n \right) = \bigcap_{n=1}^{\infty} (X \cup X_n)$.

   $\left(x \in \bigcap_{n=1}^{\infty} X_n \text{ means that } x \in X_n \text{ for every } n = 1, 2, 3, \ldots \right)$

(6) $X \cap \left(\bigcup_{n=1}^{\infty} X_n \right) = \bigcup_{n=1}^{\infty} (X \cap X_n)$.

   $\left(x \in \bigcup_{n=1}^{\infty} X_n \text{ means that } x \in X_n \text{ for some } n = 1, 2, 3, \ldots \right)$

(7) $X - \left(\bigcap_{n=1}^{\infty} X_n \right) = \bigcup_{n=1}^{\infty} (X - X_n)$.

(8) $X - \left(\bigcup_{n=1}^{\infty} X_n \right) = \bigcap_{n=1}^{\infty} (X - X_n)$.

Solution to (1) (The strategy is to use Definition 1.1.1 (7).)

First, we will show that $X \cup (Y \cap Z) \subset (X \cup Y) \cap (X \cup Z)$. Let $x \in X \cup (Y \cap Z)$. Then $x \in X$ or $x \in (Y \cap Z)$.

Case 1: Suppose $x \in X$. Then $x \in (X \cup Y)$ and $x \in (X \cup Z)$. So $x \in (X \cup Y) \cap (X \cup Z)$.

Case 2: Suppose $x \in (Y \cap Z)$. Then $x \in Y$ and $x \in Z$. Hence, $x \in (X \cup Y)$ and $x \in (X \cup Z)$. So $x \in (X \cup Y) \cap (X \cup Z)$.

From Cases 1 and 2, we have shown that $X \cup (Y \cap Z) \subset (X \cup Y) \cap (X \cup Z)$.

Conversely, we will show that $(X \cup Y) \cap (X \cup Z) \subset X \cup (Y \cap Z)$. Let $x \in (X \cup Y) \cap (X \cup Z)$. Then $x \in (X \cup Y)$ and $x \in (X \cup Z)$. Hence, $(x \in X$ or $x \in Y)$ and $(x \in X$ or $x \in Z)$. So $x \in X$ or $(x \in Y$ and $x \in Z)$. Thus $x \in X \cup (Y \cap Z)$.

Therefore, we have shown that $X \cup (Y \cap Z) = (X \cup Y) \cap (X \cup Z)$.

Solution to (5) First, we will show that $X \cup \left(\bigcap\limits_{n=1}^{\infty} X_n \right) \subset \bigcap\limits_{n=1}^{\infty} (X \cup X_n)$. Let $x \in X \cup \left(\bigcap\limits_{n=1}^{\infty} X_n \right)$. Then $x \in X$ or $x \in \left(\bigcap\limits_{n=1}^{\infty} X_n \right)$.

(Case 1): Suppose $x \in X$. Then $x \in X \cup X_n$ for every $n = 1, 2, 3, \ldots$. So $x \in \bigcap\limits_{n=1}^{\infty} (X \cup X_n)$.

(Case 2): Suppose $x \in \bigcap\limits_{n=1}^{\infty} X_n$. Then $x \in X_n$ for every $n = 1, 2, 3, \ldots$. Hence, $x \in X \cup X_n$ for every $n = 1, 2, 3, \ldots$. So $x \in \bigcap\limits_{n=1}^{\infty} (X \cup X_n)$.

From Cases 1 and 2, we have shown that $X \cup \left(\bigcap\limits_{n=1}^{\infty} X_n \right) \subset \bigcap\limits_{n=1}^{\infty} (X \cup X_n)$.

Conversely, we will show that $\bigcap\limits_{n=1}^{\infty} (X \cup X_n) \subset X \cup \left(\bigcap\limits_{n=1}^{\infty} X_n \right)$. Let $x \in \bigcap\limits_{n=1}^{\infty} (X \cup X_n)$. Then $x \in X \cup X_n$ for every $n = 1, 2, 3, \ldots$. Hence, $(x \in X$ or $x \in X_n)$ for every $n = 1, 2, 3, \ldots$. So $x \in X$ or $(x \in X_n$ for every $n = 1, 2, 3, \ldots)$. Thus $x \in X$ or $x \in \bigcap\limits_{n=1}^{\infty} X_n$.

That is, $x \in X \cup \left(\bigcap\limits_{n=1}^{\infty} X_n \right)$.

Therefore, we have shown that $X \cup \left(\bigcap\limits_{n=1}^{\infty} X_n \right) = \bigcap\limits_{n=1}^{\infty} (X \cup X_n)$.

Problem 1.1.6 Let $X, Y \subset \mathbb{R}$. Prove the following statements if they are true. If not, give counterexamples to explain why they are not.

(1) $X - Y = X \cap (\mathbb{R} - Y)$.

(2) $X \cup (\mathbb{R} - X) = \mathbb{R}$.   (2′) $X \cup (Y - X) = Y$.

(3) $X \cap (\mathbb{R} - X) = \emptyset$.   (3′) $X \cap (Y - X) = \emptyset$.

(4) $\mathbb{R} - (\mathbb{R} - X) = \emptyset$.   (4′) $Y - (Y - X) = \emptyset$.

Remark 1.1.4 (*Russell's Paradox*)

What is the definition of a set? Even though Georg Cantor is credited for initiating set theory, this question tormented him. One way to answer this question is to not define the word "set". It is unsettling, but "set" is an **undefined** word, except to say that if S is a set, then $S \notin S$. This is to avoid the following difficulty:

Suppose a set S can have S as an element. Then the collection[1] A of all sets is a set even though $A \in A$. So, we can split A into two subsets, $P = \{x \in A | x \in x\}$ and $Q = \{x \in A | x \notin x\}$. Here, P is a subset of A whose elements are sets that contains itself, and Q is a subset of A whose elements are sets that do not contain itself. Then $P \neq \emptyset$ since $A \in P$. Since $\emptyset$ is a set, and since $\emptyset$ is a set with no elements (that is to say $\emptyset \notin \emptyset$), we have $\emptyset \in Q$ so that $Q \neq \emptyset$. Also, P and Q are elements of the set A such that $A = P \cup Q$. Moreover, P and Q are disjoint since no element $x \in A$ can be $x \in x$ and $x \notin x$. Since $Q \in A$, we must have $Q \in P$ or $Q \in Q$. If $Q \in P$, then $Q \in Q$ by the definition of the set P, which is in contradiction to P and Q being disjoint. If $Q \in Q$, then $Q \notin Q$ by the definition of the set Q, which is contradiction to the definition of the set Q. We obtained these contradictions because we allowed a set S can contain itself as an element. Therefore, we assume that if S is a set, then S **cannot** contain itself as an element, i.e., $S \notin S$.

This is an example of proof by contradiction. This is a perfectly good method to prove a theorem. However, very often, some proofs by contradiction can be re-written without the use of a contradiction. If that is the case, I consider the proof not edited enough. In the same way that you edit an essay for an English class many times before you consider it to be finished, get in the habit of revisiting your proof more than once! This practice of editing improves your understanding of mathematics. Let me put it differently. Studying mathematics is not all that different from studying other subjects. You learn from reading and from writing. And if you cannot express your thoughts in writing, it is not a thought at all.

Definition 1.1.5 If S is a set, then S itself cannot be an element of S, i.e., $S \notin S$.

In particular, **there is no set of all sets**, for otherwise, it would be a set that contains itself.

---

[1] I am hesitant to use "set" here since A is not a set. So I substituted "set" with "collection".

1.2   Counting

There are two fundamental **counting principles**:

Let $m, n \geq 0$ be integers. Suppose sets X and Y are disjoint sets having n and m elements, respectively. Then

(Principle 1) there are $m + n$ ways to select one element from X **or** Y, and

(Principle 2) there are mn ways to select one element from X **and** another one element from Y.

Definition 1.2.1 $0! = 1$ by definition. $1! = 1, 2! = 1 \cdot 2, 3! = 3 \cdot 2!, 4! = 4 \cdot 3!, \ldots$ So $1! = 1, 2! = 2, 3! = 6, 4! = 24$. If $n > 0$ is an integer, then $n! = 1 \cdot 2 \cdot \ldots \cdot n$. We read "$n!$" as "n factorial". For example, we read "$3!$" as "three <u>factorial</u>".

Problem 1.2.1 By making sure to identify which counting principle is used, show that the number of ways to line up n people is $n!$.

Problem 1.2.1 There are 27 violin players and 18 piano players in a room. In the room, there are 10 people who play both violin and piano, and there are 12 people who play neither instrument.

(1) How many people are there in the room all together?
(2) How many people are there in the room who play violin but not piano?

Problem 1.2.2 Let $0 \leq m \leq n$ be integers. Show that the number of ways to line up m people out of n people is $n \cdot (n - 1) \cdot \ldots \cdot (n - m + 1) = \frac{n!}{(n-m)!}$.

Definition 1.2.2 Let $0 \leq m \leq n$ be integers. The number of ways to select m objects out of n objects is denoted by $\binom{n}{m}$, and $\binom{n}{m}$ is called a <u>binomial coefficient</u>.

We read $\binom{n}{m}$ as "n <u>choose</u> m". Note that $\binom{n}{0} = 1$ since there is only one way to select zero elements from n objects, and $\binom{n}{n} = 1$ since there is only one way to select all from n elements.

(The number of ways to rearrange m elements in a set of n elements is sometimes called the <u>permutation</u> of n <u>choose</u> m. Hence, from Problem 1.2.2, the permutation of n choose m is $\frac{n!}{(n-m)!}$. If X is a finite set with n elements, then number of permutations on X is $n!$.)

The number $\binom{n}{m}$ is sometimes called the <u>combination</u> of n <u>choose</u> m.)

Problem 1.2.3 Let $0 \le m \le n$ be integers. The point of this exercise is to use the above definition and **not** to use Problem 1.2.5.

(1) Show that $\binom{n}{m} = \binom{n}{n-m}$.

(2) Prove that $\binom{n}{m} + \binom{n}{m-1} = \binom{n+1}{m}$.

The next theorem is an important theorem which we will use later in this textbook.

Theorem 1.2.1 (*Binomial Theorem*)

Let $0 \le m \le n$ be integers. Let a, b be real numbers. Then $(a+b)^n =$
$$\sum_{m=0}^{n} \binom{n}{m} a^m b^{n-m}.$$

(The point of the proof below is to use only Definition 1.2.1 and the relevant counting principle. We will ask to use mathematical induction to prove this theorem again in Sect. 1.4.)

Proof Let $0 \le m \le n$. By writing $(a+b)^n = (a+b)_1 \cdot (a+b)_2 \cdot \cdots \cdot (a+b)_n$, we identify the i-th term of the multiplication $(a+b)^n$ with the subscript in $(a+b)_i$ for each $i = 1, 2, \ldots, n$.

This multiplication $(a+b)_1 \cdot (a+b)_2 \cdot \cdots \cdot (a+b)_n$ is usually done by distributive rule by first multiplying $(a+b)_1 \cdot (a+b)_2$, and then multiplying $(a+b)_3$ to it, and so on.

Instead of doing this, we know that the result of multiplying $(a+b)_1 \cdot (a+b)_2 \cdot \cdots \cdot (a+b)_n$ is the sum of the terms of the form $a^m b^{n-m}$. So we count number of the terms $a^m b^{n-m}$ in this multiplication. Let $0 \le m \le n$ be fixed integers. In order to obtain one term of $a^m b^{n-m}$ in this multiplication, we have to choose m many elements from the set $\{(a+b)_1, (a+b)_2, \ldots, (a+b)_n\}$ that a are selected, then we know that b are selected from the remaining $n - m$ terms in $\{(a+b)_1, (a+b)_2, \ldots, (a+b)_n\}$. Hence, the coefficient of $a^m b^{n-m}$ is the number of ways we can select m elements from the set $\{(a+b)_1, (a+b)_2, \ldots, (a+b)_n\}$ of n elements, and this is given to be $\binom{n}{m}$ by Definition 1.2.2. That is, the coefficient of $a^m b^{n-m}$ is $\binom{n}{m}$.

Hence, this proves that $(a + b)^n = \sum_{m=0}^{n} \binom{n}{m} a^m b^{n-m}$.

Problem 1.2.4 (1) Let $X = \{1, 2, 3\}$. Write all the subsets of X, and convince yourself that there are 8 subsets of X.

(2) Use the binomial theorem to show that there are 2^n subsets in a set having $n \geq 0$ elements.

If X is a finite set (a set with finite number of elements), then we use $|X|$ to denote the number of elements in X. Hence, there are $2^{|X|}$ many subsets of X. Because of this, we sometimes denote the set of all subsets of X by 2^X, and 2^X is said to be the <u>power set</u> of X.

(Hint, suppose a set X has n elements. Let m be a fixed integer such that $0 \leq m \leq n$. Count the number of subsets of X having exactly m many elements, and from there, conclude that the number of subsets of X is $\sum_{m=0}^{n} \binom{n}{m}$. Then use the binomial theorem to show

$$\sum_{m=0}^{n} \binom{n}{m} = 2^n.)$$

Problem 1.2.5 Let $0 \leq m \leq n$ be integers.

(1) Show that the number of ways to line up m people out of n people is $\binom{n}{m} \cdot m!$.

(2) From Problem 1.2.2 and (1), deduce that $\binom{n}{m} = \frac{n!}{m! \cdot (n-m)!}$.

1.3   Functions

The concept of a function is old. At the time of Fermat (1607–1665), Newton (1642–1726) and Leibniz (1646–1716), the need for the concept of a function was getting clearer, but it was still a vague concept. The word "function" was probably first introduced by Leibniz, or perhaps by Leonhard Euler (1707–1783). The importance of the concept "function" was starting to emerge, yet remained vague. We still use this vague concept of a function when we first study calculus. This ambiguous definition works for the most part in studying calculus. However, this ambiguous definition of a function becomes an obstacle when we want to prove theorems involving functions. Since we want to prove theorems involving functions, we have to define a function precisely using **sets**. In order to do this, we start with the following definition.

Definition 1.3.1 Let X, Y be nonempty sets. Let $a, x \in X$ and $b, y \in Y$. We say that $(a, b) = (x, y)$ if $a = x$ and $b = y$. The pair (x, y) is called an <u>ordered pair</u> since (x, y) is not the same as (y, x). The collection of ordered pairs is denoted by $X \times Y = \{(x, y) : x \in X \text{ and } y \in Y\}$, and it is called the cross product of X and Y. The set $X \times X$ is denoted by X^2 for short. So $\mathbb{R} \times \mathbb{R} = \mathbb{R}^2$.

From the context of the subject, the distinction between the ordered pair (a, b) and the open interval (a, b) should be clear. For example, $(1, -2)$ makes sense as an ordered pair, but it has no meaning as an interval.

(The xy-plane is identified with the set $\mathbb{R}^2$. The set $[0, 1] \times [0, 1]$ is a unit square in the xy-plane.)

Definition 1.3.2 Let X, Y be nonempty sets. A function f from X into Y, denoted by $f : X \rightarrow Y$, is a **subset** of $X \times Y$ with the following two properties:

(1) for every $x \in X$, there exists a $y \in Y$ such that $(x, y) \in f$., and

(2) if $(x, y) \in f$ and $(x, z) \in f$ for some $x \in X$, then $y = z$.

If $(x, y) \in f$, it is customary to write $y = f(x)$, and $f(x)$ is said to be the value of x under f. You might like to think that, for every $x \in X$, the element x has a unique partner $f(x)$ in the set Y. The set X is the domain of f, and the set Y is the range of f. (We do not use the word "co-domain".)

(3) Let A be a non-empty subset of X. Let $g : A \rightarrow Y$ be a function defined by $g(x) = f(x)$ for all $x \in A$. Then we write $f|_A$ for g, (that is to say $g = f|_A$) and $f|_A$ is said to be the restriction of f on A. In Chap. 5, we often want a continuous function $f : [0, 1] \rightarrow \mathbb{R}$ such that f is differentiable on the interval $(0, 1)$. In place of "such that f is differentiable on the interval $(0, 1)$", we might say that "such that $f|_{(0,1)}$ is differentiable".

First, "Let $f : X \rightarrow Y$" should be read "Let f be a function from X into Y", and "$f(x)$" is read as "f of x".

Intuitively, a function $f : X \rightarrow Y$ is a rule to choose a unique element (partner or value) in Y for each element in X. And the partner or the value of $x \in X$ in Y is denoted by $f(x) \in Y$. So, for example, we define a function $f : [0, 10] \rightarrow [-101, 101]$ by $f(x) = x^2 - 3x + 1$ for each $x \in [0, 10]$. In calculus classes, we used to say "let $f(x) = x^2 - 3x + 1$ be a function", and we did not define the domain and its range precisely. So the big difference here is that we have to specify the **domain** and **range** of a function f. And f is the name of the function $f : [0, 10] \rightarrow [-101, 101]$ defined by $f(x) = x^2 - 3x + 1$ for each $x \in [0, 10]$, and $f(x)$ is the value of the function **named** f at the point x.

Definition 1.3.3 Let X, Y be nonempty sets. Let $A \subset X$ and $B \subset Y$. Let $f : X \rightarrow Y$ be a function. The image of A under f, denoted by $f(A)$, is defined by

$$f(A) = \{f(x) : x \in A\}.$$

We read the image $f(A)$ as "f of A".

The <u>inverse image</u> of B under f, or simply the f-<u>inverse</u> of B, denoted by $f^{-1}(B)$, is defined by

$$f^{-1}(B) = \{x \in X : f(x) \in B\}.$$

(Here, the "f-inverse" and the "inverse function of f" are **different**. We have **not** defined the inverse function of f yet! Note that we are **not** defining the inverse function of f. We read $f^{-1}(B)$ as "f-<u>inverse</u> of B".)

If B is a set with one element, for example, $B = \{b\}$, it is customary to write $f^{-1}(b)$ in place of $f^{-1}(\{b\})$.

If $f(x)$ is a number, then $\left[f(x)\right]^{-1} = \frac{1}{f(x)}$ so that $f^{-1}(x) \neq \frac{1}{f(x)}$.

That is $\left[f(x)\right]^{-1} \neq f^{-1}(x)$.

Example 1.3.1 Let $f : (0, \infty) \to (0, \infty)$ be a function defined by $\left(x, \frac{1}{x}\right) \in f$ for every $x \in (0, \infty)$. The statement "$\left(x, \frac{1}{x}\right) \in f$ for every $x \in (0, \infty)$" is usually written by "$f(x) = \frac{1}{x}$ for every $x \in (0, \infty)$". So, this is usually written as "Let $f : (0, \infty) \to (0, \infty)$ be a function defined by $f(x) = \frac{1}{x}$ for every $x \in (0, \infty)$" and reads "Let f from $(0, \infty)$ into $(0, \infty)$ be a function ...". Then $f(\{-1, 2, 3\})$ and $f^{-1}(\{-1, 2, 3\})$ are undefined since $-1 \notin (0, \infty)$.

Let $F : (0, \infty) \to \mathbb{R}$ be a function defined by $F(x) = \frac{1}{x}$ for every $x \in (0, \infty)$. Then $F(\{-1, 2, 3\})$ is undefined, while $F^{-1}(\{-1, 2, 3\}) = \{\frac{1}{2}, \frac{1}{3}\}$.

$F([1, 10]) = \left[\frac{1}{10}, 1\right]$ and $F^{-1}([-2, 3]) = \left[\frac{1}{3}, \infty\right)$.

Let $g : (-\infty, 0) \to \mathbb{R}$ be a function defined by $g(x) = \frac{1}{x}$ for every $x \in (-\infty, 0)$. Then $g(\{-1, 2, 3\})$ is undefined, while $g^{-1}(\{-1, 2, 3\}) = \{-1\}$.

$g\left(\left[-1, -\frac{1}{10}\right]\right) = [-10, -1]$ and $g^{-1}([-2, 3]) = \left(-\infty, -\frac{1}{2}\right]$.

Let $G : \mathbb{R} - \{0\} \to \mathbb{R}$ be a function defined by $G(x) = \frac{1}{x}$ for every $x \in \mathbb{R} - \{0\}$. Both $G(\{-1, 2, 3\})$ and $G^{-1}(\{-1, 2, 3\})$ are defined. They are $G(\{-1, 2, 3\}) = \{-1, \frac{1}{2}, \frac{1}{3}\}$ and $G^{-1}(\{-1, 2, 3\}) = \{-1, \frac{1}{2}, \frac{1}{3}\}$. $G^{-1}([-2, 3]) = \left(-\infty, -\frac{1}{2}\right] \cup \left[\frac{1}{3}, \infty\right)$.

Even though the definitions of the functions f, F, g, G are very similar, they are considered to be different functions because their domains or their ranges are different.

Note that $G|_{(-\infty, 0)} = g$ and $G|_{(0, \infty)} = F$.

Example 1.3.2 Let $f : [-2, 3] \to [-3, 9]$ be a function defined by $f(x) = x^2 - 3$ for each $x \in [-2, 3]$. Then $f([-2, 1]) = [-3, 1]$ and $f^{-1}([-2, 2]) = [-2, -1] \cup \left[1, \sqrt{5}\right]$ (Fig. 1.1).

Fig. 1.1 A graph of
$y = f(x) = x^2 - 3$ on $[-2, 3]$

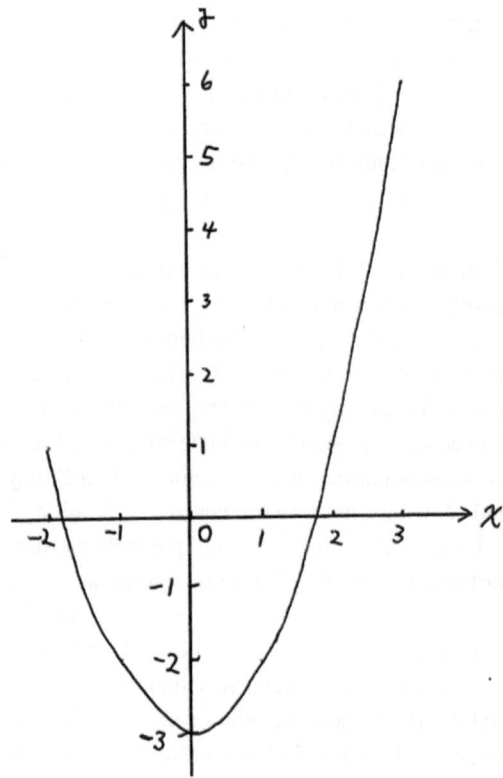

Problem 1.3.1 Let $f : (0, \infty) \to (1, \infty)$ be defined by $f(x) = x^2 + 1$ for every $x \in (0, \infty)$.
Let $F : (0, \infty) \to \mathbb{R}$ be defined by $F(x) = x^2 + 1$ for every $x \in (0, \infty)$.
Let $g : (-\infty, 0) \to \mathbb{R}$ be defined by $g(x) = x^2 + 1$ for every $x \in (-\infty, 0)$.
Let $G : \mathbb{R} \to \mathbb{R}$ be defined by $G(x) = x^2 + 1$ for every $x \in \mathbb{R}$.
Let $h : (-10, -2) \cup [4, 5] \to (0, \infty)$ be defined by $h(x) = x^2 + 1$ for every $x \in (-10, -2) \cup [4, 6]$.

(1) Explicitly write the following sets: $f(\{1, 2, 3\}), f^{-1}(\{1, 2, 3\}),$
$F(\{1, 2, 3\}), F^{-1}(\{-1, 0, 1, 2, 3\}), g(\{-1, -2, -3\}), g^{-1}(\{-1, -2, -3\}),$
$G(\{-1, 0, 1, 2, 3\}), G^{-1}(\{-1, 0, 1, 2, 3\}), h(\{-7, -3, 6\}), h^{-1}(\{-7, -3, 6, 8\}).$

(2) Use interval and set notations to express the following sets:

$f([1, 3)), f^{-1}([1, 3)), F^{-1}([-1, 3)), g([-5, 3)), g^{-1}([-5, 3)).$
$G([-5, 3)), G^{-1}([-5, 3)), h^{-1}([-1, 30]).$

  (**Selected answers:** $g([-5, 3))$ is not defined. $g^{-1}([-5, 3)) = \left[-\sqrt{3}, 0\right].$ $G([-5, 3)) =$
$[1, 25].$ $G^{-1}([-5, 3)) = \left(-\sqrt{3}, \sqrt{3}\right). h^{-1}([-1, 30]) = \left[-\sqrt{30}, -2\right) \cup [4, 5].$)

Definition 1.3.4 Let X, Y be nonempty sets. Let $f : X \to Y$ be a function.

The function f is said to be one-to-one, written "1–1" for short, if $a, x \in X$ and $f(a) = f(x)$ implies $a = x$. Equivalently, f is 1–1 if $a \neq x \in X$, then $f(a) \neq f(x)$.

The function f is said to be onto if for every $y \in Y$, there exists an $x \in X$ such that $y = f(x)$. Equivalently, the function f is onto if for every $y \in Y$, $f^{-1}(y) \neq \emptyset$. In this case, we say that the function f is X onto Y.

Example 1.3.3 Let X be **a group of adult females** looking for a male partner. Let Y be **a group of adult males** looking for a female partner. Suppose the choice of how to choose a partner in Y is given to the females in X. Then a function $g : X \to Y$ can be defined as a choice of each female in X selecting a male partner in Y with two conditions; (1) each female has exactly one choice, and (2) each female must choose one partner in Y. Hence, for each $x \in X$, $g(x)$ is the male partner in Y chosen by x. Males in Y have no right whom to choose a partner in X. So some male in Y may be chosen by more than one female in X, and some may have no partner in X. The set $g^{-1}(y)$ is the set of females who chose $y \in Y$. Let $x \in X$. Then $g(x)$ is the unique male partner of x, and $g^{-1}(g(x))$ is the set of all female partners of $g(x)$. So $g^{-1}(g(x))$ is non-empty since $x \in g^{-1}(g(x))$. The number of elements in $g^{-1}(g(x))$ is at least one. If g is 1–1, then the number of elements in $g^{-1}(g(x))$ is one, and $g^{-1}(g(x)) = \{x\}$. If g is not 1–1, then $g^{-1}(g(x))$ has at least two elements for some $x \in X$. If g is onto, then every male $y \in Y$ has at least one partner in X, i.e., $g^{-1}(y) \neq \emptyset$, and $g^{-1}(y)$ is the collection of all female partners. If g is not onto, then $g^{-1}(y) = \emptyset$ for some $y \in Y$. If g is 1–1 and onto, then each male in Y has exactly one partner in X under g. Once again, each female in X has exactly one partner in Y under this relation g because g is a function.

Example 1.3.4 (1) Let $f : (0, \infty) \to (1, \infty)$ be defined by $f(x) = x^2 + 1$ for every $x \in (0, \infty)$. Prove f is 1–1 and onto.

(2) Let $G : \mathbb{R} \to \mathbb{R}$ be defined by $G(x) = x^2 + 1$ for every $x \in \mathbb{R}$. Show G is not 1–1. Also, show that G is not onto.

Proof of (1) First, we will prove that f is 1–1. Suppose $f(x) = f(a)$ for some $x, a \in (0, \infty)$. Then $x^2 + 1 = a^2 + 1$ so that $x^2 - a^2 = (x + a)(x - a) = 0$. Hence, $x = -a$ or $x = a$. However, if $x = -a$, then $a < 0$ since $x > 0$. Since $a > 0$, we cannot have $x = -a$. Therefore, $x = a$, and this proves that f is 1–1.

Next, we prove that f is onto. Let $y \in (1, \infty)$. Then $y - 1 > 0$ so that $\sqrt{y - 1} \in (0, \infty)$. Moreover, $f\left(\sqrt{y - 1}\right) = \left(\sqrt{y - 1}\right)^2 + 1 = y - 1 + 1 = y$. Therefore, f is onto.

Proof of (2): The function G is not 1–1 since $1 \neq -1 \in \mathbb{R}$ yet $G(1) = 2 = G(-1)$.

The function G is not onto since $G(x) = x^2 + 1 \geq 1$ for every $x \in \mathbb{R}$.

In other words, G is not onto since $0 \in \mathbb{R}$ yet $G^{-1}(0) = \emptyset$.

Problem 1.3.2

(1) Let $G : \mathbb{R} - \{0\} \to \mathbb{R}$ be a function defined by $G(x) = \frac{1}{x}$ for every $x \in \mathbb{R} - \{0\}$. Is the function G 1–1 and onto? Prove your claim.

(2) Let $F : \mathbb{R} - \{0\} \to (0, \infty)$ a function defined by $F(x) = \frac{1}{x^2}$ for every $x \in \mathbb{R} - \{0\}$. Is the function F 1–1 and onto? Prove your claim.

(3) Let $f : (-\infty, -1) \to (0, \infty)$ be a function defined by $F(x) = \frac{1}{x^2}$ for every $x \in (-\infty, -1)$. Is the function f 1–1 and onto? Prove your claim.

Problem 1.3.3 Let $f : X \to Y$ be a function. Prove statements (1) and (2), and give counterexamples to (3) and (4).

(1) The function f is 1–1 if, and only if $f^{-1}(y)$ is either the empty set or a set with one element for every $y \in Y$.

(2) The function f is onto if, and only if $f(X) = Y$.

(3) The function f is onto if, and only if $f^{-1}(Y) = X$.

(4) The function f is 1–1 if, and only if $f(X) = Y$.

Problem 1.3.4 Let X, Y be nonempty sets. Let $A, B \subset X$ and $P, Q \subset Y$. Let $f : X \to Y$ be a function. Prove or give counterexamples to the following statements.

(1) $f(A \cup B) \subset f(A) \cup f(B)$.   (1') $f(A) \cup f(B) \subset f(A \cup B)$.

(2) $f(A \cap B) \subset f(A) \cap f(B)$.   (2') $f(A) \cap f(B) \subset f(A \cap B)$.

(3) $f(A - B) \subset f(A) - f(B)$.   (3') $f(A) - f(B) \subset f(A - B)$.

(4) $f^{-1}(P \cup Q) \subset f^{-1}(P) \cup f^{-1}(Q)$.

(4') $f^{-1}(P) \cup f^{-1}(Q) \subset f^{-1}(P \cup Q)$.

(5) $f^{-1}(P \cap Q) \subset f^{-1}(P) \cap f^{-1}(Q)$.

(5') $f^{-1}(P) \cap f^{-1}(Q) \subset f^{-1}(P \cap Q)$.

(6) $f^{-1}(P - Q) \subset f^{-1}(P) - f^{-1}(Q)$.

(6') $f^{-1}(P) - f^{-1}(Q) \subset f^{-1}(P - Q)$.

(7) If f is 1–1 and onto, then all the above statements are true. You should be able to prove this.

(8) Can you make any interesting observations from (1)–(6) and (1')–(6')?
     (Hint: (4)–(6) and (4')–(6') are nice.)

(9) Construct a 1–1 function from (0,1) onto $(1, \infty)$.

(A **note** on (2'): Here is a typical **wrong** proof of $f(A) \cap f(B) \subset f(A \cap B)$.

"Proof: Let $f(x) \in f(A) \cap f(B)$. Then $f(x) \in f(A)$ and $f(x) \in f(B)$. Hence, $x \in A$ and $x \in B$ so that $x \in A \cap B$. Thus, $f(x) \in f(A \cap B)$. Therefore, $f(A) \cap f(B) \subset f(A \cap B)$."

What is wrong with this "Proof"? The mistake is in the first sentence "Let $f(x) \in f(A) \cap f(B)$". By writing this way, this person is assuming that there is an x, **without** telling what x is, such that $f(x) \in f(A) \cap f(B)$. In other words, the element x is **not** quantified. It should be "$y \in f(A) \cap f(B)$" in place of "$f(x) \in f(A) \cap f(B)$". Hence, the wrong **subject** $f(x)$ was chosen in place of y. So, let us correct the "Proof".

An attempt to "prove" (2'): Let $y \in f(A) \cap f(B)$. Then $y \in f(A)$ and $y \in f(B)$. Hence, there are $a \in A$ and $b \in B$ such that $y = f(a)$ and $y = f(b)$. However, we do not know that $a = b$. So, we are stuck! Therefore, we cannot conclude that $f(A) \cap f(B) \subset f(A \cap B)$.

At this point, you would want to look for a counterexample to this statement.)

Proof for (4): Let $x \in f^{-1}(P \cup Q)$. Then $f(x) \in P \cup Q$ so that $f(x) \in P$ or $f(x) \in Q$.

Case 1: Suppose $f(x) \in P$. Then $x \in f^{-1}(P)$.

Case 2: Suppose $f(x) \in Q$. Then $x \in f^{-1}(Q)$.

Hence, $x \in f^{-1}(P)$ or $x \in f^{-1}(Q)$. Therefore, $x \in f^{-1}(P) \cup f^{-1}(Q)$. This proves that $f^{-1}(P \cup Q) \subset f^{-1}(P) \cup f^{-1}(Q)$.

Definition 1.3.5 Let X, Y be nonempty sets. Let $f : X \to Y$ be a 1–1 and onto function. Then for every $y \in Y$, there exists a unique element $x \in X$ such that $f(x) = y$. Hence, $f^{-1}(y) = \{x\}$. (We need brackets "$\{\}$" since $f^{-1}(y)$ is a set. At this moment, we are treating $f^{-1}(y)$ as the f-**inverse** of y.) Hence, when $f : X \to Y$ is a 1–1 and onto function, we **define a new function** $f^{-1} : Y \to X$ defined by, for every $y \in Y$, $f^{-1}(y)$ is the unique element $x \in X$ such that $f(x) = y$, and we write $f^{-1}(y) = x$. We call $f^{-1} : Y \to X$ the <u>inverse function</u> of f. So, because of this definition, the notation "$f^{-1}(b) = a$" makes sense, and it only makes sense when the function f is 1–1 and onto.

The notation "f^{-1}" is used in two ways; one as "f-inverse" and the other is "the inverse function of f". The "inverse function of f " and the "f-inverse" are **two different** things. If f is not 1–1 and onto, then f^{-1} is always f-inverse. Only when f is 1–1 and onto, f^{-1} is the inverse function of f (and it can be f-inverse).

Example 1.3.5 Let X be a group of adult females looking for male partners. Let Y be a group of adult males looking for female partners. Let $g : X \to Y$ be a 1–1 and onto function. (See Example 1.3.3.) Then for every female $x \in X$, $g(x)$ is her male partner in Y. But now, $g^{-1} : Y \to X$ is a function, so for every male $y \in Y$, $g^{-1}(y)$ is his unique female partner in X. Hence, the number of elements in X is exactly the number of elements in Y.

Example 1.3.6 (1) In Example 1.3.1, f is 1–1 and onto, and F, g, G are all 1–1 but not onto.

(2) Let $g : \mathbb{R} \to \mathbb{R}$ be defined by $g(x) = -3x + 5$ for every $x \in \mathbb{R}$. Then g is 1–1 and onto. By letting $y = -3x + 5$, and solving it for x, we have $x = -\frac{y-5}{3}$. Hence, $g^{-1}(y) = -\frac{y-5}{3}$ for every $y \in \mathbb{R}$.

In an elementary math class, after you solved $y = -3x + 5$ for x to obtain $x = -\frac{y-5}{3}$, you were told to interchange x and y. And the answer to this problem was $g^{-1}(x) = -\frac{x-5}{3}$. This was because the domain of any function **always** being a subset of the x-axis. This is one more bad habit of a function not being defined precisely. Please **do not do this**.

(3) Let $f : \{3, 4, 5, \ldots\} \to \{\frac{1}{n} : n \in \mathbb{N}\}$ be a function defined by $f(x) = \frac{1}{x-2}$ for every $x \in \{3, 4, 5, \ldots\}$. Then f is 1–1 and onto. So, $f^{-1} : \{\frac{1}{n} :\in \mathbb{N}\} \to \{3, 4, 5, \ldots\}$ defined by $f^{-1}(y) = \frac{1}{y} + 2$ for every $y \in \{\frac{1}{n} : n \in \mathbb{N}\}$ is the inverse function of f. The f-inverse of $\{\frac{1}{10}, \frac{1}{100}\}$ is $f^{-1}(\{\frac{1}{10}, \frac{1}{100}\}) = \{12, 102\}$.

(4) Let $f : (0, \infty) \to (0, \infty)$ be defined by $f(x) = \frac{1}{x}$ for every $x \in (0, \infty)$. Then f is 1–1 and onto. And $f^{-1}(y) = \frac{1}{y}$ for every $y \in (0, \infty)$.

(5) Let $F : (-\infty, 0] \to [0, \infty)$ be defined by $F(x) = x^2$ for every $x \in (-\infty, 0]$. Then F is 1–1 and onto. And $F^{-1}(y) = -\sqrt{y}$ for every $y \in [0, \infty)$.

(6) Let $g : (0, \infty) \to \mathbb{R}$ be defined by $g(x) = \ln x$ for every $x \in \mathbb{R}$. Then g is 1–1 and onto. The function $g^{-1} : \mathbb{R} \to (0, \infty)$ is defined by $g^{-1}(y) = e^y$ for every $y \in \mathbb{R}$.

(7) Let $f : (-1, 1) \to \mathbb{R}$ be a function defined by $f(x) = \frac{x}{1-x^2}$ for all $x \in (-1, 1)$.

First, we will prove that f is a 1–1 function. Let $-1 < a < b < 1$. Then

$$f(b) - f(a) = \frac{b}{1-b^2} - \frac{a}{1-a^2} = \frac{b(1-a^2)-a(1-b^2)}{(1-b^2)(1-a^2)} = \frac{(b-a)-ba^2+ab^2}{(1-b^2)(1-a^2)} = \frac{(b-a)-ab(b-a)}{(1-b^2)(1-a^2)} = \frac{(b-a)(1-ab)}{(1-b^2)(1-a^2)}.$$

Since $-1 < a < b < 1$, we have $b - a > 0$, $1 - ab > 0$, $1 - b^2 > 0$, and $1 - a^2 > 0$ so that $\frac{(b-a)(1-ab)}{(1-b^2)(1-a^2)} > 0$. Hence, $f(b) > f(a)$. Therefore, f is 1–1.

Second, we will prove that f is an onto function. Let $y \in \mathbb{R}$. Since $f(0) = 0$, assume that $y \neq 0$. Let $x = \frac{-1+\sqrt{1+4y^2}}{2y}$. We have to show that $x \in (-1, 1)$.

(Case 1) Suppose $y > 0$. Since $y > 0$, we have $x > 0$. We want to show that $x < 1$. On the contrary, suppose $x \geq 1$. Then $\frac{-1+\sqrt{1+4y^2}}{2y} \geq 1$, or $\sqrt{1 + 4y^2} \geq (2y + 1)$. Squaring both sides, we have $1 + 4y^2 \geq 4y^2 + 4y + 1$. This implies that $y \leq 0$, which is a contradiction to $y > 0$. Hence, we have $x \in (-1, 1)$.

(Case 2) Suppose $y < 0$. Since $y < 0$, we have $x < 0$. We want to show that $x > -1$. On the contrary, suppose $x \leq -1$. Then $\frac{-1+\sqrt{1+4y^2}}{2y} \leq -1$. By multiplying $y < 0$ to the both sides of this inequality, we have $-1 + \sqrt{1 + 4y^2} \geq -2y$. (Please note the change in the inequality direction.) Hence, $\sqrt{1 + 4y^2} \geq (-2y + 1)$. Note that $-2y + 1 > 0$. So, by squaring both sides, we have $1 + 4y^2 \geq 4y^2 - 4y + 1$. This implies that $0 \geq -4y$, which is a contradiction to $y < 0$. Hence, $x \in (-1, 1)$.

Moreover, $f(x) = f\left(\dfrac{-1+\sqrt{1+4y^2}}{2y}\right) = \dfrac{\frac{-1+\sqrt{1+4y^2}}{2y}}{1 - \left(\dfrac{-1+\sqrt{1+4y^2}}{2y}\right)^2} = y$. We leave it to the reader to

verify the last part of this simplification. Therefore, f is 1–1 and onto.

Remark 1.3.1 Let me give a brief **review of trigonometry** (triangle geometry). Draw a circle of the radius $r > 0$ centered at the origin O, and let us call this circle by ω. Let Q be the intersection of ω with the positive x-axis. Next, draw a ray emanating from the origin O, and let P be the intersection of this ray with ω. (See Figs. 1.2 and 1.3.) The arc-length between Q and P on this circle is denoted by $\overset{\frown}{QP}$ with a little twist. If we measure the arc length $\overset{\frown}{QP}$, from Q to P in the counter-clockwise direction, then $\overset{\frown}{QP}$ is a **positive** number. If we measure the arc length $\overset{\frown}{QP}$, from Q to P in the clockwise direction, then $\overset{\frown}{QP}$ is a **negative** number. The angle POQ is denoted by $\angle POQ$, and the angle measurement of it is also denoted by $\angle POQ$, and it is defined to be.

$$\angle POQ = \frac{\overset{\frown}{QP}}{r}.$$

In Fig. 1.2, $r = 5$, $\overset{\frown}{QP} = 12.5$, so $\angle POQ = \frac{12.5}{5} = 2.5$.

In Fig. 1.3, $r = 5$, $\overset{\frown}{QP} = -10.7$, so $\angle POQ = \frac{-10.7}{5} = -2.14$.

This angle measurement of $\angle POQ$ is in radian. An angle measurement is often denoted by θ **radian**, or simply θ. So if we let $\angle POQ = \theta$ radian, then

$$\theta = \frac{\overset{\frown}{QP}}{r},$$

Fig. 1.2 $\angle POQ = \frac{12.5}{5} = 2.5$

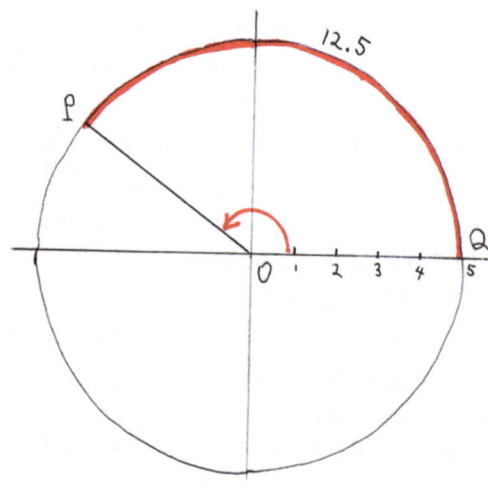

Fig. 1.3
$\angle POQ = \frac{-10.7}{5} = -2.14$

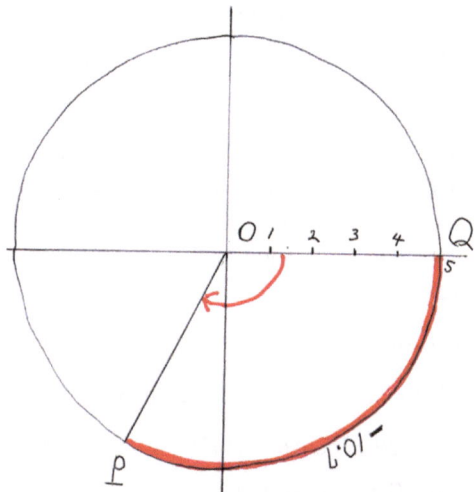

and θ is considered to be a **number.**[2]

Hence, if the angle measurement $\angle POQ = \theta$ is positive, then the positive arc length $\widehat{QP}$ is given by

$$\widehat{QP} = r\theta$$

In particular, if $P = Q$, and if $\widehat{QP}$ is the measurement by tracing the circle ω in the positive direction once around ω from Q to P, then $\widehat{QP}$ is the circumference of ω, and we denote $\angle POQ$ by 2π. This is the definition of the number π (**pi**) radian. Hence, the circumference of a circle of radius r is given by $2\pi r$ by the definition of π.

Let the Cartesian coordinate of the point P to be (x, y). The coordinates of the point Q is $(r, 0)$. Then <u>cosine</u> of θ, denoted by $\cos \theta$, and <u>sine</u> of θ, denoted by $\sin \theta$, are defined by

$$\cos \theta = \frac{x}{r} \text{ and } \sin \theta = \frac{y}{r}.$$

(Whoever originally defined cosine and sine functions, they should have called cosine the sine function, and sine the cosine function because x comes before y in alphabetical order.)

---

[2] Often a trig textbook emphasizes that if $\angle POQ$ is measured to be a degrees, $a°$, then $\angle POQ = \theta = \frac{a}{360}$ is used for the definition of $\angle POQ = \theta$ radian. It is unfortunate, and misleading.

If we measure the angle $\angle POQ$ in degrees, the degree measurement is **not** considered to be a number. So, for example, if 60 <u>degrees</u> is an angle measurement, then it always has to have the **unit** "degrees" or "°" **accompanied**, and hence, 60° is **not** a number. In other words, "60" is a number, but 60 without "degrees" have different meanings. On the other hand, θ **radian** is often simply θ because it is a number. Note that $\cos 60 \neq \cos 60°$. I try to avoid degree measurement of an angle.

So, $x = r \cos\theta$ and $y = r \sin\theta$.

Since $\left(\frac{x}{r}\right)^2 + \left(\frac{y}{r}\right)^2 = 1$ is the Pythagorean theorem on the right triangle OPQ, we have

(1) $\cos^2\theta + \sin^2\theta = 1$.

(Since $\cos^2\theta = (\cos\theta)^2$, $\cos^2\theta$ should be read as "cosine θ **square**" and not "cosine square θ" since it means the multiplication $\cos\theta \cdot \cos\theta$ here. The composite function $\cos(\cos\theta)$ is also written $\cos^2\theta$. If this is the case, $\cos^2\theta$ should be read "cosine square θ.)Here are some additional important trig identities. Let α and β be numbers. Then we have

(2) $\cos(\alpha + \beta) = \cos\alpha \cdot \cos\beta - \sin\alpha \cdot \sin\beta$

By letting $\alpha = \beta = \theta$, (2) becomes.

(2') $\cos 2\theta = \cos^2\theta - \sin^2\theta$.

By $\frac{(1)+(2')}{2}$ and $\frac{(1)-(2')}{2}$ as operations, we have

$\cos^2\theta = \frac{1+\cos 2\theta}{2}$ and $\sin^2\theta = \frac{1-\cos 2\theta}{2}$.

(These identities are useful in integrating $\int \cos^2\theta d\theta$ and $\int \sin^2\theta d\theta$.)

(3) $\sin(\alpha + \beta) = \sin\alpha \cdot \cos\beta + \cos\alpha \cdot \sin\beta$.

By letting $\alpha = \beta = \theta$, (3) becomes

(3') $\sin 2\theta = 2 \sin\theta \cos\theta$.

So, cosine and sine are **functions** $\cos : \mathbb{R} \to [-1, 1]$ and $\sin : \mathbb{R} \to [-1, 1]$, and they are onto but **not** 1–1. For example, $\cos^{-1} 1 = \{\pm 2k\pi : k = 0, 1, 2, \ldots\}$ and $\sin^{-1} 1 = \{\frac{\pi}{2} \pm 2k\pi : k = 0, 1, 2, \ldots\}$. Therefore, the inverse functions of $\cos : \mathbb{R} \to [-1, 1]$ and $\sin : \mathbb{R} \to [-1, 1]$ **do not exist**.

But it is convenient to have "**inverse trig functions**". When we talk about "inverse trig functions", we are **not** talking about inverses of $\cos : \mathbb{R} \to [-1, 1]$ and $\sin : \mathbb{R} \to [-1, 1]$ as they are defined above. We are talking about **different** cosine and sine functions by restricting their domains, and **abusively** we let

$\cos = \cos|_{[0,\pi]} : [0, \pi] \to [-1, 1]$ and $\sin = \sin|_{[-\frac{\pi}{2}, \frac{\pi}{2}]} : [-\frac{\pi}{2}, \frac{\pi}{2}] \to [-1, 1]$ when we talk about inverses of these functions. They are now 1–1 and onto, and their inverses are $\cos^{-1} : [-1, 1] \to [0, \pi]$ and $\sin^{-1} : [-1, 1] \to [-\frac{\pi}{2}, \frac{\pi}{2}]$. By making the notations short, we write $\left[\cos|_{[0,\pi]}\right]^{-1}$ by $\cos^{-1}$, and $\left[\sin|_{[-\frac{\pi}{2}, \frac{\pi}{2}]}\right]^{-1}$ by $\sin^{-1}$. We read $\cos^{-1} x$ as "cosine inverse of x". Similarly, $\sin^{-1} x$ is "sine inverse of x". The functions $\cos^{-1} x$ and $\sin^{-1} x$ are also denoted also by $\arccos x$ and $\arcsin x$, but we do not use these notations in this textbook.

Example 1.3.7 We use a compass and a ruler to sketch the angles $\sin^{-1}(-0.8)$. Let $\theta = \sin^{-1}(-0.8)$ so that $\sin\theta = -0.8 = \frac{y}{r}$ (here, we are using the preceding remark's notations).

Let us draw a semicircle with a radius $10cm$ centered at the origin O as in the figure below. Let $Q = (10, 0)$ be the intersection of the semicircle and the positive x-axis. By using $1cm$ as the unit on the xy-plane, we want to find the point $P = (x, y)$ **on the circle** having a radius of $10cm$ such that $\angle POQ = \sin^{-1}(-0.8)$. Since $\angle POQ = \theta$, we have $\sin\theta = -0.8 = \frac{y}{r}$ and $-\frac{\pi}{2} \le \theta \le \frac{\pi}{2}$. Then P is the intersection of the semicircle on the right-side of the y-axis and the horizontal line $y = -8$ since $-\frac{8}{10} = -0.8$. In other words, the y-coordinate of P is -8 and the x-coordinate of P is positive. (See Fig. 1.4.) Then the angle θ is the negative for the angle $\angle POQ$. By using a thin piece of wire to follow the contour of the semicircle, you can measure the arc length $\overset{\frown}{QP}$ (negative number) in centimeters. You may measure it to be about $-9.5cm$. By doing this, you can numerically approximate $\sin^{-1}(-0.8)$ to be about $\frac{-9.5}{10}$, or $\sin^{-1}(-0.8) \approx -0.95$. The unit is **radian**, not degrees. Using my calculator, I get $\sin^{-1}(-0.8) \approx -0.927$. From Fig. 1.4, by reading the x-coordinate of P, which is approximately 6.1, we can tell that $\cos(\sin^{-1}(-0.8)) \approx \frac{6.1}{10} = 0.61$. Using the Eq. (1), and $\sin(\sin^{-1}(-0.8)) = -0.8$, we have $\cos(\sin^{-1}(-0.8)) = \sqrt{1 - 0.64} = 0.6$.

Problem 1.3.5 Draw circles of radius of $10cm$, and use a compass and a ruler to approximate the following values: $\sin 1$, and $\cos 2$. Angles are measured in radians.

Use circles of radius $10cm$ and a ruler to sketch the following angles: $\sin^{-1}(0.8)$, $\sin^{-1}(-0.7)$, $\cos^{-1}(0.6)$, $\cos^{-1}(-0.3)$. Then approximate these angles using a piece of thin wire and a ruler in radian measurements.

No protractors, please. Use calculators to check your answers. Your answers should be close to the values on your calculators. (We rarely use degree in angle measurements in mathematics.)

Fig. 1.4 The unit 10 does not represent $10cm$ here. The angle $\sin^{-1}(-0.8) \approx -0.95$

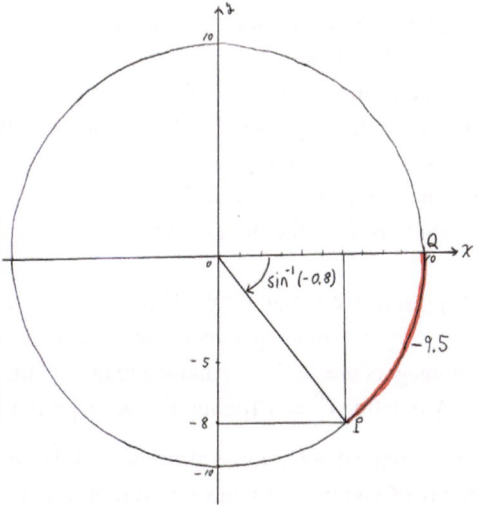

1.4   Mathematical Induction

The numbers 1, 2, 3, ... are called <u>natural numbers</u>. Let $\mathbb{N}$ be the set of all natural numbers. Some mathematicians define natural numbers using set theory from scratch. But we accept them as a basic and natural starting point and do not question their basic operations. Note that we do not consider zero, 0, as a natural number.

Let m, n, and l be elements in $\mathbb{N}$. Then we assume the following:

(N1) We know how to add n and m, meaning $m + n$ is in $\mathbb{N}$, and $m + n = n + m$. (We read "$m + n = n + m$" **as** "m *plus* n is n *plus* m" or "m *plus* n is equal to n *plus* m". For example, we write the element $1 + 1 = 2$, $1 + 2 = 2 + 1 = 3$ and so on.) And the parenthesis in $(l + m) + n$ is to perform the addition $l + m$ first, and then add n afterward. We have $l + (n + m) = (l + m) + n$. It means that the order of additions of three natural numbers does not end up with different numbers so that writing $l + m + n$ makes sense.

(N2) For two natural numbers, we can tell $n < m$ (reads "n is less than m"), or $n > m$ (n is larger than m), or $n = m$, and only one of these three holds. And if $n > m$, we know how to perform subtraction $n - m$ and it is a natural number.

(N3) We know how to multiply m and n, meaning mn is in $\mathbb{N}$, and $mn = nm$. (Sometime we write $m \cdot n$ for mn as in $2 \cdot 3 = 6$ to avoid confusion, for example, to distinguish $2 \cdot 3$ from 23.) And $(mn)l = m(nl)$. So writing mnl for the multiplication of three numbers makes sense. Note that $m \cdot 1 = m$. In particular, $1 \cdot 1 = 1$.

(N4) $m(n + l) = mn + ml$ (<u>distributive rule</u>).

(In a way, it is more important to use this property is in reverse order, $mn + ml = m(n + l)$, a **factorization**. And we often consider a factorization as a simplification. That is, a simplification of $mn + ml$ is $m(n + l)$.)

(N5) Under the usual inequality "$>$" in (N2), if A is a nonempty subset of $\mathbb{N}$, we can always find the smallest element in A. In particular, the smallest element of $\mathbb{N}$ is 1,

(N6) If S is a subset of $\mathbb{N}$ such that 1 is in S, and if an integer n in S implies that $n + 1$ is in S, then $S = \mathbb{N}$.

The property (N5) is called the <u>well ordering principle</u>. The property **(N6)** is called the <u>principle of mathematical induction</u> or simply <u>mathematical induction</u>. The properties (N5) and (N6) are equivalent. We do not prove this here since we want to concentrate on the use of mathematical induction. One use of mathematical induction is to use in definitions as in the next definition.

Definition 1.4.1 Let $r \in \mathbb{R}$. Then $r^1 = r$.

If $n \geq 0$ is an integer such that r^n is defined, then $r^{n+1} = r^n \cdot r$. Hence, r^n is defined for all integers $n = 1, 2, \ldots$, and we read "r^n" the "n-th <u>power</u> of r".

We **define** $r^0 = 1$ for any $r \in \mathbb{R}$. In particular, $0^0 = 1$.

Mathematical induction is also a powerful tool to prove theorems involving natural numbers. Many of you probably have used it before. Unfortunately, many of the proofs of theorems using mathematical induction are **flawed** in their wordings. Let me explain this by mimicking a typical flawed attempt in Example 1.4.2 after explaining mathematical induction in the next example.

Example 1.4.1 In proving a statement using mathematical induction, here is a procedure we should follow:

Let $P_1, P_2, P_3, \ldots$ be a list of statements. We wish to prove that P_n is true for all integers $n \geq 1$.

In order to prove this, we show that

(I_1) the statement P_1 is true,

and then we show that

(I_2) **if** n **is an integer** such that the statement P_n is true, then the statement P_{n+1} is true.

Note that in (I_1), P_1 is shown to be true. So in (I_2), the integer n is replaced by 1, and it implies P_2 is true. Now, in (I_2), the integer n is replaced by 2, and it implies P_3 is true. By continuing this process, we can conclude that P_n is true for all $n \in \mathbb{N}$. This is the principle of mathematical induction. That is, if these two (I_1) and (I_2) are shown, then P_n is true for all $n \in \mathbb{N}$ by mathematical induction.

Let us give an example of a **common mistake**. What is wrong with the next example?

Example 1.4.2 Prove $\sum_{k=1}^{n} k = \frac{1}{2}n(n+1)$ for all natural numbers n.

Wrong Proof: Since $1 = \frac{1}{2} \cdot 1(1+1)$, we see that $\sum_{k=1}^{n} k = \frac{1}{2}n(n+1)$ holds when $n = 1$.

For the induction step, suppose $\sum_{k=1}^{n} k = \frac{1}{2}n(n+1)$ is true. We will prove that $\sum_{k=1}^{n+1} k = \frac{1}{2}(n+1)(n+2)$ is true. ...

This continues, but let me **stop** here since the flaws are in the above two sentences.

Criticism of Example 1.4.2.

The first statement "Since $1 = \frac{1}{2} \cdot 1(1+1)$, $\sum_{k=1}^{n} k = \frac{1}{2}n(n+1)$ holds when $n = 1$" seems to suggest the assertion $\sum_{k=1}^{n} k = \frac{1}{2}n(n+1)$ is **"trivially"** true when $n = 1$. It seems to miss the point of this being a proof (an explanation).

The second statement "For the induction step, suppose $\sum_{k=1}^{n} k = \frac{1}{2}n(n+1)$ is true" is very problematic. First, the letter n is **not** quantified. Secondly, it seems to "suppose $\sum_{k=1}^{n} k = \frac{1}{2}n(n+1)$ for **all** natural numbers n". If it were, it became pointless to continue further. If it were assuming "$\sum_{k=1}^{n} k = \frac{1}{2}n(n+1)$ for **some** natural numbers n", then this would defeat the machinery of mathematical induction, and it is not going to work. So, this sentence "suppose $\sum_{k=1}^{n} k = \frac{1}{2}n(n+1)$ is true" cannot be fixed in this form because it has a **wrong subject** "$\sum_{k=1}^{n} k = \frac{1}{2}n(n+1)$". As you see in ($I_2$) of Example 1.4.1, the subject should be n. The problem is stemming from focusing on the **wrong subject**.

The next example is a correction of the proof of Example 1.4.2. Please pay attention to the underlined **subjects** in the next example in each paragraph. Subjects are underlined.

Example 1.4.3 Prove that $\sum_{k=1}^{n} k = \frac{1}{2}n(n+1)$ for **every** natural number $n = 1, 2, 3, \ldots.$

Proof Suppose $n = 1$. Then $\sum_{k=1}^{n} k = \sum_{k=1}^{1} k = 1$ and $\frac{1}{2}n(n+1) = \frac{1}{2}1(1+1) = 1$. Hence, $\sum_{k=1}^{n} k = \frac{1}{2}n(n+1)$ holds when $n = 1$.

Suppose n is a natural number such that $\sum_{k=1}^{n} k = \frac{1}{2}n(n+1)$ holds. By replacing n by $(n+1)$, we will show that $\sum_{k=1}^{n+1} k = \frac{1}{2}(n+1)((n+1)+1)$ holds. But

$$\sum_{k=1}^{n+1} k = \sum_{k=1}^{n} k + (n+1)$$

$$= \frac{1}{2}n(n+1) + (n+1) \text{ since } \sum_{k=1}^{n} k = \frac{1}{2}n(n+1)$$

$$= (n+1)\left(\frac{1}{2}n + 1\right)$$

$$= \frac{1}{2}(n+1)(n+2)$$

$$= \frac{1}{2}(n+1)((n+1)+1).$$

So, $\sum_{k=1}^{n+1} k = \frac{1}{2}(n+1)((n+1)+1)$ holds.

Therefore, by the mathematical induction, $\sum_{k=1}^{n} k = \frac{1}{2}n(n+1)$ holds for every natural number $n = 1, 2, 3, \ldots.$

Remark 1.4.2 Unfortunately, many mathematics teachers write an inductive proof like the one in Example 1.4.2. So do we have to write an inductive proof like the one in Example 1.4.3? **YES!** Each paragraph can be expressed in a different style, but **the underlined**

subjects in Example 1.4.3 cannot be changed. **If a wrong subject is used in a proof, the proof is simply wrong.**

Problem 1.4.1 Use mathematical induction to prove the following statements:

(1) $\sum_{k=1}^{n} k^2 = \frac{1}{6}n(n+1)(2n+1)$ for every natural number n.

(2) $\sum_{k=1}^{n} k^3 = \left\{\frac{1}{2}n(n+1)\right\}^2$ for every natural number n.

Remark 1.4.3 We can modify **(N6)** "If S is a subset of $\mathbb{N}$ such that 1 is in S, and if an integer n in S implies $n+1$ is in S, then $S = \mathbb{N}$" as follow:

 (N6') Let n be an **integer**. If S is a subset of $\mathbb{Z}$ such that n is in S, and if an integer k in S implies $k+1$ is in S, then $S = \{n, n+1, n+2, \ldots\}$.

 (N6") Let n be an **integer**. If S is a subset of $\mathbb{Z}$ such that n is in S, and if "k is an integer such that $i \in S$ for every $n \le i \le k$" implies $k+1$ is in S, then $S = \{n, n+1, n+2, \ldots\}$.

Problem 1.4.2 Let $x \ne t$ be non-zero numbers. Use mathematical induction to prove that $\sum_{k=0}^{n} t^{n-k}x^k = \frac{x^{n+1}-t^{n+1}}{x-t}$ for all integers $n = 0, 1, 2, 3, \ldots$. ($0^0 = 1$ by definition.)

This can be seen as a factorization of the polynomial

$$x^{n+1} - t^{n+1} = (x - t)\left(\sum_{k=0}^{n} t^{n-k}x^k\right).$$

Theorem 1.4.1 (geometric sum) Let $r \ne 1$ be a number. Then

$$\sum_{k=0}^{n} r^k = \frac{r^{n+1}-1}{r-1} \text{ for all integers } n = 0, 1, 2, 3, \ldots.$$

Proof This is a special case of Problem 1.4.2 by replacing t by 1.

Remark 1.4.4 Even though the statement in Problem 1.4.2 is more general than the one in Theorem 1.4.1, Theorem 1.4.1 is used more frequently. It is fundamental to the geometric series in Chapter 3 and to the power series in Chapter 7. If $r = 1$, then $\sum_{k=0}^{n} r^k = \sum_{k=0}^{n} 1 = n+1$ for every natural number n. Note that $\sum_{k=0}^{n} 1 \ne 1$. This is a common mistake.

Problem 1.4.3 Suppose a_n is a positive number for all natural numbers n. Use the mathe-

matical induction to prove that $\frac{1}{2^n} \sum\limits_{k=1}^{2^n} a_k \geq \left[\prod\limits_{k=1}^{2^n} a_k \right]^{\frac{1}{2^n}}$ for every natural number n. (Here,

$\prod\limits_{k=1}^{2^n} a_k = a_1 \cdot a_2 \cdot a_3 \cdot \cdots \cdot a_{2^n}$.)

Remark 1.4.5 As you might have expected, Problem 1.4.3 can be improved to "Suppose

a_n is a positive number for every natural number n. Then $\frac{1}{n} \sum\limits_{k=1}^{n} a_k \geq \left[\prod\limits_{k=1}^{n} a_k \right]^{\frac{1}{n}}$ for every

natural number n." This generalized statement of Problem 1.4.3 is difficult to prove using mathematical induction. (Try to prove this inequality for $n = 3$ to see the difficulty.) By the way, the inequality has a name. It is called <u>Arithmetic Mean—Geometric Mean Inequality</u>, or, more formally, the Inequality of Arithmetic and Geometric Means. I believe that this inequality was discovered by Isaac Newton.

Problem 1.4.4

(1) Prove the **binomial theorem** (See Theorem 1.2.1 for its statement) using mathematical induction.
(2) Let Y be a nonempty set with finite number of elements. We use $|Y|$ to denote the number of elements in Y. Use mathematical induction to prove that there are $2^{|Y|}$ many subsets of Y. (See Problem 1.2.4.)

Example 1.4.4 This is an application of Theorem 1.4.1. Part (1) is a plan for a retirement. Part (2) can be applied to plan a purchase of a house. Hence, this example is not related to analysis and can be ignored. However, this could be an important lesson for your life. (Example 5.2.5 is another similar problem.)

(1) If you deposit $500 on every first day of the month for 40 years into an account that earns the annual interest of 6% compounded monthly (this means $\frac{6}{12} = \frac{1}{2}\%$ of interest added to the principal each month, and this rate is denoted by i, and $i = \frac{0.06}{12} = 0.005$ in this case), how much will you have accumulated on the last day of the deposit? Think this is what you want to have for your retirement plan! In the mathematics of finance, the amount you will accumulate in the future after n monthly <u>payments</u> (PMT) is called the <u>future value</u> (FV).

In order to solve this, first, use Theorem 1.4.1 to **show** that i, n, FV, PMT are related by

$$FV = PMT \cdot \frac{(1+i)^n - 1}{i}.$$

In order to evaluate the numerical solution, you want to calculate FV by letting $n = 40 \times 12 = 480$ for working 40 years, $i = \frac{0.06}{12}$, and $PMT = \$500$. As you will see after calculation, you will have over a million dollars, $\$1,000,000$, at the time of future retirement.

(2) On the other hand, suppose you borrow $\$1,000,000$ for the purchase of your house at an annual interest rate of 6% compounded monthly. You are to amortize (pay back) the loan by making equal monthly payments for 30 years (12 months times 30 years $= 360$ payments).

(2a) First, we want to find your monthly payment.

The amount you borrow is called a present value (PV). In this problem, $PV = \$1,000,000$ is the present value since it is the number at this starting time. You want to find monthly payment (PMT). In order to solve this, you want to find the relation between PV, PMT, i and n, where i is the monthly interest rate and n is the number of remaining payments. In this problem, $PV = 1,000,000$, $i = \frac{0.06}{12}$, and $n = 360$. We can **show** that PV, PMT, i, n are related by

$$PV = PMT \cdot \frac{1 - (1 + i)^{-n}}{i}.$$

Hence, $PMT = PV \cdot \frac{i}{1-(1+i)^{-n}} . PMT = PV \cdot \frac{i}{1-(1+i)^{-n}}.$

For this problem, you want to find out PMT when $n = 30 \times 12 = 360$, $i = \frac{0.06}{12}$, and $PV = \$1,000,000$. Then you will see that $PMT = \$5995.51$.

(2b) Secondly, if, after 20 years of payments, you decide you want to pay back the remainder in one lump sum, how much do you owe to the bank at that time? (This situation can arise when you want to sell this house and to buy a different house.) You want to find out the amount of money you owe to the bank after 20 years. For this, the above equation in (2a) is again useful. Interpret PV to be the amount you presently owe after making $20 \times 12 = 240$ monthly payments of $PV = \$5995.51$. Hence, $360 - 240 = 120$ many payments remains. You let $PMT = \$5995.51$, $i = \frac{0.06}{12}$. However, n becomes $n = 120$ in this case. Hence, you have to pay

$PV = PMT \cdot \frac{1-(1+i)^{-n}}{i} = \$540,036.30$ to the bank in a lump sum.

Example 1.4.4 **(2).** can be viewed differently as that upon retirement, you want the bank to turn the million dollars you saved into an interest accruing fund to be dispersed back to you in equal monthly payments for the remainder of your life expectancy (which banks are willing to do), let's say 30 years at an annual 6% interest rate compounded monthly. Then you would receive the payment of $PMT = \$5995.51$ a month from the bank. But then, after 20 years, you suddenly die. Then the fund can be paid to your heir in a lump sum. This is the present value of the remaining 10 years' payment of your retirement fund, which is $PV = PMT \cdot \frac{1-(1+i)^{-n}}{i} = \$540,036.30$, where $PMT = \$5995.51$, $i = \frac{0.06}{12}$ and $n = 120$.

(Please note that these are all approximations since the interest rate does not remain constant in the real life. But these are good approximations since the annual interest rate of 6% is a very modest rate. Most likely, you will earn more than the annual interest rate of 6% compounded monthly.)

Real Numbers

2.1   Introduction

What is a real number? This seems so trivial yet it is very difficult to answer. And it is so fundamental to mathematics that we feel we should have a simple answer to this question, yet the answer to this is more difficult than you think. As for me, this is the most difficult chapter to write.

As we have seen in Sect. 1.4 on mathematical induction, the natural numbers (all positive integers) are the basis of our discussion and the basis for the construction of real numbers. This question "What is a real number?" did not arise until late nineteenth century when Georg Cantor was thinking about Set Theory and the cardinality of sets (see Theorem 2.4.2). Richard Dedekind, a friend of G. Cantor, constructed real numbers using what people now call the Dedekind cut. I do not think Cantor and Dedekind collaborated on their research, but they were close friends and they corresponded frequently with each other. You may wish to read *Essays on the Theory of Numbers* [1] by Richard Dedekind, The Open Court Publishing Company, 1948. It is also available from Dover Publisher.

Here, since the introduction of the Dedekind cut will take too much time and it is beyond the scope of this textbook and beyond the scope of the interest of most people, we introduce the real numbers by first constructing integers from natural numbers, then rational numbers, and using rational numbers to "construct" real numbers.

We will talk about the historical development of real numbers as we proceed in this chapter.

© The Author(s), under exclusive license to Springer Nature Switzerland AG 2025   31
H. Katsuura, *Introduction to Analysis*, Synthesis Lectures on Mathematics & Statistics,
https://doi.org/10.1007/978-3-031-67954-4_2

2.2   Zero and Integers

Natural numbers, as we have introduced in Sect. 1.4 have been known for a long time. The history of natural numbers is very old, probably as old as human civilizations. Knowing some of the history of real numbers might make this chapter more inviting and easier to understand.

First, the notation used to denote natural numbers by ancient civilizations was not like what we use today. It is difficult to explain their number systems and writing of their numbers since we hardly use them any longer. But we are slightly familiar with an ancient way to write numbers because Roman numerals are still in use. So, let us use Roman numerals to explain what the ancient Romans had to cope with before 1200 A.D., which was around the time when the new number system was introduced in Europe. The following is an example of Roman numerals.

$$I = 1, II = 2, III = 3, IV = 4, V = 5, VI = 6, VII = 7, VIII = 8, IX = 9, X = 10,$$

$$XI = 11, XII = 12, XIII = 13, XIV = 14, XV = 15,$$
$$XVI = 16, XVII = 17, XVIII = 18, XIX = 19, XX = 20,$$

$$XXIX = 29, XXX = 30, XXXIX = 39, XL = 40, XLIX = 49,$$
$$L = 50, LIX = 59, LX = 60,$$

$$LXIX = 69, LXX = 70, LXXIX = 79, LXXX = 80,$$
$$LXXXIX = 89, XC = 90, XCIX = 99,$$

$$C = 100, CC = 200, CCLXXXVII = 288, CCC = 300,$$
$$CD = 400, D = 500,$$

(Some people refer to a \$100 bill as a C note.)

$$DC = 600, DCLXXXVIII = 688, DCC = 700, DCCC = 800,$$
$$CM = 900, M = 1000.$$

As this example shows, it is not obvious how we can write 571 or 905 using Roman numerals, let alone, any numbers larger than one thousand. It takes a highly skilled and educated person to read and write numbers using Roman numerals.

Now, let us write some simple arithmetic, as they might have done in their time.

$$XVII + XIX = XXXVI \text{ (translation : } 17 + 19 = 36)$$

We cannot perform $XVII - XIX$ since XVII is less than XIX.
(Negative numbers did not exist until relatively recently.)

$$XVII - IX = VIII \ (17 - 9 = 8)$$

$$IV \times VIII = XXXII \ (4 \text{ times } 8 \text{ is } 32)$$

Firstly, it is also interesting to note that the notations $+, -, \times, =$ we used here did not exist in Roman times. These symbols might have been introduced during the Renaissance. So, for example, they wrote "XVII plus XIX is XXXVI" for "$XVII + XIX = XXXVI$".

Secondly, I have computed these numbers by first translating Roman numerals into our number system and translating them back to Roman numerals. But the ancient Romans did not have our number system. So, they had to calculate these arithmetic operations using only Roman numerals. In order to simplify computations, some of the privileged people had abacuses. The abacuses that survived till now in museums are made of cast bronze and must have been very expensive. It seems that only privileged Romans could have afforded them.

Thirdly, the biggest disadvantage of using Roman numerals is that there is a limit to the largest number that can be expressed in Roman numerals. Notice that $I = 1, V = 5, X = 10, L = 50, C = 100, D = 500, M = 1000$. In Roman numerals, as the number gets large, new letters must be assigned, and sooner or later, we run out of letters. For example, I have no idea how Romans wrote 1,000,000,000 in Roman numerals. At least, we can read it "one billion", and conceivably Romans had a word and a letter for it at that time. But what about 10^{100}, a number 1 followed by one hundred 0 s? We cannot read it in the same way we read 10^9 as one billion. While we can express any large natural numbers as we wish in terms of the current system, that was not the case for ancient Romans. As the alphabet letters ran out, there was no way to express very large numbers like 10^{100}. For most of the ancient Romans, the largest number they could imagine was limited to the largest number they could express using alphabetic letters.

The ancient Greeks had a number system similar to the ancient Romans. At the time of Archimedes (about 250 B.C.), the largest number of the type 10^n was 10,000, one myriad. So, to most Greeks at that time, one myriad was the largest natural number. Archimedes wrote the famous book, *The Sand Reckoner* [2]. One of the reasons he wanted to write this book was to convince people that there are infinitely many natural numbers. You might like to web-search the book and read Archimedes' introduction of the book. I found it to be very interesting in many ways.

At any rate, the ancient Romans and Greeks knew all of the six properties of natural numbers I have listed in Sect. 1.4 as well as having an understanding of **fractions** of two positive integers. For example, they knew that natural numbers were not adequate for dividing a loaf of bread. So they could calculate additions and multiplications of two fractions of integers as in $\frac{22}{7} + \frac{7}{5} = \frac{110+49}{35} = \frac{159}{35} = 4 + \frac{19}{35}$ and $\frac{22}{7} \times \frac{7}{5} = \frac{22}{5}$.

Now let us examine the reasons why our number system is advantageous.

Unlike Roman numerals, we can express any number using only 1, 2, 3, 4, 5, 6, 7, 8, 9, and 0. And here, I would like you to pay attention to 0, **zero**. (The number 0 is **not** a natural number in our definition because historically the concept of 0 and the symbol 0 was only introduced/invented in Southeast Asia (probably Cambodia according to [3]) at about 500 A.D., and it did not arrive to Europe until about 1200 A.D.) Because of 0, we can write any number by placing these ten digits in the same way any positive integers can be represented on a Chinese, Greek, Roman, or a Japanese **abacus**. On a Chinese abacus, the beads on each column represent a digit, and move on to the left (could be to the right) as the number gets larger. In exactly the same way, our number system also relies on the **position** of these digits to indicate the number the symbols represent. This positional notation of a number became practical by the invention of the notation for **zero**, 0.

Actually, there was a similar positional notation of numbers in ancient times in Mesopotamia (currently Iraq). The Mesopotamians did not have zero, so they left a space unfilled where they should have written 0, and they used base 60 instead of base 10. Dividing an hour into 60 min, and dividing a circle into 360 degrees are the Mesopotamian influences on our own system of measurement.

The number zero was a difficult concept for many ancient people. The reason Archimedes wrote *The Sand Reckoner* was to show the existence of infinitely many natural numbers without the use of zero. Actually, *The Sand Reckoner* is a very difficult book to read after its introductory section. It is difficult to read because there was no commonly agreed way to write large numbers at that time and, consequently, Archimedes had to **improvise** a way to write large numbers. I do not have the patience to learn his improvised way to write numbers. He could have written a better book if he had invented or devised a number system similar to ours which includes a 0, but he did not. So, the purpose of *The Sand Reckoner* was to show the existence of very large numbers, but it had hardly any impact in advancing arithmetic, while it had other unintended major influences.[1] And the downfall of ancient Greek mathematics, in my opinion, is that its mathematicians lacked a better number system that included **zero**.

---

[1] In the book *The Sand Reckoner*, Archimedes mentions an astronomer named **Aristarchus** who measured the distance from the earth to the sun in terms of the earth's radius. (See *Mathematical Methods in Science* by G. Polya, Mathematical Association of America, page 10,1977 [4].) Archimedes says that Aristarchus' measurement of the distance to the sun is about 19 times the radius of the earth. Archimedes thought that Aristarchus' measurement was too large and also unbelievable since Aristarchus was a crazy astronomer who thought the earth is rotating on its axis once a day and goes around the sun once a year. **Copernicus** read *The Sand Reckoner* and obtained the idea for writing his famous book *On the Revolution of Celestial Orbs* in 1543 [5].

Axiom 2.2 There is no largest natural number.

In multiplication, we pointed out in (N3) of Sect. 1.4 that $n \cdot 1 = n$ for any $n \in \mathbb{N}$, and 1 is called the <u>multiplicative identity</u>. However, there is no natural number $x \in \mathbb{N}$ such that $n + x = n$ for any $n \in \mathbb{N}$. So, we introduce a number 0, **zero**, such that $n + 0 = n$ for any $n \in \mathbb{N}$. The number zero, 0, is sometimes called the <u>additive identity</u>. Zero, 0, is an integer, $0 \in \mathbb{Z}$, but not a natural number, $0 \notin \mathbb{N}$.

As we said earlier, ancients knew about fractions of integers, $\frac{1}{2}, \frac{1}{3}, \frac{2}{3}, \ldots$ In general, if p and q are natural numbers, then they knew how to solve an equation $px = q$, and the answer is $x = \frac{q}{p}$.

It is natural to ask to find a number x such that $2 + x = 9$, and the answer is $x = 9 - 2 = 7$. It is something Archimedes had no problem answering. But finding a number x such that $9 + x = 2$ is something Archimedes could not calculate because without the concept of zero, it is impossible to conceive of <u>negative numbers</u> or <u>additive inverses</u>. By adding 0 to the set of the natural numbers, it becomes natural to introduce negative integers. Archimedes probably felt the need for something like these numbers, but he could not pinpoint the **subject** to his problem. He could not break the wall of 0 to reach the negative number $x = 2 - 9 = -7$. But, with the addition of 0, mathematicians knew exactly what to do. We expanded the set $\mathbb{N}$ of <u>natural numbers</u> 1, 2, 3, ... to the set $\mathbb{Z}$ of integers.

$$\cdots, -3, -2, -1, 0, 1, 2, 3 \ldots$$

If n is a natural number, and the solution x to the equation $n + x = 0$ is said to be the additive inverse of n, we denote the solution x by $-n$, and we call $-n$ the <u>additive inverse</u> of n. It is natural to order these numbers as in

$$\cdots < -3 < -2 < -1 < 0 < 1 < 2 < 3 < \cdots$$

by placing these numbers on the <u>number line</u>. We can think of $2 - 9$ as an addition $2 + (-9) = 2 - 9 = -7$ as in moving 9 units to the left of 2 on the number line to reach -7. And $n + 0 = n$ for every $n \in \mathbb{Z}$. In particular, $0 + 0 = 0$.

The properties of natural numbers (N1)–(N4) in Sect. 1.4 naturally extend to the following:

Let $l, m, n \in \mathbb{Z}$.

(Z1) We know how to add m and n, and $m+n$ is in $\mathbb{Z}$, and $m+n = n+m$. The addition $(l + m) + n$ means that the term inside of the parentheses $(l + m)$ is added first. And we have $(l + m) + n = l + (m + n)$. This means that the order of additions of three natural numbers does not affect the sum so that writing $l + m + n$ makes sense.

(Z2) If two integers m and n are given, then we can say $n < m$ (reads "n is less than m"), or $n > m$ (n is larger than m), or $n = m$, and only one of these three holds.

Note that $n > m$ if, and only if, $n - m > 0$.

(Z3) We know how to multiply two integers m and n. We write m times n by $mn = m \cdot n$. It is an integer $\mathbb{Z}$, and $mn = nm$. If l is another integer, then $l(mn) = (lm)n$. So writing lmn for the multiplication of three numbers makes sense. Note that $n \cdot 1 = n$.

(Z4) Let $l, m, n \in \mathbb{Z}$. Then $l(m + n) = lm + ln$. This is called the <u>distributive</u> rule, and the important way we use it is $lm + ln = l(m + n)$, which is called a <u>factorization</u>.

(Z5) The <u>additive inverse</u> of $n \in \mathbb{Z}$ is the number $x \in \mathbb{Z}$ such that $n + x = 0$. For example, since $7 - 7 = 7 + (-7) = 0$, -7 is the additive inverse of 7. And $-(-7) = 7$, so 7 is the additive inverse of -7. If n is an **integer**, we denote the <u>additive inverse</u> of n by $-n$. So, this implies that $-(-n) = n$. (Yes, we could have proved the uniqueness of the additive identity (0) and the uniqueness of the additive inverse of n denoted by $-n$. But please allow us to use it without a proof.)

A positive number is a number larger than zero. A <u>negative</u> number is a number less than zero. As you can see from here, $-n$ may not be a negative number.

(Z6) Let $n \in \mathbb{Z}$. Then $n \cdot 0 = n \cdot (0 + 0) = n \cdot 0 + n \cdot 0$. Hence, $n \cdot 0 = 0$ for any $n \in \mathbb{Z}$. So in particular, we have $0 = n \cdot 0 = n \cdot [(-1) + 1] = n \cdot (-1) + n \cdot 1 = n \cdot (-1) + n$ so that $n \cdot (-1) = -n$ for any $n \in \mathbb{Z}$.

It was once thought 0 was discovered in India. But a recent archeological study seems to suggest that Cambodian (see [3]) introduced the concept of zero. So, it was somewhere in Southeast Asia where 0 was first discovered. The time of the discovery was before 500 A.D. On the other hand, the Mayan civilization in Central America had zero (written differently from 0), and before 378 A.D. (see Gugliotta, G. (2007). The Maya Glory and Ruin, *National Geographic*, *212*(2), 68–73 [6].) So, the zero in Maya is likely older than that of Southeast Asia, but it did not contribute to the development of our number system in Europe.

2.3   Rational Numbers

The ancient Romans and Greeks could multiply natural numbers. So, they asked about the number x such that $7x = 3$, and thought of a fraction of two natural numbers and said the answer was $x = \frac{3}{7}$. In general, they thought **all** numbers can be written in the <u>fraction</u> form $\frac{m}{n}$ for some $m, n \in \mathbb{N}$.

We have introduced integers in Sect. 2.2, so we can ask about the solution x to the equation $px = q$ when $p, q \in \mathbb{Z}$. Since $q \cdot 0 = 0$ for any $q \in \mathbb{Z}$, solution to this equation $px = q$ is $x = \frac{q}{p}$ when $p \neq 0$. We say a number of the form $\frac{q}{p}$, where $p \neq 0, q \in \mathbb{Z}$, is called a <u>rational</u> number, and we use $\mathbb{Q}$ to denote the set of all rational numbers.

We know a lot about rational numbers. Let $a, b, c \in \mathbb{Q}$.

(Q1) We know how to add a and b, $a + b$ is in Q, and $a + b = b + a$. Note that we can add two numbers at a time. So, we make sure $(a + b) + c = a + (b + c)$. This expression $(a + b) + c = a + (b + c)$ means that the order of additions of three rational numbers does result in different sums so that writing $a + b + c$ makes sense. However, we usually take it to mean addition from left to right. Note that $a + 0 = a$. That is, 0 is the <u>additive identity</u>.

(Q2) We can say $a < b$, or $a > b$, or $a = b$, and only one of these three holds. And $a > b$ if, and only if, $a - b > 0$.

(Q3) We know how to multiply a and b. The multiplication of a and b is written by ab, or $a \cdot b$. As for properties of multiplications, we have $ab = ba$, and $(ab)c = a(bc)$. So writing abc for the multiplication of three numbers makes sense. Note that $a \cdot 1 = a$ and $a \cdot 0 = 0$. The number 1 is the <u>multiplicative identity</u>.

(Q4) $a(b + c) = ab + ac$, this is called the <u>distributive rule</u>, and $ab + ac = a(b + c)$, called a <u>factorization</u>.

(Q5) If $a + b = 0$, then b is the <u>additive inverse</u> of a, and b is denoted by $-a$. And $-(-a) = a$.

(Q6) $a \cdot (-1) = -a$

(Q7) If $ab = 1$, then we write $b = \frac{1}{a} = a^{-1}$, and it is the <u>multiplicative inverse</u> of a (necessarily $a \neq 0$). So $\frac{b}{a} = ba^{-1}$. If $n \in N$, then the multiplication of a n-times is written by $a \cdot a \cdot a \cdots \cdots a = a^n$ and $a^{-n} = \left(a^{-1}\right)^n = \frac{1}{a^n} = (a^n)^{-1}$. We define $a^0 = 1$ (including $0^0 = 1$) by convention. Hence, if $m, n \in Z$, we have $a^{mn} = (a^m)^n = (a^n)^m$, $a^{m+n} = a^m a^n$, and $(ab)^n = (a)^n (b)^n$.

(Note that if $r \in Q$ and $r \neq 0$, we are **not** defining a^r yet.)

(Q8) If $a > b$ or $a = b$, we write $a \geq b$. So, if $a \leq b$ and $a \geq b$, then $a = b$.

(Q9) If $a \geq b$ and $b \geq c$, then $a \geq c$.

(Q10) If $a \geq b$, then $a + c \geq b + c$.

(Q11) If $a \geq b$ and if $c \geq 0$, then $ac \geq bc$.

(Q12) If $a \geq b > 0$, then $0 < a^{-1} \leq b^{-1}$.

Problem 2.3.1 Let $a, b \in Q$. Which of following statements are false? Give counterexamples to show if they are false.

(1) $a^2 \geq 0$.

(2) Let $a \neq 0, b \neq 0$. If $a \geq b$, then $a^{-1} \leq b^{-1}$.

(3) If $a \geq b$, then $a^2 \geq b^2$.

(The use of inequalities is central to analysis. Here, I am showing you some common mistakes we have tendencies to make.)

Remark 2.3.1 Two distinct rational numbers can be compared. Find out which is larger, $\frac{32}{7}$ or $\frac{41}{9}$? In order to see this, we have to perform subtraction. $\frac{41}{9} - \frac{32}{7} = \frac{41 \cdot 7 - 32 \cdot 9}{9 \cdot 7} = \frac{287 - 288}{63} = -\frac{1}{63}$. So $\frac{41}{9} < \frac{32}{7}$. But having zero, and having the <u>division algorithm</u>, we have $\frac{32}{7} = 4.57 \cdots$ and $\frac{41}{9} = 4.55 \cdots$. So we can conclude that $\frac{41}{9} < \frac{32}{7}$. Another advantage of having zero. But it also raises a question. **What** does the three dots "$\cdots$" in "4.57$\cdots$" mean?

2.4   Contribution of Zero to Mathematics

The number 0 and probably along with it all the knowledge discussed in Sects. 2.1–2.3 arrived in the Arab nations from Southeast Asia through India in about 800 A.D. We know this because, at about 800 A.D., Al Khwarizmi, who was a Persian (Iranian) and later moved to Baghdad, made significant advances in his work on rational numbers, and wrote a book in Arabic titled *Hisab Al-jabr w'al-muqabala* [7], and established the foundation for algebra derived from the word "*Al-jabr*". Many people praise his discovery of the solution $x = \frac{-b \pm \sqrt{b^2 - 4ac}}{2a}$ to the quadratic equation $ax^2 + bx + c = 0, a \neq 0$, supposedly written in this book. It is admirable, and I agree with that. However, to me, the most significant contribution of Al Khwarizmi is the **division algorithm**[2] (which we used in the previous section to show that $\frac{32}{7} = 4.57 \cdots$), and the subsequent decimal representations of fractional numbers. You know what the division algorithm is. Essentially, the division algorithm is a process of converting a fraction of two integers into a decimal number, as in $\frac{1}{2} = 0.5, \frac{231}{8} = 28.875, \frac{32}{7} = 4.57\cdots$, and so on. Because of our heavy use of calculators these days, most of us are rusty when it comes to using the division algorithm. It is unfortunate that we no longer pay much attention to the division algorithm. What is its significance? A key is in $\frac{32}{7} = 4.57\cdots$, or in $\frac{22}{7} = 3.142857\cdots$. What is the meaning of the dots "$\cdots$" in "3.142857$\cdots$"? It can be interpreted in two ways, either it is a sequence 3.142, 3.1428, 3.14285, 3.142857, … or it is an infinite series $3 + \frac{1}{10} + \frac{4}{10^2} + \frac{2}{10^3} + \frac{8}{10^4} + \frac{5}{10^5} + \frac{7}{10^6} + \cdots$. In other words, I think that Al Khwarizmi's work led to the notions of infinite sequences and infinite series. He was able to devise the division algorithm because of two reasons, zero and the positional notation of a number using ten digits as a consequence of the discovery of zero.

Let us go back in history to the time of Pythagoras, at about 500 B.C. His main accomplishment was the Pythagorean theorem, which states that if a, b, c are the side lengths of a right triangle where c is its hypotenuse, then $a^2 + b^2 = c^2$. This was probably known before him. Probably, his significant contribution was proving this theorem. From this, we can calculate the length of the diagonal of the unit square (the square having four sides of length 1) to be $\sqrt{2}$, a positive number whose square is 2. But then, the Pythagorean school discovered the next theorem. The number $\sqrt{2}$ is read root two, or square root two.

---

[2] The origin of the word "algebra" is "*Al-jabr*", and the word "algorithm" is derived from "Al Khwarizmi".

Theorem 2.4.1 $\sqrt{2} \notin \mathbb{Q}$. That is, $\sqrt{2}$ is not a rational number.

Proof Suppose $\sqrt{2} \in \mathbb{Q}$. Then there exist two positive integers p, q such that $\sqrt{2} = \frac{p}{q}$ and the greatest common divisor of p and q being 1. So $2 = \frac{p^2}{q^2}$, or $2q^2 = p^2$. Since 2 is a prime number, 2 divides p^2. Hence, 2 divides p, i.e., $p = 2k$ for some positive integer k. So $2q^2 = p^2 = (2k)^2 = 4k^2$, or $q^2 = 2k^2$. As before, this implies that 2 divides q. This shows that a common divisor of p and q is 2. This is a contradiction to the greatest common divisor of p and q being 1. Therefore, $\sqrt{2} \notin \mathbb{Q}$.

This is probably the proof given in 500 B.C. It uses the fact that 2 is a prime number. There are numerous other proofs of this theorem. Let me introduce you to yet another alternate beautiful proof by Richard Dedekind (1831–1916) who thought deeply about real numbers, and therefore about real analysis, in a rigorous way. The significance of the next proof is that it does not use the fact that 2 is a prime number. Instead, Dedekind uses the well ordering principle of natural numbers, illustrated below.

Proof of Theorem 2.4.1: Suppose $\sqrt{2} \in \mathbb{Q}$. Then there exist two positive integers p, q such that $\sqrt{2} = \frac{p}{q}$ and the greatest common sdivisor of p and q being 1. So $2 = \frac{p^2}{q^2}$, or $p^2 - 2q^2 = 0$. By the well-ordering principle, we can choose q to be the smallest such positive number. Note that $q^2 < 2q^2 = p^2 < 2p^2 = 4q^2$. This shows that $q < p < 2q$. Let $p' = 2q - p$ and $q' = p - q$. Then $q' = p - q < 2q - q = q$, and $p'^2 - 2q'^2 = (2q - p)^2 - 2(p - q)^2 = (4q^2 - 4pq + p^2) - 2(p^2 - 2pq + q^2) = -(p^2 - 2q^2) = 0$. This is a contradiction to the minimality of q. (For more information, see an article "Irrational Thoughts" by Harold P. Boas, *Mathematics Magazine*, Vol. 93, Feb. 2020, 23–26 [8].)

We state the next theorem without a proof.

Theorem 2.4.2 Let m and n be two positive integers. Suppose that there does not exist an integer $k > 0$ such that $k^n = m$. Then there does not exist a rational number $r > 0$ such that $r^n = m$. That is, $\sqrt[n]{m} = m^{\frac{1}{n}} \notin \mathbb{Q}$.

We denote a "positive number" x such that $x^n = m$ by $\sqrt[n]{m} = m^{\frac{1}{n}}$. We read the number $m^{\frac{1}{n}}$ the n-th root of m. This theorem says that if n is a natural number, then $m^{\frac{1}{n}}$ is either an integer or an irrational number for every natural number m.

Suppose $m^{\frac{1}{n}}$ is not an integer for some natural numbers m and n. Then **I have not established** the existence of $m^{\frac{1}{n}}$, (yes, since I have not proven this, it may be possible that $m^{\frac{1}{n}}$ does not exist) so using this notation "$m^{\frac{1}{n}}$" here is misleading. Writing $\sqrt{2} \notin \mathbb{Q}$ in Theorem 2.4.1 is misleading, too. However, since I will prove the existence of $2^{\frac{1}{2}}$ in Theorem 2.5.1, and since I will prove the existence of $m^{\frac{1}{n}}$ in Sect. 3.8 in general, I am entering them into the discussion here.

Definition 2.4.1 A real number that is not a rational number is called an <u>irrational number</u>.

Definition 2.4.1 is a strange one since we have not defined what a real number is yet, and in our discussion, we do not yet know the existence of any irrational number. However, Theorem 2.4.1 and the Pythagorean theorem give a plausible reason to believe of the possible existence of an <u>irrational</u> number,[3] namely $\sqrt{2} = 2^{\frac{1}{2}}$. This is a great moment in mathematical history. However, Pythagoras and his group spent a great deal of effort in trying to hide this theorem from the public even though their effort to hide it did not succeed for long.

Why did Pythagoras and his group try to hide that $\sqrt{2}$ is an irrational number? It was because they thought that the set of all numbers was the set of rational numbers $\mathbb{Q}$, and therefore, they did not know what to make of $\sqrt{2}$. Does $\sqrt{2}$ exist? Because they did not have 0, they did not have the division algorithm, they could not convert a fraction of integers into a decimal number, and they did not have and could not have successively approximated $\sqrt{2}$ as in $\sqrt{2} = 1.4142 \cdots$. In other words, successive approximations of $\sqrt{2}$ in terms of reduced fractions of integers is very difficult and they probably did not try to find a systematic way of doing it. So, this is the reason why I praise Al Khwarizmi. His division algorithm as well as his and other's contributions to the development of algebra laid the groundwork for future mathematics while Greek mathematics, lacking the zero, was struggling to come up with new ideas. Having decimal notation, we can successively approximate as follows:

$$1 < \sqrt{2} < 2, \; 1.4 < \sqrt{2} < 1.5, \; 1.41 < \sqrt{2} < 1.42, \; 1.412 < \sqrt{2} < 1.413,$$

and so on, one digit at a time. The Greek mathematicians could have done it by fractions as in $1 < \sqrt{2} < \frac{3}{2}$, and $\frac{5}{4} < \sqrt{2} < \frac{3}{2}$. But after this point, it is not easy to continue this process.

At about the time of Al Khwarizmi, paper-making technology had spread to Arab nations from China, and that made the dissemination of new ideas somewhat easier. Before making paper, papyrus was used in the Nile delta and was imported from Egypt into the Mediterranean region. But an Egyptian pharaoh, probably Ptolemy II, embargoed the export of papyrus around 300B.C. in order to monopolize books in the Library of Alexandria. After that sheep skins made into parchment as a substitute for papyrus to make books.

It was about the year 1200 A.D. when algebra was brought to Europe from Arab countries. The translation of *Al-jabr* was already in Europe by then. But it was made widely known mainly by a son of a wealthy merchant from Pisa, Leonard Pisano, more commonly known as Fibonacci, who traveled widely in the Mediterranean, acquired the knowledge of algebra while traveling in Arab nations, and wrote *Liber Abaci (Book of Calculation)* in 1202 A.D., which introduced the Hindu-Arabic numerals including division

---

[3] The number π is a transcendental number, and so it is necessarily an irrational number, and its existence is known before Pythagoras. But π was not known to be an irrational number until after Newton. The notation π was probably due to Euler. It was proven to be a transcendental number in 1882 by Ferdinand Lindemann.

algorithm and the decimal numbers of Al Khwarizmi–all this was only possible because of the positional notations of numbers with **zero**. You might be interested in looking into the translation of *Liber Abaci* in English published in 2002 by Springer Verlag [9]. Besides *Liber Abaci*, Fibonacci is also known for introducing the *Fibonacci numbers*.

As a side note, it is interesting to note the construction in Florence of Basilica di Santa Maria del Fiore, widely known as Duomo because of its dome roof, in Florence. The construction of Duomo started in 1296 and was completed in 1436. Because of the technical difficulty in building the dome roof, it was left without a roof for many years. Some historians consider the building of this dome to be the start of the Renaissance. Florence is a city in Italy not too far from Pisa. Is it too much to imagine that Fibonacci brought paintings and sketches of Islamic mosques in Arab countries, and these paintings influenced the design and construction of Duomo?

The Renaissance brought many books to Europe including *The Sand Reckoner* by Archimedes (287–212 B.C.), which I have mentioned earlier. The essential point of *The Sand Reckoner* is to point out the deficiencies of the number system used at that time, but Archimedes was unable to provide a useful and practical solution to cope with the deficiencies because he was unable to imagine the number zero.

The importation of zero made the new way to write numbers possible, and led to significant advances in Europe. Tycho Brahe's observations of planets spreading over 30 years was recorded in this new numerical system. Using Brahe's observations, Kepler was able to derive the three laws of planetary motion based mainly on Brahe's observational records of Mars and published them in *The Epitom of Copernican Astronomy* [10] in about 1620. One of the laws states that the path of a planet is an ellipse. Here, the length of 30 years of observation is important since it takes about 30 years for Mars to revolve once around the sun. And Galileo discovered that the path of a thrown object in the air at an angle to be a parabola, which involved calculations made easier using the new number system. (Interestingly, the ellipse, hyperbola, and a parabola, which are curves obtained by slicing a cone, and are called conic sections, were studied and named by Apollonius of Perga at about 200B.C (*Conics: Book I–IV*, Green Lion Press 2013 [11]). (Galileo also created a sensation in astronomy by his telescopic observations.) From Galileo and Kepler's work (Kepler's three laws of planetary motion), and from Fermat's idea of differential calculus, Newton developed the physics of mechanics. Newton developed the theory of the universal gravitational force, succeeding in combining Galileo's physics and Kepler's astronomy. In particular, Newton was able to deduce his universal gravitational force F given by $F = G\frac{Mm}{r^2}$ between two objects having masses M and m, and having distance r between them, where G is the gravitational constant. Newton used this formula to improve Kepler's third law of planetary motion to $T^2 = \frac{4\pi^2}{GM}R^3$, where T is the period of time it takes for a planet to move around another much bigger star/planet of the mass M with the average radius R. What is the significance of this? We can measure T and R of the earth moving about the sun, for example. So, by solving $T^2 = \frac{4\pi^2}{GM}R^3$ for M to obtain $M = \frac{4\pi^2}{GT^2}R^3$, we can approximate the mass of the sun. It is a sensational equation!

And Newton together with Leibniz discovered the fundamental theorem of calculus.

All of these achievements were possible because of zero and the subsequent developments in algebra.

All of these advances made possible not only the theory of calculus but also the mathematical analysis that we aim to introduce you to in this text. Yes, the idea of limits is used by ancient Egyptians and Greeks. Eudoxus of Cnidus (410–355 B.C.) seems to be the first person who applied the method of exhaustion to calculate the volume of a circular cone. Some authors give credit to Democritus (460–370 B.C.). But generally, people seem to credit Archimedes (290–212 B.C.) with the whole of their collective efforts. This is probably because of his fame for his study of circles, spheres, and discoveries in physics. Then came Apollonius who discovered the conic sections (ellipse, parabola, and hyperbola) at about 200 B.C. as I have mentioned earlier, and the ellipse was later used to describe the orbit of a planet by Kepler. Yet the Greek mathematicians did not advance much beyond these discoveries.

Why did Greek mathematics decline and die? Once again, it seems to me the only reason for this is that they had an inferior numerical notation without zero. The discovery of zero and improvements in writing numbers in Cambodia in the sixth century eventually spread to the Arab nations in the ninth century, making possible the discovery of algebra. When those developments finally arrived in Europe about 1200 A.D., they laid the foundation for the Renaissance in Europe, furthering the study of mathematics, astronomy, and physics. Fermat laid the foundation of differential calculus. Then Newton and Leibniz, independent of each other, made a connection between the differential and integral calculus as stated in the fundamental theorem of calculus in the seventeenth century. The eighteenth century mathematicians under the leadership of Euler pushed forward the study of mathematics in ways that made mathematicians of the nineteenth century realize the need for a formal definition of "limits". Cauchy, in particular, has contributed to this development. Then a crisis hit the mathematical society in the late nineteenth century. Georg Cantor (1845–1918) proved that there is no onto function from the set $\mathbb{N}$ of all natural numbers to the set $\mathbb{R}$ of all real numbers (see Theorem 2.4.2). Both sets $\mathbb{N}$ and $\mathbb{R}$ have infinitely many elements, but $\mathbb{R}$ has "more" elements than the set $\mathbb{N}$. In a way, Cantor showed that there are many levels of infinities. This was a great moment in mathematics, yet this caused a crisis in mathematics because until then people thought they had a thorough understanding of real numbers, but now realized how little they knew. They had to ask the question: **What is a real number?** Cantor and Dedekind were close friends, and Dedekind was the first to answer this question by establishing a set theoretic construction of real numbers using the "Dedekind cut". We do not give Dedekind's construction of real numbers in this textbook. Since the time of Cantor/Dedekind, a simpler axiomatic approach to real numbers was discovered (most likely as a result of the collective effort of many mathematicians rather than a select few), and we present it in the next section.

To conclude this brief history, we present the Cantor's idea of comparing the sets $\mathbb{N}$ and $\mathbb{R}$. We have not defined real numbers yet, but I think a reader can follow Cantor's argument since I do not think Cantor himself knew exactly what a real number was when he proved this theorem.

Theorem 2.4.3 (*Georg Cantor*) There is **NO** function from the set $\mathbb{N}$ of all natural numbers **onto** the interval $(0, 1)$ of all real numbers.

Proof Suppose there is an onto function $f : \mathbb{N} \to (0, 1)$. Since $f(n)$ is a real number, we can express $f(n)$ as $f(n) = 0.a_{n,1}a_{n,2}a_{n,3} \cdots a_{n,n}a_{n,n+1} \cdots$ for every $n = 1, 2, 3, \ldots$, where $a_{n,i} \in \{0, 1, 2, 3, 4, 5, 6, 7, 8, 9\}$ for every $n = 1, 2, 3, \ldots$, and $i = 1, 2, 3, \ldots$. For every $n = 1, 2, 3, \ldots$, let $b_n = 5$ if $a_{n,n} = 0$, and $b_n = 0$ if $a_{n,n} \neq 0$. Then $0.b_1b_2 \cdots b_n \cdots$ is a real number such that $0.b_1b_2 \cdots b_n \cdots \in (0, 1)$. But for every $n = 1, 2, 3, \ldots$, $f(n) \neq 0.b_1b_2 \cdots b_n \cdots$. In order to see this, let $n = 1, 2, 3, \ldots$. We note that $f(n) = 0.a_{n,1}a_{n,2}a_{n,3} \cdots a_{n,n}a_{n,n+1} \cdots$. But $a_{n,n} \neq b_n$. Hence, $0.a_{n,1}a_{n,2}a_{n,3} \cdots a_{n,n}a_{n,n+1} \cdots \neq 0.b_1b_2 \cdots b_n \cdots$. This shows that for every $n = 1, 2, 3, \ldots$, $f(n) \neq 0.b_1b_2 \cdots b_n \cdots$. This is a contradiction to the assumption that f is onto. Therefore, there is no function from $\mathbb{N}$ **onto** $(0, 1)$. Since $(0, 1) \subset \mathbb{R}$, there is no function from $\mathbb{N}$ **onto** $\mathbb{R}$.

The above proof is short yet created a big wave in the mathematical society at that time. It was a Tsunami of a mathematical idea.

Definition 2.4.2 A finite set is said to be <u>countable</u>. If a set S is infinite with a property that there is a 1–1 and onto function $f : \mathbb{N} \to S$, then we also say the set S is <u>countable</u>. An infinite set that is not countable is said to be <u>uncountable</u>. The number of elements in a set is called a <u>cardinal number</u>. So, an integer $n \geq 0$ is a cardinal number of a finite set. The empty set $\emptyset$ has zero, 0, cardinality. The cardinal number of the set of all natural numbers is denoted by a Hebrew letter $\aleph_0$ with subscript 0, and reads "<u>alef zero</u>". The cardinal number of $\mathbb{R}$ is called the <u>continuum</u>, and it is denoted by c. Hence, $\aleph_0$ and c are two **infinite** cardinals, yet $\aleph_0$ is strictly smaller than c. If X is a set, then we denote the cardinal number of the set X by the absolute value of X as in $|X|$. So $|\{1, 3, 7\}| = 3$, $|\mathbb{N}| = \aleph_0$, and $|\mathbb{R}| = c$.

Example 2.4.1 Give an example of a function from the interval $(-1, 1)$ onto $\mathbb{R}$.

  Answer: An answer is in Example 1.3.6(7) and its solution. Hence, $|(-1, 1)| = |\mathbb{R}| = c$, the continuum.

Example 2.4.2 A countable union of countable sets is countable. We do not prove this. However, we know that $\mathbb{Z} = \mathbb{N} \cup \{0\} \cup \{-n : n \in \mathbb{N}\}$. Hence, $\mathbb{Z}$ is countable. For a fixed natural number n, the set $\left\{ \frac{i}{n} : i \in \mathbb{Z} \right\}$ is countable. Hence, $\bigcup_{n=1}^{\infty} \left\{ \frac{i}{n} : i \in \mathbb{Z} \right\}$ is a countable union of countable sets. Therefore, the set $\mathbb{Q} = \bigcup_{n=1}^{\infty} \left\{ \frac{i}{n} : i \in \mathbb{Z} \right\}$ is countable. Hence, $|\mathbb{Z}| = |\mathbb{Q}| = \aleph_0$. The set of real numbers $\mathbb{R}$ is uncountable by Theorem 2.4.3 and Example 2.4.1.

Problem 2.4.1 For your thought, we end this section with three questions for you to ponder:

(1) Is there a 1–1 function from the closed interval [0, 1] onto $\mathbb{R}$?
(2) Is there an infinite set S that has a 1–1 function from $\mathbb{R}$ **into** S, but there is no 1–1 function from $\mathbb{R}$ **onto** S?
(3) Is there a set with the cardinality strictly between $\aleph_0$ and c? You may wish to look up "continuum hypothesis".

If you are interested in these questions, you may wish to study advanced set theory.

2.5   Real Numbers

In Sect. 2.3, we have constructed the set of rational numbers denoted by $\mathbb{Q}$. From $\mathbb{Q}$, we will construct the set of real numbers in this section.

Definition 2.5.1 Let S be a nonempty subset of $\mathbb{Q}$. We say a number $a \in \mathbb{Q}$ is said to be an upper bound of S if $s \leq a$ for every $s \in S$. If S has an upper bound, we say that S is bounded above. We say a number $b \in \mathbb{Q}$ is said to be a lower bound of S if $s \geq b$ for every $s \in S$. If S has a lower bound, we say that S is bounded below. If S is bounded above and below, then S is said to be bounded. If S is not bounded, then S is said to be unbounded.

Definition 2.5.2 If S is a nonempty subset of $\mathbb{Q}$ that is bounded above, then we "assign" or "construct" a number α, we call it a real number, such that

(i) $s \leq \alpha$ for every $s \in S$, and
(ii) if $a \in \mathbb{Q}$ is an upper bound of S, then $\alpha \leq a$.

The number α is said to be the least upper bound of S. Thus, the set $\mathbb{R}$ of all real numbers is the set of least upper bounds for all the nonempty subset of $\mathbb{Q}$ that are bounded above.

Lemma 2.5.1 $\mathbb{Q} \subset \mathbb{R}$

Proof Let $a \in \mathbb{Q}$. The set $\{a\}$ is bounded above by a. And its least upper bound is a since if $b \in \mathbb{Q}$ is an upper bound of $\{a\}$, then $a \leq b$. Hence, $a \in \mathbb{R}$. This proves $\mathbb{Q} \subset \mathbb{R}$.

Example 2.5.1 Let $S = \{1, 1.4, 1.41, 1.414, 1.414, 1.4142, 1.41421, 1.414213, \ldots\}$.
     The notation "$\cdots$" in the definition of S is ambiguous. But we meant "$\cdots$" to indicate the remaining infinitely many numbers are the numbers of the form $1.414213\cdots$ whose first 7 digits are 1.414213.) Then the numbers

$$2, 1.5, 1.42, 1.415, 1.4143, 1.41422, 1.414214$$

are all **rational** upper bounds of the set S. None of these upper bounds is the least upper bound of S. As you might have guessed, I have constructed this set S thinking $\sqrt{2}$ as the least upper bound of S. But I cannot say that $\sqrt{2}$ is the least upper bound of S because of the two reasons; the "$\cdots$" in the definition of S is not clear and $\sqrt{2}$ is only a convenient way to denote a positive number whose square is 2 **if it exists**. But the existence of $\sqrt{2}$ is **not** established yet. All we can say at this moment is that S has an upper bound, and therefore, it has the least upper bound that is a **real** number. In Theorem 2.5.1, we will make this argument precise and prove that this least upper bound is $\sqrt{2}$.

Remark 2.5.1 Definition 2.5.2 is abstract, and difficult. Please take time to digest.

We assume the following (R1)–(R10) to be the **axioms** (true statements without proofs) of $\mathbb{R}$. Let $a, b, c \in \mathbb{R}$.

(R1) We know how to add a and b, and $a + b$ is in $\mathbb{R}$, and $a + b = b + a$. And $(a + b) + c = a + (b + c)$ so that writing $a + b + c$ makes sense. Also $a + 0 = a$. The number 0 is the additive identity.

(The parenthesis in $(a + b) + c$ is to indicate that you are to compute $(a + b)$ before adding c to it. The addition $a + b + c$ does not mean to add three numbers simultaneously. Technically, addition of three numbers simultaneously is not defined. Here, $a + b + c$ means either $(a + b) + c$ or $a + (b + c)$, and whichever you do, this axiom guarantees that the results are the same.)

(R2) Exactly one of $a < b, a > b$, or $a = b$ holds.

(R3) We know how to multiply a and b, their multiplication is written by ab or $a \cdot b$, and ab is in $\mathbb{R}$, and $ab = ba$. And $(ab)c = a(bc)$. So, writing abc for the multiplication of three numbers makes sense. Also $a \cdot 1 = a$. Note that $a \cdot 0 = 0$. The number 1 is the multiplicative identity of $\mathbb{R}$.

(R4) $a(b + c) = ab + ac$ is the distributive rule, $ab + ac = a(b + c)$ is the factorization.

(R5) There exists a unique element $x \in \mathbb{R}$, called the additive inverse of a, such that $a + x = 0$. We denote this $x \in \mathbb{R}$ by $-a$.

Note that

(1) $-(-a) = a$,
(2) $a > b$ is equivalent to $a - b > 0$ and to $-a < -b$,
(3) $a \cdot (-1) = -a, (-a)b = -(ab) = -ab, (-a)(-b) = ab, (-1)(-1) = 1$

(R6) If $a \neq 0$, then there exists a unique element $x \in \mathbb{R}$, called the multiplicative inverse of a, such that $ax = 1$. We write this $x \in \mathbb{R}$ by $x = \frac{1}{a}$ or $x = a^{-1}$. So $\frac{b}{a} = ba^{-1}$. If $n \in N$, the n-time multiplication of a is denoted by a^n, and $a^{-n} = \left(a^{-1}\right)^n$. We define $a^0 = 1$ (including

$0^0 = 1$) by convention. Hence, if $m, n \in Z$, we have $a^{mn} = (a^m)^n$, $a^{m+n} = a^m a^n$, and $(ab)^n = a^n b^n$.

Note that

(1) if $ab = 0$, then $a = 0$ or $b = 0$,
(2) if $a \neq 0 \neq b$, then $(a^{-1})(b^{-1}) = (ab)^{-1}$,
(3) if $a \neq 0$, then $a^{-1} \neq 0$, and $(a^{-1})^{-1} = a$,
(4) $1^{-1} = 1$,
(5) if $a \geq b > 0$, then $0 < a^{-1} \leq b^{-1}$.
(6) if $0 < a < 1$, then $a^{-1} > 1$, and if $a > 1$, then $0 < a^{-1} < 1$.

(R7) If $a > b$ or $a = b$, we write $a \geq b$. So if $a \geq b$ and $a \leq b$, then $a = b$. If $a \geq b$ and $b \geq c$, then $a \geq c$.

Note that if $a \geq b$, then $a + c \geq b + c$.

(R8) If $a \geq b$ and if $c \geq 0$, then $ac \geq bc$.

Note that (1) if $a \geq b$ and if $c \leq 0$, then $ac \leq bc$,

(2) $a^2 \geq 0$,

Remark 2.5.2 Typically, when offering an axiomatic development of real numbers, we would be more careful in stating axioms that are independent of each other. However, our goal here is to construct real numbers and state their properties as quickly as we can to move onto study the theory of calculus and analysis. So, we have stated a mishmash of axioms, their implications, and some definition of exponential numbers in the above axiomatic development, trusting that the statements in (R1)–(R8) use properties already introduced in any algebra course you have taken, and there are no surprises.

(R9) If $a < b$, then there is a $q \in \mathbb{Q}$ such that $a < q < b$.

Next, we refine Definition 2.5.1.

Definition 2.5.3 Let S be a nonempty subset of $\mathbb{R}$. We say a number $a \in \mathbb{R}$ is said to be an upper bound of S if $s \leq a$ for every $s \in S$. If S has an upper bound, we say that S is bounded above. We say a number $b \in \mathbb{R}$ is said to be a lower bound of S if $s \geq b$ for every $s \in S$. If S has a lower bound, we say that S is bounded below.

If S is a nonempty subset of $\mathbb{R}$ that is bounded above, then a number α is said to be the least upper bound or supremum of S, or simply sup of S, if

(i) α is an upper bound of S, and
(ii) if $a \in \mathbb{R}$ is an upper bound of S, then $\alpha \leq a$.

We denote this number α by $\sup S$ if it exists. (**Yes**, $\sup S$ exists by the next axiom.)

If S is a nonempty subset of $\mathbb{R}$ that is bounded below, then a number β is said to be the greatest lower bound or infimum of S, or simply inf of S, if

(i) β is a lower bound of S, and
(ii) if $a \in \mathbb{R}$ is a lower bound of S, then $\beta \geq a$.

We denote this number β by $\inf S$ if it exists.

(R10) The Completeness Axiom: Suppose S is a nonempty subset of $\mathbb{R}$ that is bounded above. Then $\sup S$ exists. That is, $\sup S$ is a real number.

Remark 2.5.3 (R10) is the axiom that generalizes Definition 2.5.2. In most introductory analysis textbooks, they usually do not give Definitions 2.5.1 and 2.5.2 because of the redundancy. But I would rather make sure to construct the set of all real numbers first, and to state the properties (R1)–(R10). Hence, I broke convention and gave Definition 2.5.2.

The remainder of this textbook is mostly about the consequences and the applications of the Completeness Axiom.

Definition 2.5.4 Suppose S is a nonempty subset of $\mathbb{R}$. If there exists an element $a \in S$ such that $s \leq a$ for all $s \in S$, then a is said to be the maximum of S, and a is denoted by $\max S$. It is the **largest** number in S. If there exists an element $b \in S$ such that $s \geq b$ for all $s \in S$, then b is said to be the minimum of S, and b is denoted by $\min S$. It is the **smallest** number in S.

Remark 2.5.4 If $\max S$ exists, then $\max S = \sup S$. But the existence of $\sup S$ does not imply the existence of $\max S$. Similarly, if $\min S$ exists, then $\min S = \inf S$. But the existence of $\inf S$ does not imply the existence of $\min S$. The distinction between sup and max of a set is very important. The next example explains these ideas.

Example 2.5.2 Please notice the distinction between inf and min, and between sup and max. $\min[-1, 8) = -1 = \inf[-1, 8)$. Be aware that we have $\sup[-1, 8) = 8$ while $\max[-1, 8)$ does not exist since $8 \notin [-1, 8)$.

Example 2.5.3 Let S be the set defined in Example 2.5.1. Then S is bounded above. So, by the completeness axiom, $\sup S$ exists. This means $\sup S$ **is a real number**. We do not know the existence of $\max S$.

If we let $T = \{1, 1.4, 1.41, 1.414, 1.414, 1.4142, 1.41421, 1.414213\}$, then $\max T = 1.414213 = \sup T$.

Definition 2.5.5 Let S be a nonempty subset of $\mathbb{R}$. Let $c \in \mathbb{R}$. Then $cS = \{cs | s \in S\}$ and $-S = \{-s : s \in S\}$. Hence, $-S = (-1)S$.

Example 2.5.4 Let S be the set defined in Example 2.5.1. Then
   $-S = \{-1, -1.4, -1.41, -1.414, -1.414, -1.4142, -1.41421, -1.414213, \ldots\}$, and -1.414214 is a lower bound of $-S$.
   We also have $-(-S) = S$, and $-(-1.414213) = 1.414213$ is an upper bound of $-(-S)$.

The next example gives you a taste of how we use sup and inf.

Example 2.5.5 Suppose S is a nonempty subset of $\mathbb{R}$ that is bounded below. We prove that $\inf S = -\sup(-S)$.

Proof First, we prove that $\sup(-S)$ exists. Let L be a lower bound of S. Let $x \in -S$. Then there is an $s \in S$ such that $x = -s$. Since L is a lower bound of S, we have $-x = s \geq L$. Hence, $x \leq -L$. This shows that $-S$ is bounded above by $-L$. By the completeness axiom, $\sup(-S)$ exists.

   Next, we prove that $-\sup(-S)$ is a lower bound of S. Let $s \in S$. Then $-s \in -S$. Hence, $-s \leq \sup(-S)$. That is, $s \geq -\sup(-S)$. This shows that $-\sup(-S)$ is a lower bound of S.

   Finally, we show that $-\sup(-S)$ is larger than any lower bound of S. To do this, let $a \in \mathbb{R}$ be a lower bound of S. Let $x \in -S$. Then $-x \in S$ Hence, $a \leq -x$ so that $x \leq -a$. Hence, $-a$ is an upper bound of $-S$ so that $\sup(-S) \leq -a$. This shows that $a \leq -\sup(-S)$. In other words, we showed that $-\sup(-S)$ is the greatest lower bound of $-S$. Therefore, $\inf S = -\sup(-S)$.

Problem 2.5.1 Suppose S is a nonempty subset of $\mathbb{R}$ that is bounded above. Then prove that $\sup S = -\inf(-S)$.

Lemma 2.5.2 Suppose S is a nonempty subset of $\mathbb{R}$ that is bounded above. Then for every $\varepsilon > 0$, there exist $s \in S$ such that $s \geq \sup S - \varepsilon$. (Here, the letter ε is Greek, and it reads epsilon. We will often use ε to indicate a positive number. In this lemma, we are interested when ε is a small positive number.)

Proof Since S is a nonempty subset of $\mathbb{R}$ that is bounded above, $\sup S$ exists. Let $\varepsilon > 0$. Then $\sup S - \varepsilon < \sup S$. Since $\sup S$ is the smallest upper bound of S, $\sup S - \varepsilon$ is not an upper bound of S. Hence, there exists exist $s \in S$ such that $s \geq \sup S - \varepsilon$.

Remark 2.5.5 One way to use this lemma is the following: Suppose S is a nonempty subset of $\mathbb{R}$ that is bounded above. Then for every $n = 1, 2, 3, \ldots$, there exist $s_n \in S$ such that $s_n \geq \sup S - \frac{1}{n}$. So, we have a sequence $\langle s_n \rangle_{n=1}^{\infty}$ "converging" to $\sup S$. We use this idea frequently in this textbook. I put the quotation marks on "converging" to $\sup S$ since we will formally introduce the convergence of a sequence in the next chapter.

Lemma 2.5.3 Let $a, b \in \mathbb{R}$.

(1) If $a - \varepsilon < b$ for every $\varepsilon > 0$, then $a \leq b$.
(2) If $a, b > 0$, and if $a\varepsilon < b$ for every $0 < \varepsilon < 1$, then $a \leq b$.

Proof of (1): Suppose $a - \varepsilon < b$ for every $\varepsilon > 0$. We will show that $a \leq b$. On the contrary, suppose $a > b$. Then $a - b > 0$. Then $a - (a - b) < b$ by our assumption. Since $a - (a - b) = b$, $a - (a - b) < b$ implies that $b < b$. This is a contradiction. Therefore, $a \leq b$.

Proof of (2): We leave this to the readers as an exercise.

Definition 2.5.6 Suppose both A and B are nonempty subsets of real numbers. Let c be a real number. Then we define new sets $A + B, A + c, cA$, and AB. We read the <u>addition of sets $A + B$</u> "A plus B".
$$A + B = \{a + b : a \in A, b \in B\}, \ A + c = \{a + c : a \in A\}, \ cA = \{ca : a \in A\}, \text{ and}$$
$AB = \{a \cdot b : a \in A, b \in B\}$. In particular, $-A = (-1)A$.

Remark 2.5.6 Suppose both A and B are nonempty subsets of real numbers. Even though we defined $A + B$ in the above definition, the **subtraction** of two sets $A - B$ was not a part of Definition 2.5.6 since $A - B$ was already defined in Definition 1.1.1(5). Note that $A - B$ is a **set theoretic notation** defined in Definition 1.1.1 part (5). Please be careful! Note that

$$A - B = \{a \in A : a \notin B\} \quad \text{and} \quad A - B \neq \{a - b : a \in A, b \in B\}.$$

The set $\{a - b : a \in A, b \in B\}$ **has to be denoted by** $A + (-B)$.

Example 2.5.6 Let $A = \{-2, 1\}$ and $B = \{1, 2\}$. Then

$$A + B = \{1 + 1, 1 + 2, -2 + 1, -2 + 2\} = \{-1, 0, 2, 3\} \quad \text{and} \quad A \cup B = \{1, 2, -2\}.$$

This shows that $A + B \neq A \cup B$.

$$A + (-B) = \{1, -2\} + \{-1, -2\} = \{0, -3\} \cup \{-1, -4\} = \{0, -1, -3, -4\}$$

and $A - B = \{-2\}$. Hence, $A - B \neq A + (-B)$.

$$A \cap B = \{1\} \quad A + 2 = \{0, 3\}$$

$$A + (-A) = \{1 - 1, -2 - 1, 1 - (-2), -2 - (-2)\} = \{-3, 0, 3\}$$

$$AB = \{1 \cdot 1, 1 \cdot 2, -2 \cdot 1, -2 \cdot 2\} = \{-4, -2, 1, 2\}$$

Problem 2.5.2 Let $A = \{0, 1, 2\}$ and $B = \{-8, 3, 6\}$. Describe the following sets by writing all elements: $2A + 3$, $A - A$, $A - B$, AB, $A + (-A)$, and $A + (-B)$.

Problem 2.5.3 Let $A = \{\frac{1}{n} : n = 1, 2, 3, \ldots\}$. Convince yourself of the following numbers: $\sup[A + A] = 2$, $\inf[A + A] = 0$, $\sup[A + (-A)] = 1$, $\inf[A + (-A)] = -1$.

(Note that $\sup[A - A]$ and $\inf[A - A]$ do not make sense since $A - A = \emptyset$.)

Example 2.5.7 Suppose both A and B are nonempty bounded sets of **positive** real numbers. We will prove that $\sup AB = \sup A \cdot \sup B$.

Proof Notice that the existence of the number $\sup AB$ is not established yet. So we have to start by establishing this by proving that the set AB is bounded. And what is the convenient upper bound of AB?

Claim 1: We prove that $\sup AB \leq \sup A \cdot \sup B$.

Let $x \in AB$. Then there are $a \in A$ and $b \in B$ such that $x = ab$. But since $0 < a \leq \sup A$ and $0 < b \leq \sup B$, we have $x = ab \leq \sup A \cdot \sup B$. Therefore, $\sup A \cdot \sup B$ is an upper bound of AB. By the completeness axiom, $\sup AB$ exists and we have established $\sup AB \leq \sup A \cdot \sup B$ since $\sup AB$ is the smallest of all upper bound of AB.

(Note that the proof of this claim cannot start with "Let $ab \in AB$ for some $a \in A$ and $b \in B$. See Problem 1.3.4(2').)

Claim 2: We prove that $\sup AB \geq \sup A \cdot \sup B$.

Note that $\sup A > 0$, and $\sup B > 0$ since A and B are sets of positive numbers. Let $\varepsilon > 0$. Then there exists a $\delta > 0$ such that $0 < \delta < \sup A$, $0 < \delta < \sup B$, and $0 < \delta < \frac{\varepsilon}{\sup A + \sup B}$. Then there are $a \in A$ and $b \in B$ such that $a > \sup A - \delta > 0$ and $b > \sup B - \delta > 0$ because $\sup A - \delta$ and $\sup B - \delta$ are not upper bounds of A and B, respectively. Therefore,

$\sup AB \geq ab > (\sup A - \delta) \cdot (\sup B - \delta)$ since $\sup A - \delta > 0$ and $\sup B - \delta > 0$

$= \sup A \cdot \sup B - \delta(\sup A + \sup B) + \delta^2$

$> \sup A \cdot \sup B - \delta(\sup A + \sup B)$ since $\delta^2 > 0$

$> \sup A \cdot \sup B - \varepsilon$ since $\delta < \dfrac{\varepsilon}{\sup A + \sup B}$.

By Lemma 2.5.3(1), we have shown that $\sup AB \geq \sup A \cdot \sup B$.

Therefore, by Claims 1 and 2, we have $\sup AB = \sup A \cdot \sup B$.

(As you will find out in Chap. 4 when we discuss the continuity of a function, a Greek letter δ, reads delta, is often used to indicate a positive number that corresponds to an arbitrary chosen positive number ε.)

Problem 2.5.4 Suppose both A and B are nonempty bounded subsets of real numbers. Let c be a positive real number. Prove or give counterexample to the following statements:

(1) $\inf(cA) = c \cdot \inf A$
(2) $\sup(cA) = c \cdot \sup A$,

(3) $A + A = 2A$

(4) $\inf(AB) = \inf A \cdot \inf B$

(5) $\sup(AB) = \sup A \cdot \sup B$

(6) $\inf(A + B) = \inf A + \inf B$

(7) $\sup(A + B) = \sup A + \sup B$

In Theorem 2.4.1, we showed that $\sqrt{2}$ was not a rational number. We also know that $\sqrt{2}$ is the length of the diagonal of the unit square. Hence, it seems that $\sqrt{2}$ exists and it is a real number. But under our construction of real numbers, it is not obvious that $\sqrt{2}$ is a real number guaranteed by the completeness axiom. So, next, we will prove $\sqrt{2}$ is a real number by using the **completeness axiom**, R10. This may be one of the most difficult theorems proved in this textbook.

Theorem 2.5.1 The number $2^{\frac{1}{2}} = \sqrt{2}$ exists. That is $\sqrt{2} \in \mathbb{R}$.

Proof Let $a_0 = 1$. Then $a_0^2 < 2 < (a_0 + 1)^2$. Let a_1 be the integer among 0, 1, 2, ..., 9 such that $\left(a_0 + \frac{a_1}{10}\right)^2 < 2 < \left(a_0 + \frac{a_1}{10} + \frac{1}{10}\right)^2$. (So $a_1 = 4$.)

Let $n \geq 1$ be an integer such that if k is an integer such that $0 \leq k \leq n$, then $a_k \in \{0, 1, 2, \ldots, 9\}$ is defined so that

$$\left(a_0 + \frac{a_1}{10} + \cdots + \frac{a_k}{10^k}\right)^2 < 2 < \left(a_0 + \frac{a_1}{10} + \cdots + \frac{a_k}{10^k} + \frac{1}{10^k}\right)^2$$

Then we define a_{n+1} to be the integer among 0, 1, 2, ..., 9 such that $\left(a_0 + \frac{a_1}{10} + \cdots + \frac{a_n}{10^n} + \frac{a_{n+1}}{10^{n+1}}\right)^2 < 2 < \left(a_0 + \frac{a_1}{10} + \cdots + \frac{a_n}{10^n} + \frac{a_{n+1}}{10^{n+1}} + \frac{1}{10^{n+1}}\right)^2$

By mathematical induction, a_n is defined for all integer $n \geq 0$.

For simplicity, let $b_n = a_0 + \frac{a_1}{10} + \cdots + \frac{a_n}{10^n}$ for every $n = 0, 1, 2, \ldots$.

Let $B = \{b_n : n = 0, 1, 2, \ldots\}$. (Here, the idea of the use of this set B is very similar to the use of set S in Example 2.5.1 for $\sqrt{2}$.)

Claim 1: We will show that B is bounded above by 2.

Let $n > 0$ be an integer. Then by Problem 1.4.2, we have

$$b_n = a_0 + \frac{a_1}{10} + \cdots + \frac{a_n}{10^n} \leq a_0 + \frac{9}{10} + \cdots + \frac{9}{10^n} = a_0 + \frac{9}{10}\left(1 + \cdots + \frac{1}{10^{n-1}}\right)$$

$$= a_0 + \frac{9}{10} \cdot \frac{\frac{1}{10^n} - 1}{\frac{1}{10} - 1} = a_0 + \left(1 - \frac{1}{10^n}\right) < a_0 + 1 = 2$$

Hence, by the completeness axiom, $\sup B$ exists. For a shorthand notation, let $\beta = \sup B$ in order to ease writing. Since $a_0 \geq 1$, we know that $\beta \geq 1$. We will prove that $\beta^2 = 2$.

Claim 2: We will show that $\beta^2 \geq 2$.

Let $\varepsilon > 0$. Then there is an integer n such that $\frac{2(a_0+1)+1}{10^n} < \varepsilon$. Then we have

$$2 < \left(a_0 + \frac{a_1}{10} + \cdots + \frac{a_n}{10^n} + \frac{1}{10^n}\right)^2 = \left(b_n + \frac{1}{10^n}\right)^2 \le \left(\beta + \frac{1}{10^n}\right)^2 = \beta^2 + \frac{2\beta}{10^n} + \frac{1}{10^{2n}} <$$

$$\beta^2 + \frac{2(a_0 + 1)}{10^n} + \frac{1}{10^{2n}} < \beta^2 + \frac{2(a_0 + 1) + 1}{10^n} < \beta^2 + \varepsilon.$$

Therefore, we have shown that $\beta^2 > 2 - \varepsilon$ for every $\varepsilon > 0$. By Lemma 2.5.3, we have $\beta^2 \ge 2$.

Claim 3: We will show that $\beta^2 \le 2$.

Let $\varepsilon > 0$. Then there is a $\delta > 0$ such that $\delta < \beta$ (here, note that that $\beta \ge 1$) and $2\beta\delta < \varepsilon$. Since $\beta - \delta$ is not an upper bound of B, there is an integer n such that $b_n \ge \beta - \delta > 0$ (by Lemma 2.5.2). Then we have $2 > b_n^2 \ge (\beta - \delta)^2 = \beta^2 - 2\beta\delta + \delta^2 > \beta^2 - 2\beta\delta > \beta^2 - \varepsilon$. Therefore, we have shown that $2 > \beta^2 - \varepsilon$ for every $\varepsilon > 0$. By Lemma 2.5.3, we have $2 \ge \beta^2$.

By Claims 2 and 3, we have shown that $\beta^2 = 2$. That is, we have shown that $2^{\frac{1}{2}} = \sqrt{2} = \beta$, and $2^{\frac{1}{2}}$ exists.

Remark 2.5.8

(1) Compare the above proof to Example 2.5.1.

Note that $a_0 = 1, a_1 = 4, a_2 = 1, a_3 = 4, a_4 = 2, a_5 = 1$, and $a_6 = 3$, So $b_0 = 1, b_1 = 1.4, b_2 = 1.41, b_3 = 1.414, b_4 = 1.4142, b_5 = 1.41421$, and $b_6 = 1.414213$. So the set B in the above proof is to formalize the definition of the set $S = \{1, 1.4, 1.41, 1.414, 1.414, 1.4142, 1.41421, 1.414213, \ldots\}$ in Example 2.5.1 with mathematical induction.

(2) In the Claim 2 of the above proof, how do we know how to choose n such that $\frac{2(a_0+1)+1}{10^n} < \varepsilon$? By reading Claim 2 backward, you can see how I chose n.

Problem 2.5.5 Let us use the same notations as in the proof of Theorem 2.5.1. Prove that $b_n + \frac{1}{10^n}$ is an upper bound of the set B for each $n = 1, 2, 3, \ldots$.

Remark 2.5.7

(1) The sequences a_n and b_n in the proof of Theorem 2.5.1 will be used as examples in the next chapter when we study sequences. So please spend enough time to study them.
(2) Let $x > 1$ be a number. If $k \ge 2$ is an integer, the existence of $x^{\frac{1}{k}}$ will be proved in the next chapter (see Theorem 3.8.1), which is a generalization of Theorem 2.5.1.

Problem 2.5.6 Let B be a nonempty bounded subset of real numbers. Let $A \subset B$. Prove that $\inf B \le \inf A \le \sup A \le \sup B$.

Problem 2.5.7 Let $A = \left\{ \left(1 + \frac{1}{n}\right)^n : n = 1, 2, 3, \ldots \right\}$.

Prove that A is bounded below by 2 and bounded above by 3.

References

1. R. Dedekind; *Theory of Numbers*, The Open Court Publishing Co. 1948 (also available by Dover Publications, Inc.).
2. Archemedes; *The Sand Reckoner*
3. A.D. Aczel, *Finding Zero*, Palgrave Macmillan (2015)
4. G. Polya, *Mathematical Methods in Science*, Mathematical Association of America (1977)
5. N. Copernicus; *On the Revolution of Celestial Orbs* in 1543, available in English translation in Great Books of the Western World, Vol. 16, Encyclopedia Britannica, Inc. (1952)
6. G. Gugliotta, The Maya glory and ruin. Natl. Geogr. **212**(2), 68–73 (2007)
7. Al Khwarizmi; *Hisab Al-jabr w'al-muqabala,* commonly known as *"Al-jabr"*
8. H.P. Boas, Irrational thoughts. Math. Mag. **93**, 23–26 (2020)
9. L. Pisano (Fibonacci) *Liber Abaci (Book of Calculation)*, translated and reproduced in English by Springer (2002)
10. J. Kepler, *The Epitome of Copernican Astronomy*, about 1620, available in English translation in Great Books of the Western World, Vol. 16, Encyclopedia Britannica, Inc. (1952)
11. Apollonius of Perga; *Conics: Book I–IV*, originally at about 200 B.C., translated and reproduced by Green Lion Press (2013)

Sequences and Series

3

3.1 Sequences

I do not know much about the history or the origin of sequences. It might have been introduced by Archimedes, but his impact on this subject seems slight. In Sect. 2.4, I speculated that the credit should go to Al Khwarizmi for its origin. I have not seen a history of mathematics book that mentions the origin of sequences explicitly. The concept of sequences is a phenomenal idea because sequences are the only way we can express irrational numbers. I would like to give the credit to Al Khwarizmi as the first person who came up with the concept of a sequence of numbers. This is purely my speculation, but the origin of the word "algorithm" goes back to Al Khwarizmi's name, and the most well-known algorithm is the division algorithm. For example, using division algorithm and applying it to a fraction like $\frac{22}{7}$ to show that its decimal version is $3.142857\cdots$ may be the beginning of the concept of a sequence. Starting from here, I think Al Khwarizmi made the concept of irrational numbers accessible because of the establishment of a sequence.

Definition 3.1.1 A <u>sequence</u> is a **function** $a : \mathbb{N} \to \mathbb{R}$ from the set $\mathbb{N}$ of natural numbers into the set $\mathbb{R}$ of real numbers. Usually, we write a_n in place of the function value notation $a(n)$, and $\langle a_n \rangle_{n=1}^{\infty}$ in place of the function notation $a : \mathbb{N} \to \mathbb{R}$. The set $\mathbb{N}$ in the definition of the sequence is often replaced by an infinite subset of $\mathbb{N} \cup \{0\}$, and we often replace $\langle a_n \rangle_{n=1}^{\infty}$ by $\langle a_n \rangle_{n=0}^{\infty}$, or $\langle a_n \rangle_{n=10}^{\infty}$, etc. We denote a sequence by $\langle a_n \rangle$ when there is no confusion.

Example 3.1.1

(1) Let $a_n = 1$ for every $n = 1, 2, 3, \dots$. Then we can denote $\langle a_n \rangle_{n=1}^{\infty}$ by $\langle 1, 1, 1, \dots \rangle$. Note that $\{a_n\}_{n=1}^{\infty} = \{1, 1, 1, \dots\} = \{1\}$ is a **set** of one element.

© The Author(s), under exclusive license to Springer Nature Switzerland AG 2025
H. Katsuura, *Introduction to Analysis*, Synthesis Lectures on Mathematics & Statistics,
https://doi.org/10.1007/978-3-031-67954-4_3

Let $a_n = (-1)^{n+1}$ for every $n = 1, 2, 3, \ldots$. We can write $\langle a_n \rangle_{n=1}^{\infty}$ as $\langle 1, -1, 1, -1, 1 \ldots \rangle$, or $\langle (-1)^{n+1} \rangle_{n=1}^{\infty}$. **Note** that $\{a_n\}_{n=1}^{\infty} = \{(-1)^{n+1}\}_{n=1}^{\infty} = \{1, -1\}$ is a set with two elements.

Some textbooks use $\{a_n\}_{n=1}^{\infty}$ or $(a_n)_{n=1}^{\infty}$ for sequences in place of $\langle a_n \rangle_{n=1}^{\infty}$. But we insist on using our notation $\langle a_n \rangle_{n=1}^{\infty}$ for a sequence. The notation $(a_n)_{n=1}^{\infty}$ is similar to ours, and I do not object to the notation $(a_n)_{n=1}^{\infty}$ as a sequence. However, the notation $\{a_n\}_{n=1}^{\infty}$ indicates a **set**, and **it is not acceptable notation for a sequence**.

We often consider the set $\{a_n\}_{n=1}^{\infty}$ of a sequence $\langle a_n \rangle_{n=1}^{\infty}$. (See, for example, the proof of Bolzano-Weierstrass Theorem, Theorem 3.7.1.) If you keep in mind that a sequence is a **function** $a : \mathbb{N} \to \mathbb{R}$, then the distinction between $\{a_n\}_{n=1}^{\infty}$ and $\langle a_n \rangle_{n=1}^{\infty}$ becomes clear. As the above definition says, $\langle a_n \rangle_{n=1}^{\infty}$ is the function $a : \mathbb{N} \to \mathbb{R}$ itself. On the other hand, $\{a_n\}_{n=1}^{\infty}$ is the **image** of $\mathbb{N}$ under a, i.e., $a(\mathbb{N}) = \{a_n\}_{n=1}^{\infty}$. To repeat, the notation $\langle a_n \rangle_{n=1}^{\infty}$ indicates a **function** a that we call it a **sequence**, while $\{a_n\}_{n=1}^{\infty}$ indicates the image of $\mathbb{N}$ under a, which is the **set** $a(\mathbb{N})$. And $\langle a_n \rangle_{n=1}^{\infty}$ and $\{a_n\}_{n=1}^{\infty}$ are **different**. Even though some textbooks denote a sequence by $\{a_n\}_{n=1}^{\infty}$, we consider it is completely **wrong** to denote a sequence $\langle a_n \rangle_{n=1}^{\infty}$ by a set notation $\{a_n\}_{n=1}^{\infty}$ in this textbook.

(2) $\langle 1, 2, 3, \ldots \rangle$ is a sequence such that $a_n = n$ for every $n = 1, 2, 3, \ldots$. We write it as $\langle n \rangle_{n=1}^{\infty}$.

(3) $\langle \frac{1}{n} \rangle_{n=1}^{\infty} = \langle 1, \frac{1}{2}, \frac{1}{3}, \ldots \rangle$ is a very important sequence. It is called the harmonic sequence. Please compare the following two sentences:

(a) For every $\varepsilon > 0$, there is a positive integer n such that $\frac{1}{n} < \varepsilon$.

(b) There is a positive integer n such that $\frac{1}{n} < \varepsilon$ for every $\varepsilon > 0$.

The only difference between these two sentences is the placement of "for every $\varepsilon > 0$". Both are grammatically correct sentences, but (a) is a true statement, and (b) is a **false** statement. In (a), a positive number ε was chosen before a positive integer n was chosen. In (b), a positive integer n was chosen before ε was chosen, and therefore, there is **no** positive integer n such that $\frac{1}{n} < \varepsilon$ for **every** $\varepsilon > 0$. Please be careful.

(4) We said earlier that $\frac{22}{7} = 3.142857 \cdots$. The notation $3.142857 \cdots$ can be interpreted as a sequence $\langle 3.142, 3.1428, 3.14285, 3.142857, \ldots \rangle$.

(5) If $\langle a_n \rangle$ is the sequence $\langle 3.142, 3.1428, 3.14285, 3.142857, \ldots \rangle$ in part (4), we want to say that the sequence $\langle a_n \rangle$ converges to $\frac{22}{7}$. The concept of convergence is old. But the actual definition of the **convergence** of a sequence was a result of Cauchy's work from about 1820–1830. The purpose of this section is to understand this definition, which is fundamental to understanding of series and the limit of a function. For that matter, the concept of convergence is fundamental to the study of modern mathematical analysis.

(6) The numbers a_n and b_n defined in the proof of Theorem 2.5.1 define a sequence of integers $\langle a_n \rangle_{n=1}^{\infty}$ and a sequence of rational numbers $\langle b_n \rangle_{n=0}^{\infty}$. We will use these examples often in this chapter.

Definition 3.1.2 A sequence $\langle a_n \rangle_{n=1}^{\infty}$ is said to <u>converge</u> to a number L if for every $\varepsilon > 0$, there exists a positive integer M such that $|a_n - L| < \varepsilon$ for all integers $n \geq M$. If $\langle a_n \rangle_{n=1}^{\infty}$ converges to a number L, then we write "$a_n \to L$ as $n \to \infty$" or $\lim_{n \to \infty} a_n = L$. If a sequence $\langle a_n \rangle_{n=1}^{\infty}$ converges to some number, we say $\langle a_n \rangle_{n=1}^{\infty}$ <u>converges</u>. If $\langle a_n \rangle_{n=1}^{\infty}$ does not converge to L, then we write $\lim_{n \to \infty} a_n \neq L$. If a sequence $\langle a_n \rangle_{n=1}^{\infty}$ does not converge to any number, we say the sequence $\langle a_n \rangle_{n=1}^{\infty}$ <u>diverges</u>.

Remark 3.1.1

(1) The interpretation of $|a_n - L|$ should be the **distance** between a_n and L. As a matter of fact, by writing $d(a_n, L)$, reads "the distance between a_n and L", in place of $|a_n - L|$, we will extend the convergence of a sequence later to a **metric** space. So, the idea of limit is that no matter what $\varepsilon > 0$ is chosen (ε is usually close to 0), there exists a positive integer M such that the **distance** between a_n and L is smaller than ε whenever $n \geq M$. This integer M depends on the choice of ε so that M can be thought as a function of ε. The word "whenever" is often replaced by "for all" or "for every".
(2) The following is **not** the same as Definition 3.1.2, and it is a **wrong** definition of the limit of a sequence some of my past students have given: "A sequence $\langle a_n \rangle_{n=1}^{\infty}$ converges to L if there exists a positive integer M such that $|a_n - L| < \varepsilon$ for all integers $n \geq M$ for every $\varepsilon > 0$". The reason is that this integer M is no longer depending on the choice of ε. This is similar to Example 3.1.1(3).

Theorem 3.1.1 Let c be a real number. Let $a_n = c$ for every $n = 1, 2, 3, \ldots$. Then the sequence $\langle a_n \rangle_{n=1}^{\infty}$ converges to c, and we express this by $\lim_{n \to \infty} a_n = c$ or $\lim_{n \to \infty} c = c$.

Proof Let $\varepsilon > 0$. Then $|a_n - c| = |c - c| = 0 < \varepsilon$ for every $n = 1, 2, 3, \ldots$. Therefore, $\lim_{n \to \infty} a_n = c$.

Remark 3.1.2 Just because Definition 3.1.2 says "for every $\varepsilon > 0$", please note that the above proof does not start by saying "for every $\varepsilon > 0$". Because of "for every $\varepsilon > 0$" in Definition 3.1.1, we start the proof by picking an arbitrary $\varepsilon > 0$ by "Let $\varepsilon > 0$". It reads "Let epsilon be a positive number". As you will see below, the proofs of Theorem 3.1.2, Example 3.1.2 and Theorem 3.1.5 all start by "Let $\varepsilon > 0$". Please **do not start** these proofs by "For every $\varepsilon > 0$, ...". The statement "Let $\varepsilon > 0$" is setting the **subject** for the paragraph. Proofs of theorems are very much like computer programs. And mathematicians read proofs like computers read programs. If a sentence starts by "For every $\varepsilon > 0$, ...", then the number ε is only valid within the sentence. Hence, you cannot use the number ε in the next sentence expecting it to have the same meaning unless it is **again** redefined in the second sentence. And if you redefine ε in the second sentence, the number ε in the first sentence and the number ε in the second sentence are **not** the same. Thus, if you want to use ε throughout, in a paragraph you have to set the **subject** by saying "Let $\varepsilon > 0$" at the beginning. **Therefore, a proof that begins by saying "For every $\varepsilon > 0$, ..." is most likely not a valid proof.**

We restate the following axiom we stated in Sect. 2.2.

Axiom 2.2 There is no largest natural number.

Theorem 3.1.2 The sequence $\langle \frac{1}{n} \rangle_{n=1}^{\infty} = \langle 1, \frac{1}{2}, \frac{1}{3}, \ldots \rangle$ converges to 0.

Proof Let $\varepsilon > 0$. Since the set $\mathbb{N}$ of natural numbers is not bounded above, there exists a positive integer (natural number) M such that $M > \frac{1}{\varepsilon}$. Hence, if $n \geq M$ is an integer, then $\left| \frac{1}{n} - 0 \right| = \left| \frac{1}{n} \right| = \frac{1}{n} \leq \frac{1}{M} < \varepsilon$. Therefore, $\langle \frac{1}{n} \rangle_{n=1}^{\infty}$ converges to 0.

Remark 3.1.3

(1) In the above proof, it seems silly to start an equation by writing $\left| \frac{1}{n} - 0 \right| = \left| \frac{1}{n} \right|$. But it is **necessary** to write $\left| \frac{1}{n} - 0 \right|$ since we are comparing the distance between $\frac{1}{n}$ and 0. That is, $\left| \frac{1}{n} - 0 \right|$ is the **subject**.

(2) Theorem 3.1.2 is the fundamental tool to prove the convergence of sequences and series as in the next example.

Example 3.1.2 We know that $\langle \frac{3n^2 - 5n - 3}{4n^2 + n - 56} \rangle_{n=1}^{\infty}$ converges to $\frac{3}{4}$. But this is because we learned how to use Theorem 3.1.4 below in a calculus class, as shown in Example 3.1.4 below. Here, we are learning how to use the definition. That is, we only have **Definition** 3.1.2 to show that $\langle \frac{3n^2 - 5n - 3}{4n^2 + n - 56} \rangle_{n=1}^{\infty}$ converges to $\frac{3}{4}$.

(In this textbook, **we almost always take a point of view that** if we have not yet proven a theorem in this textbook, then we do not yet know that theorem–unless otherwise stated.)

We need some preparation by estimating the distance between $\frac{3n^2 - 5n - 3}{4n^2 + n - 56}$ and $\frac{3}{4}$.

$$\left| \frac{3n^2 - 5n - 3}{4n^2 + n - 56} - \frac{3}{4} \right| = \left| \frac{4(3n^2 - 5n - 3) - 3(4n^2 + n - 56)}{(4n^2 + n - 56)4} \right|$$

$$= \left| \frac{4(-5n - 3) - 3(n - 56)}{(4n^2 + n - 56)4} \right| = \left| \frac{-23n + 156}{(4n^2 + n - 56)4} \right|$$

$$= \frac{|-23n + 156|}{|4n^2 + n - 56|}$$

$$\leq \frac{1}{4} \cdot \frac{23n + 156}{|4n^2 + n - 56|} \quad \text{by the triangle inequality}$$

$$= \frac{1}{4} \cdot \frac{23n + 156}{(4n^2 + n - 56)} \quad \text{if } n \geq 56$$

(I know that $\left|4n^2 + n - 56\right| = 4n^2 + n - 56$ if $n \geq 10$ for example, but all I needed was to find a number M such that $\left|4n^2 + n - 56\right| = 4n^2 + n - 56$ if $n \geq M$. So, by paying attention to $n - 56$, I chose 56 for the number M.)

Assume $n \geq 56$. Then $4n^2 + n - 56 \geq 4n^2 + 56 - 56 = 4n^2$. Hence,

$$\frac{1}{4} \cdot \frac{23n + 156}{\left(4n^2 + n - 56\right)} \leq \frac{1}{4} \cdot \frac{23n + 156}{4n^2} < \frac{1}{4} \cdot \frac{23n + 156n}{4n^2} = \frac{1}{4} \cdot \frac{179n}{4n^2} < \frac{1}{4} \cdot \frac{200n}{4n^2}$$

$$= \frac{25}{2n} < \frac{13}{n}$$

Proof of $\lim\limits_{n \to \infty} \frac{3n^2 - 5n - 3}{4n^2 + n - 56} = \frac{3}{4}$: Let $\varepsilon > 0$. Then there exists a positive integer M such that $M > \frac{13}{\varepsilon} + 56$. Let $n \geq M$ be an integer. Then

$$\left|\frac{3n^2 - 5n - 3}{4n^2 + n - 56} - \frac{3}{4}\right| = \left|\frac{4\left(3n^2 - 5n - 3\right) - 3\left(4n^2 + n - 56\right)}{\left(4n^2 + n - 56\right)4}\right|$$

$$= \left|\frac{4(-5n - 3) - 3(n - 56)}{\left(4n^2 + n - 56\right)4}\right| = \left|\frac{1}{4} \cdot \frac{-23n + 156}{\left(4n^2 + n - 56\right)}\right|$$

$$= \frac{1}{4} \cdot \frac{|-23n + 156|}{\left|4n^2 + n - 56\right|}$$

$$\leq \frac{1}{4} \cdot \frac{23n + 156}{\left|4n^2 + n - 56\right|} \text{ by the triangle inequality}$$

$$= \frac{1}{4} \cdot \frac{23n + 156}{\left(4n^2 + n - 56\right)} \text{ since } n \geq M > \frac{13}{\varepsilon} + 56 > 56$$

Also, since $n > 56$, we have $4n^2 + n - 56 \geq 4n^2 + 56 - 56 = 4n^2$. Hence, $\frac{1}{4} \cdot \frac{23n + 156}{\left(4n^2 + n - 56\right)} \leq \frac{1}{4} \cdot \frac{23n + 156}{4n^2} < \frac{1}{4} \cdot \frac{23n + 156n}{4n^2} = \frac{179n}{4 \cdot 4n^2} < \frac{200n}{4 \cdot 4n^2} = \frac{25}{2n} < \frac{13}{n}$. Since $n \geq M > \frac{13}{\varepsilon}$, we have $\frac{13}{n} < \varepsilon$. This shows that $\left|\frac{3n^2 - 5n - 3}{4n^2 + n - 56} - \frac{3}{4}\right| < \varepsilon$ for every $n \geq M$. (More precisely, we have shown that for an arbitrary chosen $\varepsilon > 0$, there is an integer M such that $\left|\frac{3n^2 - 5n - 3}{4n^2 + n - 56} - \frac{3}{4}\right| < \varepsilon$ for every $n \geq M$.)

Therefore, we have $\lim\limits_{n \to \infty} \frac{3n^2 - 5n - 3}{4n^2 + n - 56} = \frac{3}{4}$.

Remark 3.1.4

(1) The proof in Example 3.1.2 varies depending on the preparation work. So, if I rewrite this proof with a new preparation, I will probably come up with a different proof. Your solution to this type of problem should be different from your friend's solution.

(2) We usually do not write the proof by showing the preparation as I did in Example 3.1.2. So if you wanted to see how I came up with the above proof if the preparation had not been written, you would have to read the above proof in the reverse order.

(3) The conclusion "Therefore, we have $\lim\limits_{n\to\infty} \frac{3n^2-5n-3}{4n^2+n-56} = \frac{3}{4}$." is a paragraph by itself in the proof, unless the proof is very short like the one in Theorem 3.1.1.

We take a view that the statement of a theorem or what you want to prove as in the above example to be an introduction to an essay. The "proof" is the main body of the essay and the conclusion. Some people may conclude a proof without stating the conclusion thinking it to be "obvious", but that is not recommended. After you have worked hard to write a proof, please proudly write the conclusion as in "Therefore, we have $\lim\limits_{n\to\infty} \frac{3n^2-5n-3}{4n^2+n-56} = \frac{3}{4}$." Some people end with "QED" or by " $\blacksquare$ " for an emphasis. This is just a visual clarification to indicate the end of a proof, and not a part of the proof.

Problem 3.1.1 Prove that each of the following sequences converges:

1. $\left(\frac{-3}{5n-6}\right)_{n=1}^{\infty}$.

2. $\left(\pi - \frac{1}{n}\right)_{n=1}^{\infty}$.

3. $\left(\frac{5n^2+13}{4n^2+n+6}\right)_{n=1}^{\infty}$.

4. $\left(\frac{5n^2+13}{-2n^2-n+6}\right)_{n=1}^{\infty}$.

5. $\left(\frac{5n^2+13}{n^2-17n+6}\right)_{n=1}^{\infty}$.

6. $\left(-5 - (-1)^n \frac{1}{n}\right)_{n=1}^{\infty}$.

Remark 3.1.5 The **negation** of Definition 3.1.1 is the following:

A sequence $\langle a_n \rangle_{n=0}^{\infty}$ does **not** converge to a number L if, for any positive integer M, there exist an $\varepsilon > 0$ and an integer $n \geq M$ such that $|a_n - L| \geq \varepsilon$. This statement is equivalent to Definition 3.1.1.

Example 3.1.3 We will show that $\langle \frac{1}{n} \rangle_{n=1}^{\infty} = \langle 1, \frac{1}{2}, \frac{1}{3}, \ldots \rangle$ does not converge to any number L different from 0. (Note that we have already shown that the sequence converges to 0. This does not prove that the sequence does not converge to $L \neq 0$ since we do not have Theorem 3.1.5 yet. Yes, I intentionally delayed the presentation of Theorem 3.1.5.)

Let L be a number different from 0. Then $|L| > 0$. Let M be a positive integer. Since $\lim\limits_{n\to\infty} \frac{1}{n} = 0$, there exists $n \geq M$ such that $\left|\frac{1}{n}\right| = \left|\frac{1}{n} - 0\right| < \frac{|L|}{2}$. Hence, by the triangle inequality Theorem 1.1.1, we have $\left|\frac{1}{n} - L\right| \geq \left||L| - \frac{1}{n}\right| \geq |L| - \frac{1}{n} > |L| - \frac{|L|}{2} = \frac{|L|}{2} > 0$. This proves that $\frac{1}{n}_{n=1}^{\infty} = 1, \frac{1}{2}, \frac{1}{3}, \ldots$ does not converge to the number L.

Please note that we chose ε (in Remark 3.1.5) to be $\frac{|L|}{2}$. However, we did not say "Let $\varepsilon = \frac{|L|}{2}$" since $\frac{|L|}{2}$ is a perfectly good number and simple to write. Hence, we did not rename $\frac{|L|}{2}$ to be ε. Unless we have a number expressed in a more complicated way, I prefer not to rename it. Some people may be upset if you call them by a name that is different from their given name. I respect that in mathematics, too.

Problem 3.1.2 Use only the definition of convergence to prove the following:

(1) Show that $\left(\pi + \frac{2}{n}\right)_{n=1}^{\infty}$ does not converge to 2.

(Showing $\lim_{n\to\infty} \left(\pi + \frac{2}{n}\right) = \pi$ and concluding that the sequence does not converge to 2 since $\pi \neq 2$ is not good enough for this proof at this time because we do not yet have Theorem 3.1.4 to draw on and, lacking that, you will have to negate the definition of convergence to prove this.)

(2) Show that $\langle n\rangle_{n=1}^{\infty}$ does not converge to 1,000,000. (Please do not prove this by contradiction. Saying "$\lim_{n\to\infty} n = \infty$ and therefore, $\lim_{n\to\infty} n \neq 1,000,000$" is not a proof. We do not know exactly what $\lim_{n\to\infty} n = \infty$ means yet. The notation "∞" is only a symbol without any meaning at this point. Therefore, "$\lim_{n\to\infty} n = \infty$" has no meaning at this stage.)

(3) Show that $\left((-1)^{n+1}\right)_{n=1}^{\infty}$ does not converge to -3. (Please do not prove this by contradiction.)

(4) Show that $\left((-1)^{n+1}\right)_{n=1}^{\infty}$ diverges. (This is to show that $\left((-1)^{n+1}\right)_{n=1}^{\infty}$ does not converge to any number. Again, please do not prove this by contradiction.)

(5) Show that $\left((-1)^{n+1} \cdot n\right)_{n=1}^{\infty}$ diverges. (Again, please do not prove this by contradiction.)

(6) For every positive integer n, let $a_{3n} = 0, a_{3n+1} = 1, a_{3n+2} = 2$. Use Definition 3.1.1 to prove that the sequence $\langle a_n\rangle_{n=1}^{\infty}$ diverges. Again, please do not prove this by contradiction.

Remark 3.1.6 There is nothing wrong with using contradiction in a proof. However, if there is a way to prove the statement/theorem without contradiction, it is **usually** better not to use contradiction. Problem 3.1.2 can be proven by contradiction, but with a deeper understanding of the definition of the convergence of a sequence, this problem can be proved without contradiction. Choosing to avoid the use of contradiction as your default approach is a form of patient editing. By taking the time to consider options, you will improve your presentation of your proof and achieve a deeper understanding of the subject.

Definition 3.1.3

(1) A sequence $\langle a_n \rangle_{n=1}^{\infty}$ <u>diverges to infinity</u>, and we write $\lim\limits_{n \to \infty} a_n = \infty$ if for every positive number A, there exists a positive integer M such that $a_n > A$ for all integer $n \geq M$. A sequence $\langle a_n \rangle_{n=1}^{\infty}$ <u>diverges to minus infinity</u>, and write $\lim\limits_{n \to \infty} a_n = -\infty$ if $\lim\limits_{n \to \infty} (-a_n) = \infty$.

Definition 3.1.4 A sequence $\langle a_n \rangle$ is said to be <u>bounded above</u> if the set $\{a_n\}_{n=1}^{\infty}$ is bounded above. A sequence $\langle a_n \rangle$ is said to be <u>bounded below</u> if the set $\{a_n\}_{n=1}^{\infty}$ is bounded below. If a sequence is bounded above and bounded below, then it is <u>bounded</u>. If a sequence is not bounded, it is <u>unbounded</u>.

Theorem 3.1.3 If $\langle a_n \rangle_{n=1}^{\infty}$ converges, then the set $\{a_n\}_{n=1}^{\infty}$ is bounded. In short, we say that a convergent sequence is <u>bounded</u>.

(Please see Example 3.1.1 part (1) for the distinction between $\langle a_n \rangle_{n=1}^{\infty}$ and $\{a_n\}_{n=1}^{\infty}$.)

Proof Suppose $\langle a_n \rangle_{n=1}^{\infty}$ converges to c for some real number c. Then there exists a positive integer M such that $|a_n - c| < 1$ for all integer $n \geq M$. Hence, $-|c| - 1 \leq c - 1 < a_n < c + 1 \leq |c| + 1$. Hence, $|a_n| \leq |c| + 1$ for all integers $n \geq M$. Let $A = \max\{|a_1|, |a_2|, \dots, |a_M|\}$. Therefore, $\{|a_n|\}_{n=1}^{\infty}$ is bounded above by $A + |c| + 1$. But since $-|a_n| \leq a_n \leq |a_n|$, $\{a_n\}_{n=1}^{\infty}$ is bounded below by $-(A + |c| + 1)$, and bounded above by $A + |c| + 1$. Hence, the sequence $\langle a_n \rangle_{n=1}^{\infty}$ is bounded.

Corollary 3.1.3.1

(1) If $\lim\limits_{n \to \infty} |a_n| = \infty$, then the sequence $\langle a_n \rangle_{n=1}^{\infty}$ diverges.
(2) $\lim\limits_{n \to \infty} n = \infty$

Problem 3.1.3

(1) Prove Corollary 3.1.3.1.
(2) Prove that $\lim\limits_{n \to \infty} \frac{-5n^2 + 33}{n+5} = -\infty$.

Problem 3.1.4

(1) Prove that if $\lim\limits_{n \to \infty} a_n = \infty$, then the set $\{a_n\}_{n=1}^{\infty}$ is not bounded without using Theorem 3.1.3.
(2) Give an example of a divergent sequence $\langle a_n \rangle_{n=1}^{\infty}$ such that $\lim\limits_{n \to \infty} |a_n| \neq \infty$.
(3) Is a bounded sequence convergent? Either prove it or give a counterexample.

Theorem 3.1.4 Suppose $\langle a_n \rangle_{n=1}^{\infty}$ and $\langle b_n \rangle_{n=1}^{\infty}$ are convergent. (This means $\lim_{n \to \infty} a_n$ and $\lim_{n \to \infty} b_n$ exist and they are numbers, and $\lim_{n \to \infty} a_n \neq \pm\infty$ or $\lim_{n \to \infty} b_n \neq \pm\infty$.) Then we have the following:

(1) $\lim_{n \to \infty} (a_n + b_n) = \lim_{n \to \infty} a_n + \lim_{n \to \infty} b_n$

(This says that the sequence $\langle a_n + b_n \rangle_{n=1}^{\infty}$ converges to the number $\lim_{n \to \infty} a_n + \lim_{n \to \infty} b_n$.)

(2) $\lim_{n \to \infty} (a_n \cdot b_n) = \left(\lim_{n \to \infty} a_n \right) \cdot \left(\lim_{n \to \infty} b_n \right)$

(3) If $b_n \neq 0$ for every $n = 1, 2, 3, \ldots$, and if $\lim_{n \to \infty} b_n \neq 0$, then $\lim_{n \to \infty} \frac{1}{b_n} = \frac{1}{\lim\limits_{n \to \infty} b_n}$.

(4) If $b_n \neq 0$ for every $n = 1, 2, 3, \ldots$, and if $\lim_{n \to \infty} b_n \neq 0$, then $\lim_{n \to \infty} \frac{a_n}{b_n} = \frac{\lim\limits_{n \to \infty} a_n}{\lim\limits_{n \to \infty} b_n}$.

Proofs of (1), and (2) are left to the readers as exercises.

Proof of (3): (This proof is rather difficult, but it is instructive.)

Suppose $\lim_{n \to \infty} b_n = c$ for some $c \neq 0$. Since $\lim_{n \to \infty} b_n = c$, there exists a positive integer M such that $|b_n - c| < \frac{|c|}{3}$ for all $n \geq M$.

Let $n \geq M$. Then $|b_n - c| < \frac{|c|}{3}$ implies that $-\frac{|c|}{3} < b_n - c < \frac{|c|}{3}$, or $c - \frac{|c|}{3} < b_n < c + \frac{|c|}{3}$. Let $A = \frac{2|c|}{3}$. If $c > 0$, then $|b_n| = b_n > c - \frac{|c|}{3} = \frac{2|c|}{3} = A$. If $c < 0$, then $0 > c + \frac{|c|}{3} = c - \frac{c}{3} = \frac{2c}{3}$ so that $A = \frac{2|c|}{3} = \left| c - \frac{c}{3} \right| = \left| c + \frac{|c|}{3} \right| < -b_n = b_n$ since $b_n < c + \frac{|c|}{3} < 0$. Then $|b_n| > A$. So regardless of $c > 0$ or $c < 0$, we have shown that $|b_n| > A$, or $\frac{1}{|b_n|} < \frac{1}{A}$ for all $n \geq M$.

Let $\varepsilon > 0$. Since $\lim_{n \to \infty} b_n = c$, there exists a positive integer L such that $|b_n - c| < \varepsilon A |c|$ for all integer $n \geq L$.

Let $n \geq L + M$. Then

$$\left| \frac{1}{b_n} - \frac{1}{c} \right| = \frac{|b_n - c|}{|b_n||c|}$$

$$< \frac{|b_n - c|}{A|c|} \quad \text{since } n \geq M$$

$$< \frac{\varepsilon A |c|}{A|c|} = \varepsilon \quad \text{since } n \geq L$$

Therefore, $\lim_{n \to \infty} \frac{1}{b_n} = \frac{1}{c} = \frac{1}{\lim\limits_{n \to \infty} b_n}$.

(Here, we renamed $\lim_{n \to \infty} b_n$ by c because it would be very cumbersome to write this proof without it.)

Proof of (4) is by (2) and (3).

Problem 3.1.5 Prove Theorem 3.1.4 (1) and (2).

(Hint for proving (2): $a_n b_n - ab = a_n(b_n - b) + (a_n - a)b$.)

Problem 3.1.6 Give an example of a sequence $\langle b_n \rangle_{n=1}^{\infty}$ such that $b_n \neq 0$ for every $n = 1, 2, 3, \ldots$, and $\lim\limits_{n \to \infty} \frac{1}{b_n} \neq \frac{1}{\lim\limits_{n \to \infty} b_n}$.

(An answer: Let $b_n = n$ for every $n = 1, 2, 3, \ldots$. Then $\lim\limits_{n \to \infty} \frac{1}{b_n} = 0$, but $\frac{1}{\lim\limits_{n \to \infty} b_n}$ has no meaning. Please, $\frac{1}{\infty}$ is not a number.)

Theorem 3.1.5 Let b and c be real numbers. Let $\langle a_n \rangle_{n=1}^{\infty}$ be a sequence. If $\lim\limits_{n \to \infty} a_n = b$ and $\lim\limits_{n \to \infty} a_n = c$, then $b = c$.

(In other words, if a sequence converges, it has a unique limit.)

Proof Let $\varepsilon > 0$. Since $\lim\limits_{n \to \infty} a_n = b$, there exists a positive integer L such that $|a_n - b| < \frac{\varepsilon}{2}$ for all integer $n \geq L$. Since $\lim\limits_{n \to \infty} a_n = c$, there exists a positive integer M such that $|a_n - c| < \frac{\varepsilon}{2}$ for all integer $n \geq M$. Let $n \geq L + M$. Then $|b - c| \leq |a_n - b| + |a_n - c| < \frac{\varepsilon}{2} + \frac{\varepsilon}{2} = \varepsilon$ since $n \geq L$ and $n \geq M$.

Therefore, we have shown that $|b - c| < \varepsilon$ for every $\varepsilon > 0$. That is, we have shown that $b = c$.

Remark 3.1.7 With Theorem 3.1.5 being proven, we can give an alternate solution to Problem 3.1.2

(1) as follows: Since $\lim\limits_{n \to \infty} \left(\pi + \frac{2}{n} \right) = \pi$ (by Theorem 3.1.4 (1)) and $\pi \neq 2$, we have $\lim\limits_{n \to \infty} \left(\pi + \frac{2}{n} \right) \neq 2$ by the uniqueness of the limit of convergent sequence as shown in Problem 3.1.1.

Problem 3.1.7

(1) Let $k \geq 2$ be an integer. Prove that $\lim\limits_{n \to \infty} \frac{1}{n^k} = 0$.

(2) If $\lim\limits_{n \to \infty} |a_n| = \infty$, and if $a_n \neq 0$ for every $n = 1, 2, 3, \ldots$, prove that $\lim\limits_{n \to \infty} \frac{1}{a_n} = 0$.

Example 3.1.4 In Example 3.1.2, we have shown $\lim\limits_{n \to \infty} \frac{3n^2 - 5n - 3}{4n^2 + n - 56} = \frac{3}{4}$ using the definition of limits. Now, since we have Theorem 3.1.4, we can prove this as follow: $\lim\limits_{n \to \infty} \frac{3n^2 - 5n - 3}{4n^2 + n - 56} = \frac{\lim\limits_{n \to \infty} (3n^2 - 5n - 3) \cdot \frac{1}{n^2}}{\lim\limits_{n \to \infty} (4n^2 + n - 56) \cdot \frac{1}{n^2}} = \frac{\lim\limits_{n \to \infty} \left(3 - 5 \cdot \frac{1}{n} - 3 \cdot \frac{1}{n^2} \right)}{\lim\limits_{n \to \infty} \left(4 + \frac{1}{n} - 56 \cdot \frac{1}{n^2} \right)} = \frac{3}{4}$ since $\lim\limits_{n \to \infty} \frac{1}{n} = 0 = \lim\limits_{n \to \infty} \frac{1}{n^2}$.

Problem 3.1.8 Let S be a nonempty subset of real numbers that is bounded above. By the completeness axiom, $\sup S$ exists. Let $\sup S = \alpha$. Then for every $n = 1, 2, 3, \ldots$, there exists an element $a_n \in S$ such that $a_n > \alpha - \frac{1}{n}$ since $\alpha - \frac{1}{n}$ is not an upper bound of S. Prove that this sequence $\langle a_n \rangle_{n=1}^{\infty}$ converges to α by Definition 3.1.1.

3.2 Important Theorems

Here, we give three additional important theorems fundamental to this chapter for sequences and series introduced in the next section.

Definition 3.2.1 A sequence $\langle a_n \rangle$ such that $a_n \leq a_{n+1}$ for every $n = 1, 2, 3, \ldots$ is said to be monotonically increasing or simply increasing. If $a_n \geq a_{n+1}$ for every $n = 1, 2, 3, \ldots$, then the sequence is said to be monotonically decreasing or simply decreasing. If a sequence is either increasing or decreasing, then the sequence is said to be monotonic or a monotone sequence.

Example 3.2.1 The sequence of rational numbers $\langle b_n \rangle$ in the proof of Theorem 2.5.1 is an increasing sequence.

In Theorem 3.1.3, we proved that a convergent sequence is bounded.

Problem 3.2.1

(1) Prove that a sequence a_n is bounded if, and only if, there exists two positive integers M and N such that $|a_n| \leq M$ for every integer $n \geq N$.
(2) By negating the part (1), complete the following sentence:

A sequence $\langle a_n \rangle$ is not bounded if $\cdots$.

(3) Give an example of unbounded sequence.
(4) Give an example of a decreasing sequence that is not bounded below.
(5) Prove that a decreasing sequence is bounded above.

Theorem 3.2.1

(1) If a monotonically increasing sequence is bounded above, then the sequence converges.
(2) If a monotonically decreasing sequence is bounded below, then the sequence converges.

Proof of (1): Suppose $\langle a_n \rangle$ is a monotonically increasing sequence that is bounded above. Then $\{a_n\}_{n=1}^{\infty}$ is bounded above. By **the completeness axiom**, $\sup\{a_n\}_{n=1}^{\infty}$ exists. Let $\sup\{a_n\}_{n=1}^{\infty} = \alpha$. We will show that $\lim_{n\to\infty} a_n = \alpha$. Let $\varepsilon > 0$. Since $\alpha - \varepsilon$ is not an upper bound of $\{a_n\}_{n=1}^{\infty}$, there exists a positive integer M such that $a_M > \alpha - \varepsilon$. Let $n \geq M$ be an integer. Since $\langle a_n \rangle$ is a monotonically increasing sequence, we have $a_1 \leq a_2 \leq a_3 \leq \cdots$. In particular, we have $a_M \leq a_n$. Hence, $a_n > \alpha - \varepsilon$. But since α is an upper bound of $\{a_n\}_{n=1}^{\infty}$, we have $a_n \leq \alpha < \alpha + \varepsilon$. Thus we have $\alpha - \varepsilon < a_n < \alpha + \varepsilon$, or $|a_n - \alpha| < \varepsilon$.

Therefore, we have shown $\lim_{n\to\infty} a_n = \alpha$, i.e., a_n converges.

Proof of (2) is left to the readers.

Remark 3.2.1

(1) Compare Theorem 3.2.1 to Problem 3.1.8.
(2) The entire Chap. 2 can be thought of as a preparation to proving Theorem 3.2.1, which is probably be the most important theorem in this chapter. As you will see, many convergence tests of infinite series will be based on Theorem 3.2.1.

Theorem 3.2.2 (The sandwich theorem) Suppose $\langle a_n \rangle$, $\langle b_n \rangle$, and $\langle c_n \rangle$ are sequences. Suppose $a_n \leq b_n \leq c_n$ for every $n = 1, 2, 3, \ldots$.

(1) Suppose $\lim_{n\to\infty} a_n$ and $\lim_{n\to\infty} b_n$ exist. Then $\lim_{n\to\infty} a_n \leq \lim_{n\to\infty} b_n$.
(2) If $\lim_{n\to\infty} a_n = \alpha = \lim_{n\to\infty} c_n$ for some $\alpha \in \mathbb{R}$, then $\lim_{n\to\infty} b_n = \alpha$.
(3) If $\lim_{n\to\infty} a_n = \infty$, then $\lim_{n\to\infty} b_n = \infty$.
(4) If $\lim_{n\to\infty} c_n = -\infty$, then $\lim_{n\to\infty} b_n = -\infty$.

Proof of (1): For simplicity, let $\lim_{n\to\infty} a_n = \alpha$ and $\lim_{n\to\infty} b_n = \beta$. Suppose, on the contrary, that $\alpha > \beta$. Then $\alpha - \beta > 0$.

Since $\lim_{n\to\infty} a_n = \alpha$, there is an integer $M > 0$ such that $|a_n - \alpha| < \frac{\alpha-\beta}{2}$, or $-\frac{\alpha-\beta}{2} < a_n - \alpha < \frac{\alpha-\beta}{2}$ for all $n \geq M$. The left side $-\frac{\alpha-\beta}{2} < a_n - \alpha$ implies that $\frac{\alpha+\beta}{2} < a_n$ for all $n \geq M$.

Since $\lim\limits_{n\to\infty} b_n = \beta$, there is an integer $N > 0$ such that $|b_n - \beta| < \frac{\alpha-\beta}{2}$, or $-\frac{\alpha-\beta}{2} < b_n - \beta < \frac{\alpha-\beta}{2}$ for all $n \geq N$. The right side $b_n - \beta < \frac{\alpha-\beta}{2}$ implies that $b_n < \frac{\alpha+\beta}{2}$ for all $n \geq N$.

Hence, let $n = N + M$. Then we have $b_n < \frac{\alpha+\beta}{2} < a_n$. This is a contradiction to $a_n \leq b_n$ for every $n = 1, 2, 3, \ldots$. Therefore, we must have $\lim\limits_{n\to\infty} a_n \leq \lim\limits_{n\to\infty} b_n$.

Proof of (2)[1]: Let $\varepsilon > 0$. Since $\lim\limits_{n\to\infty} a_n = \alpha = \lim\limits_{n\to\infty} c_n$, there exists two positive integers M_a and M_c such that $|a_n - \alpha| < \frac{\varepsilon}{3}$ for all $n \geq M_a$ and $|c_n - \alpha| < \frac{\varepsilon}{3}$ for all $n \geq M_c$. Let n be an integer such that $n \geq M_a + M_c$. Then $|a_n - \alpha| < \frac{\varepsilon}{3}$ and $|c_n - \alpha| < \frac{\varepsilon}{3}$. Hence,

$$
\begin{aligned}
|b_n - \alpha| &= |b_n - a_n + a_n - \alpha| \leq |b_n - a_n| + |a_n - \alpha| \\
&\leq |c_n - a_n| + |a_n - \alpha| \quad \text{since } a_n \leq b_n \leq c_n \\
&= |c_n - \alpha + \alpha - a_n| + |a_n - \alpha| \\
&\leq |c_n - \alpha| + |\alpha - a_n| + |a_n - \alpha| \\
&= |c_n - \alpha| + 2|a_n - \alpha| < \frac{\varepsilon}{3} + 2 \cdot \frac{\varepsilon}{3} = \varepsilon
\end{aligned}
$$

Therefore, $\lim\limits_{n\to\infty} b_n = \alpha$.

Problem 3.2.2

(1) Prove Theorem 3.2.1(2).
(2) Prove Theorem 3.2.2 (3) and (4).
(3) Prove that the sequence of integers $\langle a_n \rangle$ in the proof of Theorem 2.5.1 is not monotonically increasing.

Theorem 3.2.3

(1) If $0 < r < 1$, then $\lim\limits_{n\to\infty} r^n = 0$.
(2) If $r > 1$, then $\lim\limits_{n\to\infty} r^n = \infty$.

Proof of (1): Since $r > r^2 > r^3 > r^4 > \cdots > 0$, the sequence r^n is decreasing and bounded below. Hence, it converges, say $\lim\limits_{n\to\infty} r^n = \alpha$. By Theorem 3.2.2, we have $0 \leq \alpha \leq 1$. By Theorem 3.1.4, we have $\lim\limits_{n\to\infty} r^{n+1} = r \cdot \lim\limits_{n\to\infty} r^n = r\alpha$. But we also have $\lim\limits_{n\to\infty} r^{n+1} = \lim\limits_{n\to\infty} r^n = \alpha$. So $r\alpha = \alpha$ or $(r-1)\alpha = 0$. Since $r \neq 1$, we must have $\alpha = 0$, i.e., $\lim\limits_{n\to\infty} r^n = 0$.

---

[1] Here, the existence of $\lim\limits_{n\to\infty} b_n$ is not assumed. Part (2) says that if $\lim\limits_{n\to\infty} a_n = \alpha = \lim\limits_{n\to\infty} c_n$, then $\lim\limits_{n\to\infty} b_n$ exists and $\lim\limits_{n\to\infty} b_n = \alpha$. It seems that part (1) of this theorem implies this part (2). But this was not clear to me. So I proved part (2) without part (1).

Proof of (2) is left as an exercise.

Problem 3.2.3

(1) If $0 < r < 1$, prove that the sequence $\langle r^n \rangle_{n=1}^{\infty}$ is decreasing using mathematical induction.

(2) If $-1 < r < 0$, prove that $\lim\limits_{n \to \infty} r^n = 0$.

(3) If $|r| > 1$, then what can you say about $\lim\limits_{n \to \infty} r^n$?

(4) Let $r > 1$, and assume that $r^{\frac{1}{n}}$ exists for every integer $n > 0$. What can you say about $\lim\limits_{n \to \infty} r^{\frac{1}{n}}$? (The answer will be given in Lemma 3.9.2. But think about this without seeing this lemma.)

(5) If $0 < r < 1$, then what can you say about $\lim\limits_{n \to \infty} r^{\frac{1}{n}}$?

(6) Prove Theorem 3.2.3(2).

3.3   Infinite Series

An important and an interesting way to construct a new **sequence** from an existing one is by infinite series as in the next definition.

Definition 3.3.1 Let $\langle a_n \rangle_{n=1}^{\infty}$ be a sequence. For every $n = 1, 2, 3, \ldots$, we denote $a_1 + a_2 + \cdots + a_n$ by $\sum_{k=1}^{n} a_k$. We call the **sequence** $\langle \sum_{k=1}^{n} a_k \rangle_{n=1}^{\infty}$ an <u>infinite series</u> or a <u>series</u>, and we denote this sequence $\langle \sum_{k=1}^{n} a_k \rangle_{n=1}^{\infty}$ by $\sum_{n=1}^{\infty} a_n$, i.e.,

$$\left\langle \sum_{k=1}^{n} a_k \right\rangle_{n=1}^{\infty} = \sum_{n=1}^{\infty} a_n.$$

The finite sum $\sum_{k=1}^{n} a_k$ is called the <u>n-th partial sum</u> of the infinite series $\sum_{n=1}^{\infty} a_n$. If the sequence $\langle \sum_{k=1}^{n} a_k \rangle_{n=1}^{\infty}$ **converges to** s, we also denote its **limit** s by $\sum_{n=1}^{\infty} a_n$, i.e.,

$$\lim_{n \to \infty} \sum_{k=1}^{n} a_k = s = \sum_{n=1}^{\infty} a_n$$

In this case, we say $\sum_{n=1}^{\infty} a_n$ <u>exists</u> or $\sum_{n=1}^{\infty} a_n$ <u>converges</u>. If the sequence $\langle \sum_{k=1}^{n} a_k \rangle_{n=1}^{\infty}$ diverges, we say the infinite series $\sum_{n=1}^{\infty} a_n$ <u>diverges</u>. In this case, we also say that $\sum_{n=1}^{\infty} a_n$ <u>does not exist</u>.

Hence, the notation $\sum_{n=1}^{\infty} a_n$ has **two** meanings, one is to denote the sequence $\langle \sum_{k=1}^{n} a_k \rangle_{n=1}^{\infty}$, and the other is to denote the limit of the sequence $\lim\limits_{n \to \infty} \sum_{k=1}^{n} a_k = \sum_{n=1}^{\infty} a_n$ if it exists.

Remark 3.3.1 By writing $\sum_{n=1}^{\infty} a_n$, it appears that we are talking about summation of infinite terms. But **this is completely false** because $\sum_{n=1}^{\infty} a_n = \lim_{n\to\infty} \sum_{k=1}^{n} a_k$ and it is the limit of the sequence of **finite** n-th partial sums $\langle \sum_{k=1}^{n} a_k \rangle_{n=1}^{\infty}$.

Unfortunately, since many people erroneously think $\sum_{n=1}^{\infty} a_n$ as a summation of infinite terms, the limit, $\lim_{n\to\infty} \sum_{k=1}^{n} a_k$, is said to be the <u>sum</u> of the series. Hence, in place of "the sum $\sum_{n=1}^{\infty} a_n$", I often say "the **limit** $\sum_{n=1}^{\infty} a_n$" since I think of $\lim_{n\to\infty} \sum_{k=1}^{n} a_k$ for $\sum_{n=1}^{\infty} a_n$ in this textbook. I always think of $\sum_{n=1}^{\infty} a_n$ as the sequence $\langle \sum_{k=1}^{n} a_k \rangle_{n=1}^{\infty}$.

Theorem 3.3.1

(1) If $\sum_{n=1}^{\infty} a_n$ converges, then $\lim_{n\to\infty} a_n = 0$.

(2) **(The divergence test)** If $\lim_{n\to\infty} a_n \neq 0$, then $\sum_{n=1}^{\infty} a_n$ diverges.

Proof of (1): For every $n = 1, 2, 3, \ldots$, let $s_n = \sum_{k=1}^{n} a_k$. Then

$\lim_{n\to\infty} a_n = \lim_{n\to\infty} (s_n - s_{n-1}) = \lim_{n\to\infty} s_n - \lim_{n\to\infty} s_{n-1} = 0$ by Theorem 3.1.4(3) because both $\lim_{n\to\infty} s_n$ and $\lim_{n\to\infty} s_{n-1}$ exist and $\lim_{n\to\infty} s_n = \lim_{n\to\infty} s_{n-1}$.

(2) is a contraposition of (1), and therefore, (2) is true.

It is a good idea to apply the divergence test **first** to test convergence of a series.

Theorem 3.3.2

(1) **(The geometric series)** If $-1 < r < 1$, then

$$\sum_{n=0}^{\infty} r^n = 1 + r + r^3 + \cdots = \frac{1}{1-r}. \text{ (\textbf{Note} that this infinite series starts from } n = 0.)$$

(2) If $|r| \geq 1$, then $\sum_{n=0}^{\infty} r^n$ diverges.

Proof

(1) From Theorem 1.4.1, we have $\sum_{k=0}^{n} r^k = \frac{r^{n+1}-1}{r-1}$ for every $n \in \mathbb{N}$. Theorem 3.2.3(1) says that $\lim_{n\to\infty} r^n = 0$. Hence, by Theorem 3.1.4, we have $\sum_{n=0}^{\infty} r^n = \lim_{n\to\infty} \sum_{n=0}^{\infty} r^k =$

$\lim_{n\to\infty} \frac{r^{n+1}-1}{r-1} = \frac{\left(\lim_{n\to\infty} r^{n+1}\right)-1}{r-1} = \frac{-1}{r-1} = \frac{1}{1-r}$.

Proof of (2): Since $\lim_{n\to\infty} |r^n| = \lim_{n\to\infty} |r|^n \geq \lim_{n\to\infty} 1^n = 1$ by the sandwich theorem, we have $\lim_{n\to\infty} r^n \neq 0$. By the divergence test, $\sum_{n=0}^{\infty} r^n$ diverges.

Example 3.3.1 I cannot emphasize enough how useful the above two theorems are.

(1) $\sum_{n=0}^{\infty} \left(\frac{3}{2}\right)^n$ diverges since $\lim_{n\to\infty} \left(\frac{3}{2}\right)^n \neq 0$. $\sum_{n=0}^{\infty} (-1)^n$ diverges since $\lim_{n\to\infty} (-1)^n \neq 0$.

(2) $\sum_{n=0}^{\infty} \frac{1}{2^n} = \sum_{n=0}^{\infty} \left(\frac{1}{2}\right)^n = 1 + \frac{1}{2} + \left(\frac{1}{2}\right)^2 + \left(\frac{1}{2}\right)^3 + \cdots = \frac{1}{1-\frac{1}{2}} = 2$.

(3) $1.999\cdots = 1 + \sum_{n=1}^{\infty} \frac{9}{10^n} = 1 + \frac{9}{10} \sum_{n=0}^{\infty} \left(\frac{1}{10}\right)^n = 1 + \frac{9}{10} \cdot \frac{1}{1-\frac{1}{10}} = 2$.

The number 2 has two representations!

(4) Let $\langle a_n \rangle$ be the sequence of integers defined in the proof of Theorem 2.5.1. Then we prove that $\langle a_n \rangle$ diverges.

Please keep in mind that $\{a_n\}_{n=1}^{\infty} \subset \{0, 1, 2, \ldots, 9\}$. Suppose a_n converges, say $\lim_{n\to\infty} a_n = L$ for some number L. Then there exists an integer $M > 0$ such that $|a_n - L| < \frac{1}{2}$, or $L - \frac{1}{2} < a_n < L + \frac{1}{2}$ for all $n \geq M$. Since the interval $\left(L - \frac{1}{2}, L + \frac{1}{2}\right)$ has the length 1, it contains exactly one integer. Since a_n is an integer, "$L - \frac{1}{2} < a_n < L + \frac{1}{2}$ for all $n \geq M$" implies that or $a_n = a_M$ for all $n \geq M$, and a_M is an integer.

Let $n > M$. Let b_n be the number defined in the proof of Theorem 2.5.1. Then $b_n = \sum_{k=0}^{n} \frac{a_k}{10^k} = \sum_{k=0}^{M-1} \frac{a_k}{10^k} + \sum_{k=M}^{n} \frac{a_k}{10^k} = \sum_{k=0}^{M-1} \frac{a_k}{10^k} + \sum_{k=M}^{n} \frac{a_k}{10^k}$. Now, $\lim_{n\to\infty} \sum_{k=M}^{n} \frac{a_M}{10^k} = \sum_{k=M}^{\infty} \frac{a_M}{10^k} \frac{a_M}{10^k} = \frac{a_M}{10^M} \sum_{k=0}^{\infty} \frac{a_M}{10^k} \frac{1}{10^k} = \frac{a_M}{10^M} \cdot \frac{1}{1-\frac{1}{10}} = \frac{a_M}{10^M} \cdot \frac{10}{9} = \frac{a_M}{9 \cdot 10^{M-1}}$. Therefore, $\sqrt{2} = \lim_{n\to\infty} b_n = \sum_{k=0}^{M-1} \frac{a_k}{10^k} + \frac{a_M}{9 \cdot 10^{M-1}}$. This shows that $\sqrt{2}$ is a rational number, which is a contradiction to $\sqrt{2}$ being irrational. Therefore, $\langle a_n \rangle$ diverges.

As an application of Theorem 3.1.4, we have the following.

Theorem 3.3.3 Suppose $\sum_{n=1}^{\infty} a_n$ and $\sum_{n=1}^{\infty} b_n$ converge.

(1) Then $\sum_{n=1}^{\infty} (a_n + b_n) = \sum_{n=1}^{\infty} a_n + \sum_{n=1}^{\infty} b_n$.
(2) If c is a constant, then $\sum_{n=1}^{\infty} c \cdot a_n = c \sum_{n=1}^{\infty} a_n$.

Proof is left to the readers since this theorem is a consequence of Theorem 3.1.4.

Problem 3.3.1 Find the following **limits** if they exist. (Colloquially, we often say, "Find the following **sums** if they exist". See Remark 3.3.1.):

(1) $\sum_{n=1}^{\infty} \frac{3}{(-2)^n}$
(2) $\sum_{n=3}^{\infty} \frac{4}{3^n}$
(3) $\sum_{n=2}^{\infty} \left(-\frac{3}{2}\right)^n$
(4) $1.090909\cdots = 1.\overline{09}$. The bar in $1.\overline{09}$ is to indicate "09" is repeated indefinitely. Hence, $1.\overline{09} = 1 + \sum_{n=1}^{\infty} \frac{9}{100^n}$. Convert this number as a fraction of two integers.

(5) Use Theorem 3.3.2 to show that $3.\overline{14285} = \frac{22}{7}$.

Any numbers that repeat itself in decimal representations as in this and (4) are all rational numbers by geometric series.

(6) $\sum_{n=3}^{\infty} \frac{n+4}{n-2}$

(7) Let $r < 0.$ $\sum_{n=1}^{\infty} \frac{1}{n^r}$.

Example 3.3.2 Since $\frac{1}{k(k+1)} = \frac{1}{k} - \frac{1}{k+1}$ (a partial fraction) for any integer $k > 0$, we have

$$\sum_{k=1}^{n} \frac{1}{k(k+1)} = \sum_{k=1}^{n} \left(\frac{1}{k} - \frac{1}{k+1} \right) = \left(\frac{1}{1} - \frac{1}{2} \right) + \left(\frac{1}{2} - \frac{1}{3} \right) + v + \left(\frac{1}{n} - \frac{1}{n+1} \right).$$

$$= 1 \left(-\frac{1}{2} + \frac{1}{2} \right) + \left(-\frac{1}{3} + \frac{1}{3} \right) + v + \left(-\frac{1}{n} + \frac{1}{n} \right) + \frac{1}{n+1} = 1 - \frac{1}{n+1}$$

Hence, $\sum_{n=1}^{\infty} \frac{1}{n(n+1)} = \lim_{n\to\infty} \sum_{k=1}^{n} \frac{1}{k(k+1)} = \lim_{n\to\infty} \left(1 - \frac{1}{n+1} \right) = 1.$

This type of series is called a telescoping series. It resembles the following integration you might have learned when studying the techniques of integrations:

$$\int \frac{1}{x(x+1)} dx = \int \frac{1}{x} dx - \int \frac{1}{x+1} dx = \ln|x| - \ln|x+1| + C$$

Problem 3.3.2

(1) Find the sum $\sum_{n=1}^{\infty} \frac{1}{n(n+2)}$

(2) Find the sum $\sum_{n=3}^{\infty} \frac{1}{n(n-1)}$.

The next theorem seems almost trivial, but it is a very useful theorem.

Theorem 3.3.4 (Comparison Test) Suppose $\langle a_n \rangle_{n=1}^{\infty}$ and $\langle b_n \rangle_{n=1}^{\infty}$ are sequences of **positive** terms such that $a_n \leq b_n$ for every $n = 1, 2, 3, \ldots$.

(1) If $\sum_{n=1}^{\infty} b_n$ converges, then $\sum_{n=1}^{\infty} a_n$ converges and $\sum_{n=1}^{\infty} a_n \leq \sum_{n=1}^{\infty} b_n$.

(2) If $\sum_{n=1}^{\infty} a_n = \infty$, then $\sum_{n=1}^{\infty} b_n = \infty$.

Proof of (1): Since $\langle a_n \rangle_{n=1}^{\infty}$ is a sequence of **positive** terms, $\left(\sum_{k=1}^{n} a_k \right)_{n=1}^{\infty}$ is increasing. Suppose $\sum_{n=1}^{\infty} b_n$ converges. Then $\sum_{n=1}^{\infty} b_n$ is an upper bound of the sequence $\left(\sum_{k=1}^{n} a_k \right)_{n=1}^{\infty}$. Therefore, $\sum_{n=1}^{\infty} a_n$ converges by Theorem 3.2.1.

Proof of (2) is left to the readers.

Remark 3.3.2 Theorem 3.2.1 and Theorem 3.3.4 are essentially **identical**. As you have learned in Calculus courses, there are several other convergence tests for infinite series.

And we will talk about some of them later. But for now, let us learn how to use the above comparison test.

Theorem 3.3.5

(1) (Nicole Oresme, 1323–1382) $\sum_{n=1}^{\infty} \frac{1}{n}$ diverges. More precisely, $\sum_{n=1}^{\infty} \frac{1}{n} = \infty$.

The sequence $\frac{1}{n}$ is called the <u>harmonic sequence</u>, and $\sum_{n=1}^{\infty} \frac{1}{n}$ is called the <u>harmonic series</u>.

(2) $\sum_{n=1}^{\infty} \frac{1}{n^2}$ converges, and $\sum_{n=1}^{\infty} \frac{1}{n^2} < 2$.

Proof of (1): Let n be a positive integer. Then

$$\sum_{k=1}^{2^n} \frac{1}{k} = 1 + \frac{1}{2} + \frac{1}{3} + \cdots + \frac{1}{8} + \frac{1}{9} + \cdots + \frac{1}{2^n - 1} + \frac{1}{2^n}$$

$$= 1 + \frac{1}{2} + \left(\frac{1}{3} + \frac{1}{4}\right) + \left(\frac{1}{5} + \cdots + \frac{1}{8}\right) + \cdots + \left(\frac{1}{2^{n-1}} \cdots + \frac{1}{2^n - 1} + \frac{1}{2^n}\right)$$

$$> 1 + \frac{1}{2} + \left(2 \cdot \frac{1}{4}\right) + \left(4 \cdot \frac{1}{8}\right) + \cdots + \left(2^{n-1} \cdot \frac{1}{2^n}\right)$$

$$= 1 + \underbrace{\frac{1}{2} + \frac{1}{2} + \cdots + \frac{1}{2}}_{n \text{ terms}} = 1 + n \cdot \frac{1}{2}.$$

Since $\lim_{n \to \infty} \left(1 + n \cdot \frac{1}{2}\right) = \infty$, this shows that $\sum_{n=1}^{\infty} \frac{1}{n}$ by the comparison test.

Proof of (2): $\frac{1}{n^2} < \frac{1}{n(n-1)}$ for every $n = 2, 3, \ldots$. So, we have $\sum_{n=1}^{\infty} \frac{1}{n^2} = 1 + \sum_{n=2}^{\infty} \frac{1}{n^2} < 1 + \sum_{n=2}^{\infty} \frac{1}{n(n-1)} = 1 + \sum_{n=1}^{\infty} \frac{1}{n(n+1)} = 1 + 1 = 2$ since $\sum_{n=1}^{\infty} \frac{1}{n(n+1)}$ by Example 3.3.2. Therefore, $\sum_{n=1}^{\infty} \frac{1}{n^2}$ converges by the comparison test.

Problem 3.3.3 Test the convergence of the following:

(1) $\sum_{n=1}^{\infty} \frac{1}{n(n^2+2)}$

(2) $\sum_{n=1}^{\infty} \frac{1}{n^2+2}$

(3) $\sum_{n=2}^{\infty} \frac{1}{\sqrt{n}}$

(4) $\sum_{n=2}^{\infty} \frac{1}{5n+1}$

(5) $\sum_{n=2}^{\infty} \frac{1}{\sqrt{n(n-1)}}$

(6) $\sum_{n=2}^{\infty} \frac{n+3}{n(n-1)}$

(7) $\sum_{n=3}^{\infty} \frac{1}{n(n^2-1)}$

(8) $\sum_{n=1}^{\infty} \frac{1}{n(2^n-1)}$

(9) $\sum_{n=1}^{\infty} \frac{\sqrt{n}}{n(2^n-1)}$

(10) $\sum_{n=1}^{\infty} \frac{n^n}{n!}$

(11) $\sum_{n=1}^{\infty} \frac{n!}{n^n}$

(12) Let $\langle a_n \rangle_{n=1}^{\infty}$ be a sequence of integers in the set $\{1, 2, \ldots, 100\}$. $\sum_{n=1}^{\infty} \frac{a_n}{3^n}$.

Remark 3.3.3 By defining $\zeta(z) = \sum_{n=1}^{\infty} \frac{1}{n^z}$ as a function of $z \in \mathbb{R}$, we read $\zeta(z)$ "the Riemann zeta function of z", or call it simply the zeta function. (Often, z is a complex number, but here we are only considering when z as a real number.) The zeta function is an important function in number theory. Euler lived before Reimann, and it was Euler who established that $\zeta(2) = \frac{\pi^2}{6}$, $\zeta(4) = \frac{\pi^4}{90}$, etc. These are Euler's identities, his first important discovery that made him famous. Later, Riemann made his famous conjecture, called the Riemann hypothesis, related to this zeta function, and thus his name is attached to this function. By the comparison test and from above Theorem 3.3.5, we know that $\zeta(z) = \sum_{n=1}^{\infty} \frac{1}{n^z}$ diverges if $z \leq 1$, and $\zeta(z) = \sum_{n=1}^{\infty} \frac{1}{n^z}$ converges if $z \geq 2$. What about when $1 < z < 2$? We answer this question in the next theorem.

In the next theorem, we will prove that $\zeta(z) = \sum_{n=1}^{\infty} \frac{1}{n^z}$ converges if $z > 1$. As mentioned earlier, in general we only prove theorems in this book by using definitions, theorems and proofs previously introduced in the text, but I do not know how to do this without integrations, which we have not yet covered in this textbook. Likewise, the proof assumes n^z to be defined for all real numbers $z > 1$. The exact definition of n^z when z is an irrational number has to wait until Definition 3.9.1 is introduced. However, integrations are covered in basic calculus courses so I know you will be able to follow the proof.

Theorem 3.3.6 If $z > 1$, then $\zeta(z) = \sum_{n=1}^{\infty} \frac{1}{n^z}$ converges.

Proof Note that $\frac{1}{n^z}$ represent the area of a rectangle with the base 1 and the height $\frac{1}{n^z}$. So $\zeta(z) = \sum_{k=1}^{\infty} \frac{1}{k^z}$ can be interpreted as the area of the n rectangles of the base length 1 and the height $\frac{1}{k^z}$, $k = 1, 2, \ldots, n$. Note that $\int_{k-1}^{k} \frac{1}{x^z} dx$ represents the area under the curve $y = \frac{1}{x^z}$ over the interval $[k-1, k]$, and it contains the rectangle with the base 1 and the height $\frac{1}{k^z}$ for each $k = 2, 3, \ldots, n$. Hence, $\frac{1}{k^z} < \int_{k-1}^{k} \frac{1}{x^z} dx$ for each $k \geq 2$. Thus if $n \geq 2$, we have

$$\sum_{k=1}^{\infty} \frac{1}{k^z} = 1 + \sum_{k=1}^{\infty} \frac{1}{k^z} \leq 1 + \sum_{k=2}^{n} \left(\int_{k-1}^{k} \frac{1}{x^z} dx \right)$$

$$= 1 + \int_{1}^{n} \frac{1}{x^z} dx = 1 - \frac{1}{z-1} \left[\frac{1}{x^{z-1}} \right]_{x=1}^{x=n} = 1 - \frac{1}{z-1} \left[\frac{1}{n^{z-1}} - 1 \right] < 2$$

We have shown that $\sum_{k=1}^{n} \frac{1}{k^z}$ is bounded above by 2 for every $n = 2, 3, \ldots$. Therefore, $\sum_{n=1}^{\infty} \frac{1}{n^z}$ converges if $z > 1$. (Figs. 3.1 and 3.2)

Problem 3.3.4 Use the figure below to prove that $\displaystyle\sum_{k=1}^{n} \frac{1}{k} > \ln n$ for every positive integer n.
(This also shows that the series $\sum_{n=1}^{\infty} \frac{1}{n}$ diverges since.

$$\lim_{n\to\infty} \ln n = \lim_{n\to\infty} \ln e^n = \lim_{n\to\infty} n\cdot\ln e = \lim_{n\to\infty} n = \infty\Big)$$

Remark 3.3.4 In relation to the divergence of the harmonic series, Euler proved that if $\langle p_n\rangle_{n=1}^{\infty} = \langle 2, 3, 5, 7, 11, 13, \ldots\rangle$ represent the increasing sequence of all prime numbers, then $\sum_{n=1}^{\infty} \frac{1}{p_n} = \infty$. It can be proven that $\sum_{k=1}^{n} \frac{1}{p_k} > \ln(\ln(n))$ for every positive integer n.

Problem 3.3.4 Test the convergence of the following series. If the limits exist, it is better to find the limits (sums) than simply saying "converges". If the sums can be found, find them.

Fig. 3.1 $\sum_{k=1}^{n} \frac{1}{k^2} \le$
$1 + \sum_{k=2}^{n} \left(\int_{k-1}^{k} \frac{1}{x^2} dx\right)$

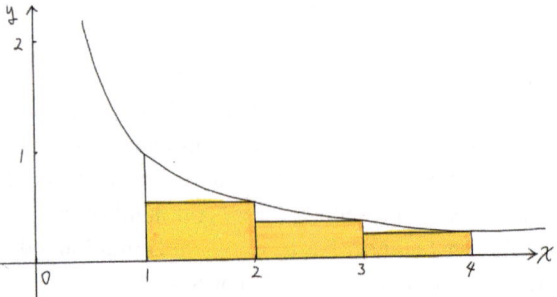

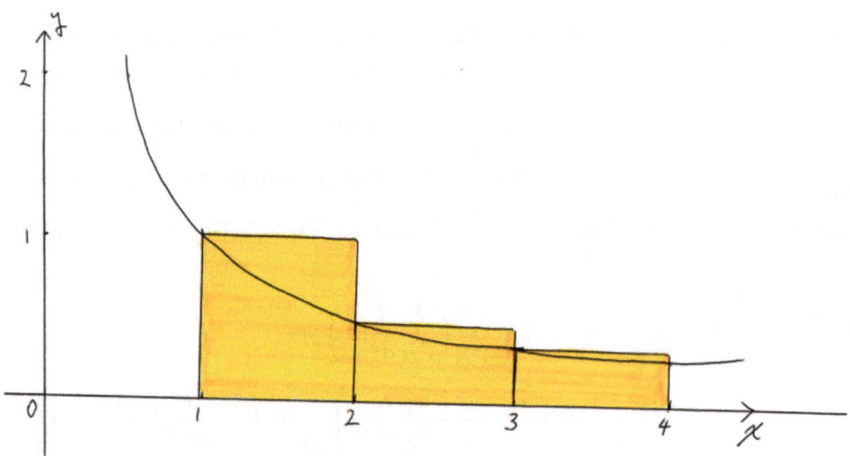

Fig. 3.2 $\sum_{k=1}^{n} \frac{1}{k} \ge \sum_{k=1}^{n} \left(\int_{k-1}^{k} \frac{1}{x} dx\right)$

(1) $\sum_{n=1}^{\infty} \frac{1}{3^n}$

(2) $\sum_{n=1}^{\infty} \frac{1}{3^n - 100}$

(3) $\sum_{n=1}^{\infty} \frac{1}{(-1)^n}$

(4) $\sum_{n=1}^{\infty} \frac{1}{(-3)^n}$

(5) $\sum_{n=1}^{\infty} \frac{43}{3^n + 100}$

(6) $\sum_{n=1}^{\infty} \frac{2^{n+3}}{3^n}$

(7) $\sum_{n=1}^{\infty} \frac{2^{n+3}}{(-3)^n}$

(8) $\sum_{n=1}^{\infty} \frac{1}{n\sqrt{n}}$

(9) $\sum_{n=1}^{\infty} \frac{n - \sqrt{n}}{n + \sqrt{n}}$

(10) $\sum_{n=1}^{\infty} \frac{n - \sqrt{n}}{n^3 + \sqrt{n}}$.

3.4   Alternating Series

If $\langle a_n \rangle_{n=1}^{\infty}$ is a **positive** terms sequence, the series $\sum_{n=1}^{\infty} (-1)^{n+1} a_n = a_1 - a_2 + a_3 - a_4 + \cdots$ is called an <u>alternating series</u>. The convergence of an alternating series is easier to determine than an arbitrary series as seen in the following theorem.

Theorem 3.4.1 (Alternating series test) If $\langle a_n \rangle_{n=1}^{\infty}$ is a monotonically **decreasing** sequence of **positive** numbers converging to 0, then the alternating series $\sum_{n=1}^{\infty} (-1)^{n+1} a_n = a_1 - a_2 + a_3 - a_4 + \cdots$ converges.

Proof Let n be a positive integer. Then we have

$$\sum_{k=1}^{2n} (-1)^{k+1} a_k = a_1 - a_2 + a_3 - a_4 + \cdots - a_{2n}$$
$$= (a_1 - a_2) + (a_3 - a_4) + a_5 - \cdots$$
$$+ (a_{2n-1} - a_{2n}).$$

Since $\langle (a_1 - a_2), (a_3 - a_4), \ldots, (a_{2n-1} - a_{2n}), \ldots \rangle$ is a positive term sequence, the sequence $\left\langle \sum_{k=1}^{2n} (-1)^{k+1} a_k \right\rangle_{n=1}^{\infty}$ is increasing. But

$$\sum_{k=1}^{2n} (-1)^{k+1} a_k = a_1 - a_2 + a_3 - a_4 + \cdots - a_{2n}$$
$$= a_1 + (-a_2 + a_3) + (-a_4 + a_5) + \cdots$$
$$+ (-a_{2n-2} + a_{2n-1}) - a_{2n} \leq a_1$$

since $(-a_2 + a_3), (-a_4 + a_5), \cdots, (-a_{2n-2} + a_{2n-1})$ are all negative numbers.

Since the increasing sequence $\left\langle \sum_{k=1}^{2n} (-1)^{k+1} a_k \right\rangle_{n=1}^{\infty}$ is bounded above by a_1, $\left\langle \sum_{k=1}^{2n} (-1)^{k+1} a_k \right\rangle_{n=1}^{\infty}$ converges. This almost proves the theorem. We have to note that

$$\lim_{n\to\infty} \sum_{k=1}^{2n} (-1)^{k+1} a_k = \lim_{n\to\infty} \sum_{k=1}^{2n} (-1)^{k+1} a_k + \lim_{n\to\infty} a_{n+1} = \lim_{n\to\infty} \sum_{k=1}^{2n} (-1)^{k+1} a_k$$

since $\lim_{n\to\infty} a_{n+1} = 0$. Hence, $\lim_{n\to\infty} \sum_{k=1}^{2n} (-1)^{k+1} a_k = \lim_{n\to\infty} \sum_{k=1}^{2n} (-1)^{k+1} a_k$.

Therefore, $\sum_{n=1}^{\infty} (-1)^{n+1} a_n$ converges.

Problem 3.4.1 Test the convergence of the following series:

(1) $\sum_{n=1}^{\infty} (-1)^{n+1} \frac{1}{n}$

(2) $\sum_{n=1}^{\infty} (-1)^{n+1} \frac{\sqrt{n-1}}{\sqrt{n+1}}$

(3) $\sum_{n=1}^{\infty} (-1)^{n+1} \left(1 + \frac{1}{\sqrt{n}}\right)$

(4) $\sum_{n=1}^{\infty} (-1)^{n} \frac{1}{n}$

(5) $\sum_{n=1}^{\infty} (-1)^{n+1} \frac{\sqrt{n-1}}{\sqrt{n^2+1}}$

(6) $\sum_{n=1}^{\infty} (-1)^{n+1} \left(\frac{1}{\sqrt{n+1}} - \frac{1}{\sqrt{n}}\right)$

Let us take a closer look at the alternating series in the above Theorem 3.4.1. Let $\sum_{n=1}^{\infty} (-1)^{n+1} a_n = \alpha$. Let n be a positive integer. Let $S_n = \sum_{k=1}^{n} (-1)^{k+1} a_k$ Then we have

$$S_{2n+1} = \sum_{k=1}^{2n+1} (-1)^{k+1} a_k = a_1 - a_2 + a_3 - a_4 + \cdots - a_{2n} + a_{2n+1}$$
$$= a_1 + (-a_2 + a_3) + (-a_4 + a_5) + \cdots + (-a_{2n} + a_{2n+1}).$$

Since $\langle (-a_2 + a_3), (-a_4 + a_5), \ldots, (-a_{2n} + a_{2n+1}), \ldots \rangle$ is a negative terms sequence, the sequence $\left\langle \sum_{k=1}^{2n+1} (-1)^{k+1} \right\rangle_{n=1}^{\infty} = \langle S_{2n+1} \rangle_{n=1}^{\infty}$ is decreasing. But

$$S_{2n} = \sum_{k=1}^{2n} (-1)^{k+1} a_k = (a_1 - a_2) + (a_3 - a_4) + \cdots + (a_{2n-1} - a_{2n})$$

Since the terms of the sequence $\langle (a_1 - a_2), (a_3 - a_4), \ldots, (a_{2n-1} - a_{2n}) \ldots \rangle$ are positive, the sequence $\left\langle \sum_{k=1}^{2n+1} (-1)^{k+1} \right\rangle_{n=1}^{\infty} = \langle S_{2n+1} \rangle_{n=1}^{\infty}$ is increasing.

Hence, we have $S_2 < S_4 < S_6 < \cdots < S_{2n} < \alpha < S_{2n+1} < \cdots < S_5 < S_3 < S_1$, and $S_{2n+1} - S_{2n} = a_{2n+1}$. In particular, we have the following corollary.

Corollary 3.4.2 Let $\langle a^n \rangle_{n=1}^{\infty}$ be a monotonically decreasing positive terms sequence converging to 0. Let $\sum_{n=1}^{\infty} (-1)^{n+1} a_n = \alpha$. For every positive integer n, we have

$$\left| \sum_{k=1}^{n} (-1)^{k+1} a_k - \alpha \right| \leq a_{n+1}$$

Example 3.4.1 The series $\sum_{n=1}^{\infty}(-1)^{n+1}\frac{1}{n}$ is called <u>the alternating harmonic series</u>, and it converges by Theorem 3.4.1. This is a special series since we can find its limit.

$$\sum_{k=1}^{2n}(-1)^{k+1}\frac{1}{k} = 1 - \frac{1}{2} + \frac{1}{3} - \frac{1}{4} + \cdots - \frac{1}{2n}$$

$$= \left(1 + \frac{1}{2} + \frac{1}{3} + \frac{1}{4} + \cdots + \frac{1}{2n}\right) - 2\left(\frac{1}{2} + \frac{1}{4} + \frac{1}{6} + \cdots + \frac{1}{2n}\right)$$

$$= \left(1 + \frac{1}{2} + \frac{1}{3} + \frac{1}{4} + \cdots + \frac{1}{2n}\right) - \left(1 + \frac{1}{2} + \frac{1}{3} + \cdots + \frac{1}{n}\right)$$

$$= \frac{1}{n+1} + \frac{1}{n+2} + \frac{1}{n+3} + \cdots + \frac{1}{2n}.$$

As in Theorem 3.3.6, we have

$$\int_{n+1}^{2n}\frac{1}{x}dx < \frac{1}{n+1} + \frac{1}{n+2} + \frac{1}{n+3} + \cdots + \frac{1}{2n} < \int_{n}^{2n-1}\frac{1}{x}dx$$

But

$$\int_{n+1}^{2n}\frac{1}{x}dx = \ln 2n - \ln(n+1) = \ln\frac{2n}{n+1} = \ln\frac{2}{1+\frac{1}{n}}$$

and

$$\int_{n}^{2n-1}\frac{1}{x}dx = \ln(2n-1) - \ln n = \ln\left(2 - \frac{1}{n}\right)$$

This shows that $\lim_{n\to\infty}\int_{n+1}^{2n}\frac{1}{x}dx = \ln 2 = \lim_{n\to\infty}\int_{n}^{2n-1}\frac{1}{x}dx$ so that

$$\sum_{n=1}^{\infty}(-1)^{n+1}\frac{1}{n} = \ln 2$$

This limit is credited to Leibniz while the method we used here is not his. See Remark 3.4.1 (1) below. We will come back to this in Sect. 7.4.

Problem 3.4.2

(1) Following a similar idea in Example 3.4.1, prove that.

$$\sum_{n=0}^{\infty}\left(\frac{1}{3n+1} + \frac{1}{3n+2} - \frac{2}{3n+3}\right) = \left(\frac{1}{1} + \frac{1}{2} - \frac{2}{3}\right) + \left(\frac{1}{4} + \frac{1}{5} - \frac{2}{6}\right) + \cdots = \ln 3$$

(2) Prove that $\sum_{n=0}^{\infty}\left(\frac{1}{3n+1} + \frac{1}{3n+2} - \frac{2}{3n+3}\right) = \left(\frac{1}{1} + \frac{1}{2} - \frac{1}{3}\right) + \left(\frac{1}{4} + \frac{1}{5} - \frac{1}{6}\right) + \cdots$ diverges.

Remark 3.4.1

(1) Leibniz claimed that $1 - \frac{1}{3} + \frac{1}{5} - \frac{1}{7} + \frac{1}{9} - \frac{1}{11} + \frac{1}{13} - \frac{1}{15} + \cdots = \frac{\pi}{4}$, but he did not prove it. Abel proved it. (See Example 7.5.4) This and $\sum_{n=1}^{\infty} (-1)^{n+1} \frac{1}{n} = \ln 2$ are called Leibniz's identities.

(2) In the paper Euler's Miracle, (Euleriana Volume 1, Issue 2, Article 5, 2021 [1]), William Dunham explains the following Euler's series:

$$1 + \frac{1}{2} + \frac{1}{3} + \frac{1}{4} - \frac{1}{5} + \frac{1}{6} + \frac{1}{7} + \frac{1}{8} + \frac{1}{9} - \frac{1}{10} + \frac{1}{11} + \frac{1}{12} - \frac{1}{13}$$
$$+ \frac{1}{14} - \frac{1}{15} + \frac{1}{16} - \frac{1}{17} + \frac{1}{18} + \frac{1}{19} - \frac{1}{20} + \frac{1}{21} + \frac{1}{22} + \frac{1}{23} + \frac{1}{24}$$
$$+ \frac{1}{25} - \frac{1}{26} + \frac{1}{27} + \frac{1}{28} - \frac{1}{29} - \frac{1}{30} + \frac{1}{31} + \cdots = \pi.$$

(3) Regarding Example 3.4.1 and Example 3.4.1, if $k \geq 2$ is an integer, we can prove that

$$\sum_{n=0}^{\infty} \left(\frac{1}{kn+1} + \frac{1}{kn+2} + \cdots + \frac{1}{kn+(k-1)} - \frac{k-1}{kn+k} \right) = \ln k.$$ (This is due to A. Kheyfits, Problem 623, The College Mathematics Journal, Vol. 29, 1998 [2]) and (C. Kicey and S. Goel, *A Series for* $\ln k$, The American Mathematical Monthly, Vol. 105(6), 1998, 552–554 [3]. The problem and the article appeared almost simultaneously, independent of each other.)

(4) $\left(\frac{1}{1} - \frac{2}{3} + \frac{1}{5} \right) + \left(\frac{1}{7} - \frac{2}{9} + \frac{1}{11} \right) + \cdots = \frac{\ln 3}{2}$

$\left(\frac{1}{1} - \frac{3}{4} + \frac{1}{7} + \frac{1}{10} \right) + \left(\frac{1}{13} - \frac{3}{16} + \frac{1}{19} + \frac{1}{22} \right) + \cdots = \frac{\ln 4}{3}$, etc. See *A New Infinite Series Representation of* $\ln k$ by H. Katsuura, The American Mathematical Monthly, Vol.122(4), 2015, page 376 [4].

Problem 3.4.3 Let $a_n = \begin{cases} 1 \text{ if } n \text{ is odd} \\ 2 \text{ if } n \text{ is even} \end{cases}$. Prove that the alternating series $\sum_{n=1}^{\infty} (-1)^{n+1} \frac{a_n}{n}$ diverges. Note that the sequence $\left\langle \frac{a_n}{n} \right\rangle_{n=1}^{\infty}$ is not monotonic.

3.5   Series of Mixed Terms

The comparison test is for the series of **positive** terms, and so far, we only dealt with positive term series or alternating series. In this section, we deal with series of **mixed** terms. That is, the sequence $\langle a_n \rangle$ may not be all positive numbers in this section. Some or all terms in $\langle a_n \rangle$ may be negative numbers.

Theorem 3.5.1 If $\sum_{n=1}^{\infty} |a_n|$ converges, then $\sum_{n=1}^{\infty} a_n$ converges.

Proof Since $\sum_{n=1}^{\infty} (|a_n| + a_n)$ is a series of positive term, and since $\sum_{n=1}^{\infty} (|a_n| + a_n) \leq$ $2 \sum_{n=1}^{\infty} |a_n|$, we know that $\sum_{n=1}^{\infty} (|a_n| + a_n)$ converges by the comparison test. Hence, by Theorem 3.3.3, we have $\sum_{n=1}^{\infty} a_n = \sum_{n=1}^{\infty} (|a_n| + a_n) - \sum_{n=1}^{\infty} a_n$ and, therefore, $\sum_{n=1}^{\infty} a_n$ converges.

Definition 3.5.1 If $\sum_{n=1}^{\infty} |a_n|$ converges, then we say that the series $\sum_{n=1}^{\infty} a_n$ converges absolutely. So, the above theorem can be re-stated as follows: If a series converges absolutely, then it converges.

Theorem 3.5.2 (The ratio test) If $a_n \neq 0$ for all integer n, and $\lim_{n \to \infty} \frac{|a_{n+1}|}{|a_n|} < 1$, then $\sum_{n=1}^{\infty} a_n$ converges absolutely.

Proof Let $\lim_{n \to \infty} \frac{|a_{n+1}|}{|a_n|} = \alpha < 1$, and let r be a number such that $\alpha < r < 1$. Since $r - \alpha > 0$, there exists a positive integer M such that $\left| \frac{|a_{n+1}|}{|a_n|} - \alpha \right| < r - \alpha$ for any integer $n \geq M$. Let $n \geq M$. Then $-(r - \alpha) < \left(\frac{|a_{n+1}|}{|a_n|} - \alpha \right) < r - \alpha$. In particular, we have $0 < \frac{|a_{n+1}|}{|a_n|} < r$, or $|a_{n+1}| < r|a_n|$. Applying this inductively, we have

$$|a_{M+1}| < r|a_M|$$

$$|a_{M+2}| < r|a_{M+1}| < r^2|a_M|$$

$$|a_{M+3}| < r|a_{M+2}| < r^3|a_M|$$

and so on.

We have $|a_{M+n}| < r^n|a_M|$ for every $n = 1, 2, 3, \ldots$. Hence, we have

$$\sum_{n=1}^{\infty} |a_n| = \sum_{k=1}^{M-1} a_n + \sum_{n=M}^{\infty} |a_n| < \sum_{k=1}^{M-1} |a_n| + \sum_{n=M}^{\infty} |a_M|$$

$$< \sum_{k=1}^{M-1} |a_n| + |a_M| \sum_{n=0}^{\infty} r^n = \sum_{k=1}^{M-1} |a_n| + |a_M| \frac{1}{1-r}.$$

Note that $\sum_{k=1}^{M-1} a_n$ is a finite sum and it is a positive number. So $\sum_{n=1}^{\infty} a_n$ is bounded above by $\sum_{k=1}^{M-1} |a_n| + |a_M| \frac{1}{1-r}$. Therefore, $\sum_{n=1}^{\infty} a_n$ converges absolutely by the comparison test.

Remark 3.5.1

(1) Many Calculus textbooks also list "the root test" next to the ratio test. It is similar to the ratio test. But I do not write it here because the ratio test is easier to apply than the root test and the exponential power of a real number is not fully explained yet. And I cannot think of any infinite series that the root test applies but not the ratio test.

(2) If $\lim\limits_{n\to\infty} \frac{|a_{n+1}|}{|a_n|} = 1$, then $\sum_{n=1}^{\infty} a_n$ may or may not converge. Apply the ratio test to $\sum_{n=1}^{\infty} \frac{1}{n}$, $\sum_{n=1}^{\infty} \frac{(-1)^{n+1}}{n}$ and $\sum_{n=1}^{\infty} \frac{1}{n^2}$ to convince yourself. The series $\sum_{n=1}^{\infty} \frac{1}{n}$ diverges while $\sum_{n=1}^{\infty} \frac{(-1)^{n+1}}{n}$ and $\sum_{n=1}^{\infty} \frac{1}{n^2}$ converge.

(3) The ratio test works for some problems in Sect. 3.3. But since applications of the comparison test for convergence not only helps you understand the infinite series better but it is also aesthetically more pleasing, I postponed the introduction of the ratio test until now. Please learn to use the ratio test as a last resort.

(4) The **main purpose/use** of the ratio test is its application to power series as given in the next example. We will talk more about this in Chap. 7.

Example 3.5.1 Let us find all the x-values so that a power series $\sum_{n=0}^{\infty} \frac{(2x)^n}{5n+1}$ converges absolutely. Let $a_n = \frac{(2x)^n}{5n+1}$. Then $\frac{a_{n+1}}{a_n} = \frac{(2x)^{n+1}}{5(n+1)+1} \cdot \frac{5n+1}{(2x)^n} = (2x) \cdot \frac{5n+1}{5(n+1)+1} = (2x) \cdot \frac{5+\frac{1}{n}}{5+\frac{6}{n}}$.

Hence, $\lim\limits_{n\to\infty} \left| \frac{a_{n+1}}{a_n} \right| = |2x|$. Since $|2x| < 1$ implies $|x| < \frac{1}{2}$, the power series converges absolutely when $|x| < \frac{1}{2}$, or when $x \in \left(-\frac{1}{2}, \frac{1}{2}\right)$. This interval $\left(-\frac{1}{2}, \frac{1}{2}\right)$ is called the <u>interval of absolute convergence</u> of this power series by the ratio test. Unfortunately, the ratio test does not tell us about the convergence when $x = -\frac{1}{2}$ or when $x = \frac{1}{2}$. These two cases needed to be dealt with separately. I leave these to the readers as in the next problem.

Problem 3.5.1

(1) Show that the power series $\sum_{n=0}^{\infty} \frac{(2x)^n}{5n+1}$ diverges when $x = \frac{1}{2}$.

(2) Show that the power series $\sum_{n=0}^{\infty} \frac{(2x)^n}{5n+1}$ converges when $x = -\frac{1}{2}$.

(3) Show that the power series $\sum_{n=0}^{\infty} \frac{(2x)^n}{5n+1}$ diverges for all $x \in \left(-\infty, -\frac{1}{2}\right) \cup \left(\frac{1}{2}, \infty\right)$.

Because of (1)–(3) together with Example 3.5.1, the interval $[-\frac{1}{2}, \frac{1}{2})$ is said to be the <u>interval of convergence</u> for the series $\sum_{n=0}^{\infty} \frac{(2x)^n}{5n+1}$.

Problem 3.5.2

(1) Find two divergent series $\sum_{n=1}^{\infty} a_n$ and $\sum_{n=1}^{\infty} b_n$ such that $\sum_{n=1}^{\infty} (a_n + b_n)$ converges.

(2) Suppose two series $\sum_{n=1}^{\infty} a_n$ and $\sum_{n=1}^{\infty} b_n$ converge absolutely. Prove that $\sum_{n=1}^{\infty} (a_n + b_n)$ converges absolutely, and $\sum_{n=1}^{\infty} (a_n + b_n) = \sum_{n=1}^{\infty} a_n + \sum_{n=1}^{\infty} b_n$.

(3) Find two series $\sum_{n=1}^{\infty} a_n$ and $\sum_{n=1}^{\infty} b_n$ that converge absolutely, but $\sum_{n=1}^{\infty} (a_n \cdot b_n) \neq \left(\sum_{n=1}^{\infty} a_n\right) \cdot \left(\sum_{n=1}^{\infty} b_n\right)$.

3.6   Numbers in Bases 2 or 3; Cantor Set

Why do we care to think about the number representations in bases 2 or 3? This is related to our earlier question; what is a real number? In order to understand real numbers better, thinking of number representations other than base 10 is one way to accomplish this. Another purpose of this section is to introduce the Cantor set.

Example 3.6.1

(1) Note that $0.\bar{9} = \sum_{n=1}^{\infty} \frac{9}{10^n} = \frac{9}{10} \sum_{n=0}^{\infty} \frac{1}{10^n} = \frac{9}{10} \cdot \frac{1}{1-\frac{1}{10}} = 1$. Any number s in the interval

$[0,1]$ can be written as $s = \sum_{n=1}^{\infty} \frac{s_n}{10^n} = 0.s_1 s_2 s_3 \cdots$ for some sequence $\langle s_n \rangle_{n=1}^{\infty}$ from

elements in the set $\{0,1,2,3,4,5,6,7,8,9\}$. For example, $\frac{5}{10} + \frac{6}{10^2} + \frac{3}{10^3} = 0.563$. And if m is a positive integer, there exists an integer n and a finite sequence $\langle t_k \rangle_{k=0}^{n}$ from the set $\{0,1,2,3,4,5,6,7,8,9\}$ such that $m = t_0 + t_1 \cdot 10 + t_2 \cdot 10^2 + \cdots + t_n \cdot 10^n$. We usually write $m = t_n t_{n-1} \cdots t_1 t_0$. For example, $5 + 6 \cdot 10 + 3 \cdot 10^2$ is usually written 365. So

$m + s = t_0 + t_1 \cdot 10 + t_2 \cdot 10^2 + \cdots + t_n \cdot 10^n + \sum_{n=1}^{\infty} \frac{s_n}{10^n} = t_n t_{n-1} \cdots t_1 t_0 . s_1 s_2 s_3 \cdots$. For

example, $365.563 = 5 + 6 \cdot 10 + 3 \cdot 10^2 + \frac{5}{10} + \frac{6}{10^2} + \frac{3}{10^3}$.

Because of these, our number system is said to be of **base** 10, and a number in base 10 is called a <u>decimal number</u>.

(2) Any number s in the interval $[0,1]$ can be written as $S = \sum_{n=1}^{\infty} \frac{s_n}{2^n}$ for some $s_n \in \{0, 1\}$ for every integer $n = 1, 2, 3, \ldots$. Hence, $\langle s_n \rangle_{n=1}^{\infty}$ is a sequence from the set $\{0,1\}$. And if m is a positive integer, there exists an integer n and a finite sequence $\langle t_k \rangle_{k=0}^{n}$ from the set $\{0,1\}$ such that $m = t_0 + t_1 \cdot 2 + t_2 \cdot 2^2 + \cdots + t_n \cdot 2^n$. These are **base** 2 expressions of numbers s and m. A number in base 2 is called a <u>binary number</u>. Hence,

$m + s = t_0 + t_1 \cdot 2 + t_2 \cdot 2^2 + \cdots + t_n \cdot 2^n + \sum_{n=1}^{\infty} \frac{s_n}{2^n} = t_n t_{n-1} \cdots t_1 t_0 . s_1 s_2 s_3 \cdots$. The

number $t_n t_{n-1} \cdots t_1 t_0 . s_1 s_2 s_3 \cdots$ in part (1) and in here are different, and we want to distinguish between the two. So we write a base 2 number as in

$$t_n t_{n-1} \cdots t_1 t_0 . s_1 s_2 s_3 \cdots {}_{(2)}$$

The **subscript** (2) here is to indicate that the number is a binary number. For example, if 10011.0011 is in base 2, we write it $10011.0011_{(2)}$ with the subscript (2) to distinguish it from base 10 numbers. Note that $10011.0011_{(2)} \neq 10011.0011$ since

$$10011.0011_{(2)} = 2^4 + 2 + 1 + \frac{1}{2^3} + \frac{1}{2^4} = 16 + 2 + 16 + \frac{1}{8} + \frac{1}{16}$$

Another examples:

(a) $0.\overline{011}_{(2)} = \left(\frac{1}{2^2} + \frac{1}{2^3}\right) + \left(\frac{1}{2^5} + \frac{1}{2^6}\right) + \nu = \frac{11_{(2)}}{8} + \frac{11_{(2)}}{8^2} + \cdots$

$= \sum_{n=1}^{\infty} \frac{0 \cdot 2^2 + 1 \cdot \cdot 2^1 + 1 \cdot 2^0}{8^n} = \frac{3}{8} \sum_{n=1}^{\infty} \frac{1}{8^n}$

$= \frac{3}{8} \cdot \frac{1}{1 - \frac{1}{8}} = \frac{3}{7}.$

(b) $10011.0011_{(2)} = 10011.0010111\ldots_{(2)} = 10011.0010\overline{1}_{(2)}$

(c) $1.0011_{(2)} + 0.011_{(2)} = 1.1001_{(2)}$

(d) $1.0011_{(2)} - 0.011_{(2)} = 0.1101_{(2)}$

(3) Any number s in the interval $[0,1]$ can be written as $s = \sum_{n=1}^{\infty} \frac{s_n}{3^n}$ for some sequence $\langle s_n \rangle_{n=1}^{\infty}$ from the set $\{0,1,2\}$. And if m is a positive integer, there exists an integer n and a finite sequence $\langle t_k \rangle_{k=0}^{n}$ from the set $\{0,1,2\}$ such that $m = t_0 + t_1 \cdot 3 + t_2 \cdot 3^2 + \cdots + t_n \cdot 3^n$. These are **base** 3 expressions of numbers. A number in base 3 is called a ternary number. Hence, $m + s = t_0 + t_1 \cdot 3 + t_2 \cdot 3^2 + \cdots + t_n \cdot 3^n + \sum_{n=1}^{\infty} \frac{s_n}{3^n} = t_n t_{n-1} \cdots t_1 t_0.s_1 s_2 s_3 \cdots_{(3)}.$

Again, the subscript (3) here is to indicate that it is a ternary number.

For example, $2100202.001021_{(3)} = 2 \cdot 3^6 + 3^5 + 2 \cdot 3^2 + 2 + \frac{1}{3^3} + \frac{2}{3^4} + \frac{1}{3^5}.$

Another examples:

(a) $0.\overline{210}_{(3)} = \left(\frac{2}{3^1} + \frac{1}{3^2}\right) + \left(\frac{2}{3^4} + \frac{1}{2^5}\right) + \cdots = \frac{210_{(3)}}{27} + \frac{210_{(3)}}{27^2} + \cdots$

$= \sum_{n=1}^{\infty} \frac{2 \cdot 3^2 + 1 \cdot \cdot 3^1 + 0 \cdot 3^0}{27^n} = \frac{21}{27} \sum_{n=0}^{\infty} \frac{1}{27^n} = \frac{21}{27} \cdot \frac{1}{1 - \frac{1}{27}} = \frac{21}{26}.$

(b) $2100202.001021_{(3)} = 2100202.00102022\cdots_{(3)} = 2100202.0010\overline{202}_{(3)}.$

(c) $2.001021_{(3)} + 0.\overline{210}_{(3)} = 2.21200\overline{1210}_{(3)}$

(d) $2.001021_{(3)} - 0.\overline{210}_{(3)} = 1.02011\overline{0012}_{(3)}.$

Example 3.6.2

(1) A carpenter's ruler (or a tape measure) is a mixture of bases 2, 3 and 12. For example 10.75 inches is usually said to be "10 and ¾ inches" ($10 + \frac{1}{2} + \frac{1}{2^2}$ inches). (26 and $\frac{7}{32}$

inches) is 2 feet plus (2 and $\frac{7}{32} = \frac{0}{2} + \frac{0}{2^2} + \frac{1}{2^3} + \frac{1}{2^4} + \frac{1}{2^5}$) inches. And (89 inches is 7 feet and 5 inches) is equal to 2 yards 1 foot 5 inches.

(2) Measuring time is very complicated, since one year (the time it takes for the earth to rotate around the sun once) is not exactly 365 days. But if the time period is relatively short, the bases of time are 60 and 12. For example, (126,100 s) is (2101 min 40 s), which is equal to (35 h 1 min 40 s), and it is equal to (1 day 11 h 1 min 40 s).

When you read about ancient history, its events are dated in B.C. (Before Christ). The calendar in ancient times is usually based on the Mesopotamian (or Babylonian) calendar. In this calendar, one year has 360 days. So, the summer August becomes winter August in about 36 years since approximately $36 \times 5 = 180$ days "are lost" over 30 years. To correct that, Julius Caesar established the Julian calendar in about 46 B.C. in which 1 year is calculated as 365 days except for a leap year every four year which has 366 days. (Note that the year 1 A.D. is not 0 B.C.) But this was not quite right, either. In 1582 AD, Pope Gregory further corrected the Julian calendar, and established the Gregorian calendar. (I leave it to the readers to find out how the corrections were made.) However, not all countries adopted the Gregorian calendar at that time. Britain and its colonies including what would become the United States, continued to use the Julian calendar. But the U.S. adopted the Gregorian calendar after it declared independence. So, for example, George Washington's birthday was February 11 before the revolution, and it was changed to February 22 after the revolution. Even now, the Greek orthodox church keeps the Julian calendar so that their Christmas date is different from the Catholic Christmas date.

Problem 3.6.1

(1) Write integers from 1 to 20 in binary numbers.
(2) Convert 100 into the binary number.
(3) Write integers from 1 to 30 in ternary numbers.
(4) Convert 1000 into the ternary number.
(5) Convert 2741 into the ternary number.

(**Answers**: (2) $100 = 110010_{(2)}$, (5) $2741 = 10202112_{(3)}$.)

Example 3.6.3 Arithmetic in ternary numbers:

(1) $1_{(3)} + 2_{(3)} = 10_{(3)}$, $12_{(3)} + 21_{(3)} = 110_{(3)}$.

$$1.02_{(3)} + 2.21_{(3)} = 11.00_{(3)}, \quad 12.12_{(3)} + 21.22_{(3)} = 111.11_{(3)}.$$

(2) All you need to do the multiplications in base 3 is to know that $1_{(3)} \times 0_{(3)} = 0_{(3)}$, $1_{(3)} \times 1_{(3)} = 1_{(3)}$, $1_{(3)} \times 2_{(3)} = 2_{(3)}$, and $2_{(3)} \times 2_{(3)} = 11_{(3)}$. From these, we can compute the following.

$$12_{(3)} \times 2_{(3)} = 101_{(3)}, \quad 12_{(3)} \times 20_{(3)} = 1010_{(3)}.$$

Hence,

$$12_{(3)} \times 22_{(3)} = 12_{(3)} \times 2_{(3)} + 12_{(3)} \times 20_{(3)} = 101_{(3)} + 1010_{(3)} = 1111_{(3)}$$

(3) We compute $212.2_{(3)} \times 22.1_{(3)}$. For simplicity, we drop the subscripts in Fig. 3.3.

From Fig. 3.3, we have $2122_{(3)} \times 221_{(3)} = 2102202_{(3)}$.
    Hence, $212.2_{(3)} \times 22.1_{(3)} = 21022.02_{(3)}$.

(4) Compute $\frac{2122_{(3)}}{21_{(3)}} = 2122_{(3)} \div 21_{(3)}$.

Figure 3.4 is the division algorithm in base 3. From Fig. 3.4, we have

$$2122_{(3)} \div 21_{(3)} = 101.0102 \cdots_{(3)}$$

However, the remainder 1 appeared twice (see "$\leftarrow$" in Fig. 3.4), so we know that

$$2122_{(3)} \div 21_{(3)} = 101.\underbrace{0102}\underbrace{0102}\cdots_{(3)} = 101.\overline{0102}_{(3)}$$

(5) Write $0.\overline{21}_{(3)}$ as a fraction of two integers in base 3.

(We drop subscripts for simplicity in the next line.)

$$0.\overline{21} = \sum_{n=1}^{\infty} \frac{21}{100^n} = \frac{21}{100}\left(1 + \frac{1}{100} + \frac{1}{100^2} + \cdots\right) = \frac{21}{100} \cdot \frac{1}{1 - \frac{1}{100}} = \frac{21}{100} \cdot \frac{100}{22} = \frac{21}{22}$$

Hence, $0.\overline{21}_{(3)} = \frac{21_{(3)}}{22_{(3)}}$ (Figs. 3.3 and 3.4).

Problem 3.6.2 Compute the following using only base 2 or base 3, without converting them into decimal numbers.

(1) $2100202.001021_{(3)} + 1202.1220121_{(3)}$ (Answer should be in base 3.)
(2) $2100202.001021_{(3)} + 1202.1220121_{(3)}$ (Answer should be in base 3.)

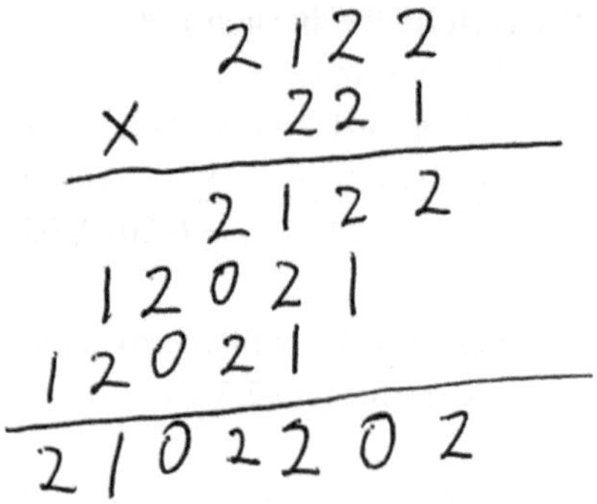

Fig. 3.3 Base 3 multiplication

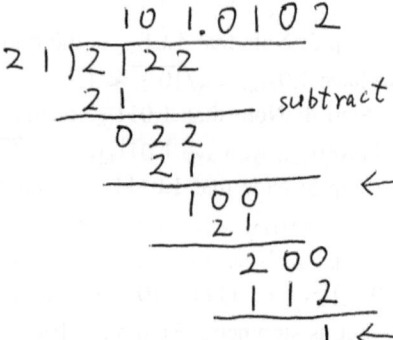

Fig. 3.4 Base 3 division
algorithm

(3) All you need to do the multiplications in base 2 is to know that $1_{(2)} \times 0_{(2)} = 0_{(2)}$ and
$1_{(2)} \times 1_{(2)} = 1_{(2)}$. Compute $10101011_{(2)} \times 1101_{(2)}$ (Answer should be in base 2).

(4) Perform the multiplication $120_{(3)} \times 20_{(3)}$ without using decimal numbers. Answer should
be in base 3.

(5) Perform the multiplication $120_{(3)} \times 2021_{(3)}$ without using decimal numbers. Answer
should be in base 3.

(6) Compute $21021_{(3)} \times 121_{(3)}$ (Answer should be in base 3.)

(7) Compute $21.021_{(3)} \times 1.21_{(3)}$ (Answer should be in base 3.)

(8) Convert $\frac{1_{(3)}}{11_{(3)}}$ into $\sum_{n=1}^{\infty} \frac{s_n}{3^n}$ for some sequence $\langle s_n \rangle_{n=1}^{\infty}$ from the set $\{0,1,2\}$. This is
asking you to perform the division algorithm in ternary numbers.

(9) Repeat (8) for $\frac{11_{(3)}}{121_{(3)}}$ and $\frac{121_{(3)}}{11_{(3)}}$.

(10) Perform the division $120_{(3)} \div 20_{(3)}$ without using decimal numbers. Answer should
be in base 3.

(11) Perform the division $120_{(3)} \div 2021_{(3)}$ only without using decimal numbers. Answer
should be in base 3.

(12) Perform the division $1.20_{(3)} \div 2.021_{(3)}$ only without using decimal numbers. Answer should be in base 3.

(Selected **Answers**: (3) $100010101111_{(2)}$; (5) $22122_{(3)}$; (7) $110.22011_{(3)}$;

(9) $\frac{11_{(3)}}{121_{(3)}} = 0.\overline{02}_{(3)} = \sum_{n=1}^{\infty} \frac{2}{3^{2n}}$, and $\frac{121_{(3)}}{11_{(3)}} = 11_{(3)}$

(12) $1.20_{(3)} \div 2.021_{(3)} = 0.2012202\ldots_{(3)}.$)

Example 3.6.4 We will approximate $\sqrt{2} = \sqrt{10_{(2)}}$ in binary numbers. (This is similar to the proof of Theorem 2.5.1.)

Step 1: $\left(1_{(2)}\right)^2 = 1_{(2)}$ and $\left(10_{(2)}\right)^2 = 100_{(2)}$, so $1_{(2)} < \sqrt{10_{(2)}} = 10_{(2)}$.

Step 2: Note that $1.1_{(2)}$ is the midpoint of $1_{(2)}$ and $10_{(2)}$. Since $1.1_{(2)}^2 = 10.01_{(2)}$, we have $1_{(2)} < \sqrt{10_{(2)}} < 1.1_{(2)}$.

Step 3: Note that $1.01_{(2)}$ is the midpoint of $1_{(2)}$ and $1.1_{(2)}$. Since $1.01_{(2)}^2 = 1.1001_{(2)}$, we have $1.01_{(2)} < \sqrt{10_{(2)}} < 1.1_{(2)}$.

Step 4: Note that $1.011_{(2)}$ is the midpoint of $1.01_{(2)}$ and $1.1_{(2)}$. Since $1.011_{(2)}^2 = 1.111001_{(2)}$, we have $1.011_{(2)} < \sqrt{10_{(2)}} < 1.1_{(2)}$.

Step 5: Note that $1.0111_{(2)}$ is the midpoint of $1.011_{(2)}$ and $1.1_{(2)}$. Since $1.0111_{(2)}^2 = 10.00000001_{(2)} > 2$, we have $1.011_{(2)} < \sqrt{10_{(2)}} < 1.0111_{(2)}$.

Step 6: Note that $1.01101_{(2)}$ is the midpoint of $1.011_{(2)}$ and $1.0111_{(2)}$. Since $1.01101_{(2)}^2 = 1.1111101001_{(2)}$, we have $1.01101_{(2)} < \sqrt{10_{(2)}} < 1.0111_{(2)}$.

Let us stop here. We have $\sqrt{10_{(2)}} = 1.01101\cdots_{(2)}$.

Problem 3.6.3 Determine $s_1, s_2, s_3, s_4 \in \{0, 1, 2\}$ such that $(1.s_1s_2s_3s_4)_{(3)} < \sqrt{2} = \sqrt{2}_{(3)} < (1.s_1s_2s_3s_4)_{(3)} + (1.00001)_{(3)}$.

Problem 3.6.4 Convert 7 and 12 into ternary numbers. Then convert $\frac{7}{12}$ (in base 10) into the ternary number.

(**Answer**: $\frac{7}{12} = \frac{21_{(3)}}{110_{(3)}} = 0.12\overline{02}_{(3)}$. Note that $0.12\overline{02}_{(3)} = \frac{1}{3} + \frac{2}{3^2} + \sum_{n=1}^{\infty} \frac{2}{3^{2+2n}}.$)

Example 3.6.5 Some numbers have two expressions. For example, $1 = 0.999\ldots = 0.\overline{9}$ and $28.3 = 28.2\overline{9}$. In order to see $28.3 = 28.2\overline{9}$, note that

$$28.2\overline{9} = 28.2 + \sum_{n=2}^{\infty} \frac{9}{10^n} = 28.2 + \frac{9}{10^2} + \frac{9}{10^3} + \cdots$$

$$= 28.2 + \frac{9}{100}\left(1 + \frac{1}{10} + \frac{1}{10^2} + \cdots\right) = 28.2 + \frac{9}{100} \cdot \frac{1}{1 - \frac{1}{10}} = 28.2 + \frac{1}{10} = 28.3.$$

Similarly, we also have $1.11_{(2)} = 1.101111\cdots_{(2)} = 1.10\overline{1}_{(2)}$, $0.22_{(3)} = 0.21\overline{2}_{(3)}$, and $122.0021_{(3)} = 122.0020222\cdots_{(3)} = 122.0020\overline{2}_{(3)}$. (Prove these.)

Problem 3.6.5

(1) A sequence s_n is defined by $s_{3n} = 0$, $s_{3n+1} = 1$, $s_{3n+2} = 2$ for all integers $n > 0$. Find the limit $\sum_{n=0}^{\infty} \frac{s_n}{3^n}$. In other words, convert this ternary number $\sum_{n=0}^{\infty} \frac{s_n}{3^n} = 0.120120\cdots_{(3)} =$ $0.12\overline{012}_{(3)} = 0.\overline{120}_{(3)}$ into a fraction of two decimal integers. By division algorithm, check your answer as in Problem 3.6.4.

(2) Also, express $0.120120\cdots_{(3)} = 0.\overline{120}_{(3)}$ in a fraction of two ternary integers, without using the answer in (1). Do not use decimal numbers.

(3) Express $0.\overline{10}_{(2)}$ in a fraction of two binary integers. Do not use decimal numbers.

(4) Express $0.\overline{212}_{(3)}$ in fraction of two ternary integers. Do not use decimal numbers.

(**Answer** for (2): All the numbers in this answer are in base 3.

$$0.120120\cdots_{(3)} = \frac{1}{10} + \frac{2}{10^2} + \frac{0}{10^{10}} + \frac{1}{10^{11}} + \frac{2}{10^{12}} + \frac{0}{10^{20}} + \cdots = \frac{120}{1000} + \frac{120}{1000^2} + \cdots$$

$$= \frac{120}{1000}\left(1 + \frac{1}{1000} + \frac{1}{1000^2} + \cdots\right) = \frac{120}{1000}\cdot\frac{1}{1 - \frac{1}{1000}} = \frac{120}{1000}\cdot\frac{1000}{222} = \frac{120_{(3)}}{222_{(3)}}.)$$

Definition 3.6.1 Let C be the set of all numbers $\sum_{n=1}^{\infty}\frac{s_n}{3^n} = 0.s_1 s_2 s_3 \cdots_{(3)}$ for some sequence $\langle s_n\rangle_{n=1}^{\infty}$ from the set $\{0,2\}$. So $s_n = 0$ or $s_n = 2$ for all integers $n > 0$. Then the set C is called the <u>Cantor (ternary) set</u>. So, $1 = 1_{(3)} = 0.222\cdots_{(3)}$, $\frac{1}{9} = 0.01_{(3)} = 0.00222\cdots_{(3)}$, and $0.2021_{(3)} = 0.202022\cdots_{(3)}$ are all points in C. But $0.12_{(3)}, 0.0212\cdots_{(3)}, 0.2121_{(3)}$ are not in C (Figs 3.5 and 3.6).

Fig. 3.5 A part of Cantor ternary set

Fig. 3.6 A magnification of the interval $[0, \frac{1}{9}]$ inside the Cantor set

Example 3.6.6

(1) $\frac{2}{3} = 0.2_{(3)} \in C$, and $\frac{1}{3} = 0.1_{(3)} = 0.0222\cdots_{(3)} = \sum_{n=2}^{\infty} \frac{2}{3^n} \in C$.
(2) $\frac{1}{9}, \frac{2}{9}, \frac{3}{9} = \frac{1}{3}, \frac{7}{9}, \frac{8}{9}, \frac{9}{9} = 1$ are all points in C.
(3) $\frac{7}{10} \in C$ since $\frac{7}{10} = \frac{21_{(3)}}{101_{(3)}} = 0.2002002\ldots_{(3)} = 0.2\overline{002}_{(3)} \in C$.

Problem 3.6.6

(1) Find all integers n such that $\frac{n}{81}$ is a point in C.
(2) Rewrite 4 in ternary integer. Use a division algorithm in base 3 to express $\frac{1}{4}$ in ternary number. This will show that $\frac{1}{4} \in C$.
(3) Show $\frac{1}{5} \notin C$. Use a similar strategy as in (2).
(4) Is $\frac{9}{10} \in C$? Prove your claim.
(5) Is $\frac{11}{12} \in C$? Prove your claim.
(6) Is $\sqrt{3} - 1 \in C$? Prove your claim.

(**Answers**: (2) $\frac{1}{4} = 0.\overline{02}_{(3)} \in C$; (3) $\frac{1}{5} = 0.012\ldots_{(3)} \notin C$; (6) Since $1.21_{(3)} < \sqrt{3} < 1.211_{(3)}$, we know that $\sqrt{3} - 1 \notin C$.)

Example 3.6.7 Suppose a point on the interval $[0, 1]$ is equally likely to be picked. Then the probability of picking a point on the Cantor set is zero. In order to see this, we will calculate the probability of picking a point from the compliment of the Cantor set C.

<u>Step 1</u>: Since all numbers in the open interval $U_1 = (0.1_{(3)}, 0.2_{(3)})$ has the form $0.1\cdots_{(3)}$, no number in U_1 is a point of C. The probability of picking a point in U_1 is given by its length, which is $\frac{1}{3}$. Note that $U_1 = (\frac{1}{3}, \frac{2}{3})$.

<u>Step 2</u>: No numbers in $U_2 = (0.01_{(3)}, 0.02_{(3)}) \cup (0.21_{(3)}, 0.22_{(3)})$ is a point of C. We can write U_1 as $U_1 = (0.10_{(3)}, 0.20_{(3)})$, so we can see that $U_1 \cap U_2 = \emptyset$. The probability of picking a point in U_2 is given by its total length, which is $2 \cdot \frac{1}{3^2}$.

Note that $U_2 = (\frac{1}{3^2}, \frac{2}{3^2}) \cup (\frac{7}{3^2}, \frac{8}{3^2})$, and $U_2 = [\frac{1}{10_{(3)}} \cdot U_1] \cup [\frac{1}{10_{(3)}} \cdot U_1 + 0.2_{(3)}]$.

(Also, note that $\frac{1}{10_{(3)}} = \frac{1}{3}$, $0.2_{(3)} = \frac{2}{3}$, and see Definition 2.5.6 for the set $\frac{1}{10_{(3)}} \cdot U_1 + 0.2_{(3)}$.)

<u>Step 3</u>: No numbers in
$$U_3 = (0.001_{(3)}, 0.002_{(3)}) \cup (0.021_{(3)}, 0.022_{(3)}) \cup (0.201_{(3)}, 0.202_{(3)}) \cup (0.221_{(3)}, 0.222_{(3)})$$
is a point of C. We can write U_2 as
$$U_2 = (0.010_{(3)}, 0.020_{(3)}) \cup (0.210_{(3)}, 0.220_{(3)})$$
so we can see that $U_2 \cap U_3 = \emptyset$. The probability of picking a point in U_3 is given by its total length, which is $2^2 \cdot \frac{1}{3^3}$.

Note that $U_3 = \left(\frac{1}{3^3}, \frac{2}{3^3}\right) \cup \left(\frac{7}{3^3}, \frac{8}{3^3}\right) \cup \left(\frac{19}{3^3}, \frac{20}{3^3}\right) \cup \left(\frac{25}{3^3}, \frac{26}{3^3}\right)$, and $U_3 = \left[\frac{1}{10_{(3)}} \cdot U_2\right] \cup$ $\left[\frac{1}{10_{(3)}} \cdot U_2 + 0.2_{(3)}\right]$.

Step n: Let $n \geq 1$ be an integer. Suppose U_n is defined with no elements U_n being a point of C and the total length of being $2^{n-1} \cdot \frac{1}{3^n}$.

Step $n+1$: Let $U_{n+1} = \left[\frac{1}{10_{(3)}} \cdot U_n\right] \cup \left[\frac{1}{10_{(3)}} \cdot U_n + 0.2_{(3)}\right]$. Then $U_n \cap U_{n+1} = \emptyset$, no points of U_{n+1} is a point of C, and the total length of U_{n+1} is $2^n \cdot \frac{1}{3^{n+1}}$.

Hence, U_n is defined for every integer $n \geq 1$. Note that $C = [0, 1] - \bigcup_{n=1}^{\infty} U_n = \bigcup_{n=1}^{\infty} ([0, 1] - U_n)$. The Cantor set C is the complement of $\bigcup_{n=1}^{\infty} U_n$, and the probability of picking a point from the **compliment** of the Cantor set is given by the geometric series

$$\frac{1}{3} + 2 \cdot \frac{1}{3^2} + 2^2 \cdot \frac{1}{3^3} + \cdots = \frac{1}{3}\left\{1 + \frac{2}{3} + \left(\frac{1}{3}\right)^2 + \cdots\right\} = \frac{1}{3} \cdot \frac{1}{1 - \frac{2}{3}} = 1$$

This shows that the probability of picking a point on the Cantor set is zero. Moreover, if we let $I_n = [0, 1] - U_n$ for all integers $n > 0$, then the Cantor set C is given by $C = \bigcap_{n=1}^{\infty} I_n$ (Fig. 3.7)

Theorem 3.6.1 Let $f : C \to [0, 1]$ be a function defined by

$f\left(\sum_{n=1}^{\infty} \frac{S_n}{3^n}\right) = \sum_{n=1}^{\infty} \frac{\left(S_n/2\right)}{2^n} = \sum_{n=1}^{\infty} \frac{S_n}{2^{n+1}}$ for every $\sum_{n=1}^{\infty} \frac{S_n}{3^n} \in C$. (This function f is called the Cantor function.) Then, this function f is **onto**.

Proof Let $y \in [0, 1]$. Then $y = \sum_{n=1}^{\infty} \frac{a_n}{2^n}$ for some $a_n \in \{0, 1\}$ for all $n = 1, 2, 3, \ldots$. Since $\sum_{n=1}^{\infty} \frac{2a_n}{3^n} \in C$ and $f\left(\sum_{n=1}^{\infty} \frac{2a_n}{3^n}\right) = \sum_{n=1}^{\infty} \frac{a_n}{2^n}$, we can see that f is onto (Fig. 3.8).

Problem 3.6.7 Show that the Cantor function is **not** one-to-one.

Remark 3.6.2 Example 3.6.7 and Theorem 3.6.1 seem contradictory to each other, but they are not. Theorem 3.6.1 shows that the cardinality of the Cantor set is greater than or equal to the interval $[0, 1]$. Since the Cantor set is a subset of $[0, 1]$, the cardinality of the interval $[0, 1]$ is greater than or equal to the Cantor set. Therefore, it is plausible to say that there is a one-to-one function from the Cantor set onto $[0, 1]$. And, yes, it is the case. (See Problem

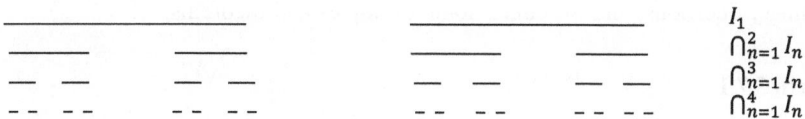

Fig. 3.7 Construction of Cantor set described in Example 3.6.7

Fig. 3.8 A graph of Cantor
function

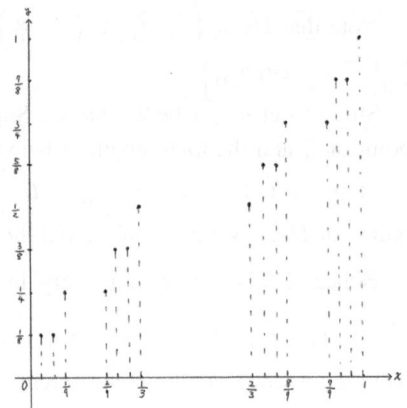

1 J on page 15 of General Topology by Stephen Willard, 1970 [5].) The cardinality of both
the Cantor set and the interval [0, 1] are the same continuum c. Cantor set is a mysterious
set with many other interesting properties.

Example 3.6.8 By looking at Fig. 3.5, it seems that all points in C are rational points. But
by Theorem 3.6.1, there must be many irrational points in C. So let's find one. By Example
3.6.4, we know that $\sqrt{2} = \sqrt{10_{(2)}} = 1.01101\cdots_{(2)}$. Hence, $\sqrt{2} - 1 = 0.01101\cdots_{(2)}$.
Let $f : C \to [0, 1]$ be the Cantor function. Then $f\left(0.02202\cdots_{(3)}\right) = 0.01101\cdots_{(2)}$, where
$0.02202\cdots_{(3)}$ is exactly the number obtained by replacing all "1" in $0.01101\cdots_{(2)} = \sqrt{2}-1$
by "2".

Suppose the number $0.02202\cdots_{(3)}$ is rational. Then the tail end of it must repeat, which
implies that the tail end of $0.01101\cdots_{(2)}$ must repeat. Hence, $0.01101\cdots_{(2)}$ is rational,
which is a contradiction to $\sqrt{2} - 1$ being irrational. Hence, the number $0.02202\cdots_{(3)}$ must
be an irrational point in C.

Analogously, if $y \in [0, 1]$ is an irrational number, then the set $f^{-1}(y)$ must be a set with
one element, and it is an irrational point in C.

3.7   Cauchy Sequence

Before introducing the definition of a Cauchy sequence, let us start by asking some
preliminary questions, and introduce some examples and theorems.

Problem 3.7.1

(1) Prove that a convergent sequence is bounded. (This is a repeat of Theorem 3.1.3.)

(2) Prove that if $\langle n_k \rangle_{k=1}^{\infty}$ is a strictly increasing sequence of positive integers, then $n_k \geq k$ for each $k = 1, 2, 3, \ldots$. Please prove this formally by mathematical induction.

Definition 3.7.1 If $\langle n_k \rangle_{k=1}^{\infty}$ is a **strictly** increasing sequence of positive integers, then the sequence $\langle a_{n_k} \rangle_{k=1}^{\infty}$ is said to be a <u>subsequence</u> of $\langle a_n \rangle_{n=1}^{\infty}$.

Example 3.7.1

(1) Let $\langle a_n \rangle_{n=1}^{\infty}$ be a sequence. Then $\langle a_n \rangle_{n=1}^{\infty}$ itself is a subsequence of $\langle a_n \rangle_{n=1}^{\infty}$.

Also, the sequences $\langle a_n \rangle_{n=100}^{\infty}$, $\langle a_{3n} \rangle_{n=1}^{\infty}$, and $\langle a_4, a_5, a_{10}, a_{123}, \ldots \rangle$ are subsequences of $\langle a_n \rangle_{n=1}^{\infty}$.

But sequences $\langle a_4, a_5, a_3, a_{13}, a_8, \ldots \rangle$ is not a subsequence of $\langle a_n \rangle_{n=1}^{\infty}$ since a_3 comes after a_5, and $\langle a_4, a_5, a_5, a_{10}, a_{10}, a_{13}, \ldots \rangle$ is **not** subsequence of $\langle a_n \rangle_{n=1}^{\infty}$ since a_5 and a_{10} are repeated.

(2) A sequence $\left(\frac{1}{3n-2} \right)_{n=1}^{\infty}$ can be thought of as a subsequence of $\langle \frac{1}{n} \rangle_{n=1}^{\infty}$ since $\left(\frac{1}{3k-2} \right)_{k=1}^{\infty} = \langle a_{n_k} \rangle_{k=1}^{\infty}$ where $a_{n_k} = \frac{1}{3k-2}$.

(3) The sequences $\langle 1, 1, 1, \ldots \rangle_{n=1}^{\infty} = \langle (-1)^{2n} \rangle_{n=1}^{\infty}$ and $\langle (-1)^n \rangle_{n=1}^{\infty}$ can be thought of as subsequences of $\langle (-1)^{n+1} \rangle_{n=1}^{\infty}$. This sequence $\langle (-1)^{n+1} \rangle_{n=1}^{\infty}$ is bounded and divergent (see Problem 3.1.2 (4)).

(4) The sequence $\langle a_n \rangle_{n=1}^{\infty}$ in the proof of Theorem 2.5.1 is a bounded divergent sequence. See Example 3.3.1(4).

Problem 3.7.2

(1) Suppose $\langle a_n \rangle_{n=1}^{\infty}$ is a convergent sequence. If $\langle a_{n_k} \rangle_{k=1}^{\infty}$ is a subsequence of $\langle a_n \rangle_{n=1}^{\infty}$, prove that $\lim\limits_{k \to \infty} a_{n_k} = \lim\limits_{n \to \infty} a_n$. (In short, we can state this by saying that a subsequence of a convergent sequence is convergent.)

(2) It is known that there is a function from the set of all natural numbers $\mathbb{N}$ onto the set of all rational numbers in $[0, 1] \cap \mathbb{Q}$. Let $f : \mathbb{N} \to [0, 1] \cap \mathbb{Q}$ be an onto function. Then the sequence $\langle f(n) \rangle_{n=1}^{\infty}$ is bounded by 0 and 1. Prove that $\langle f(n) \rangle_{n=1}^{\infty}$ is a divergent sequence. (Hint: If you assume that $\langle f(n) \rangle_{n=1}^{\infty}$ is convergent, then this will lead you to a contradiction.)

There are bounded sequences that diverge as you have seen above. We will prove next that a bounded sequence has a convergent subsequence. (The sequences $\langle 1, 1, 1, \ldots \rangle_{n=1}^{\infty} = \langle (-1)^{2n} \rangle_{n=1}^{\infty}$ and $\langle (-1)^{2n+1} \rangle_{n=1}^{\infty}$ are convergent subsequences of $\langle (-1)^{n+1} \rangle_{n=1}^{\infty}$ while $\langle (-1)^{n+1} \rangle_{n=1}^{\infty}$ diverges. However, the existence of a convergent subsequence of $\langle f(n) \rangle_{n=1}^{\infty}$ in Problem 3.7.2(2) is not clear.) In Example 3.6.4, we approximated $\sqrt{2} = \sqrt{10_{(2)}}$ with binary numbers. The proof of the next theorem is very similar to the argument in Example 3.6.4 and Theorem 2.5.1. As in the proof of Theorem 2.5.1, we will use mathematical induction to prove Theorem 3.7.1. The next theorem is named after two mathematicians, Bolzano and Weierstrass. it will be used to prove the convergence of a Cauchy sequence in Theorem 3.7.2 below.

Theorem 3.7.1 (Bolzano-Weierstrass Theorem) A bounded sequence has a convergent subsequence.

Proof Let $\langle a_n \rangle_{n=1}^{\infty}$ be a bounded sequence. We will construct a convergent subsequence of $\langle a_n \rangle_{n=1}^{\infty}$.

Step 1: Let b_1 and c_1 be the lower and upper bounds of the set $\{a_n\}_{n=1}^{\infty}$. Let $n_1 = 1$. Then $b_1 \leq a_n \leq c_1$ for infinitely many $n \in \{n_1 + 1, n_1 + 2, n_1 + 3, \ldots\}$.

Step k: Suppose k is a positive integer such that $b_k, c_k,$ and n_k are defined so that $b_k \leq a_n \leq c_k$ for **infinitely** many $n \in \{n_k + 1, n_k + 2, n_k + 3, \ldots\}$.

Step (k + 1): Suppose $b_k \leq a_n \leq \frac{b_k+c_k}{2}$ for **infinitely** many $n \in \{n_k + 1, n_k + 2, n_k + 3, \ldots\}$. We let n_{k+1} to be the smallest integer in $\{n_k + 1, n_k + 2, n_k + 3, \ldots\}$ such that $b_k \leq a_{n_{k+1}} \leq \frac{b_k+c_k}{2}$. Let $b_{k+1} = b_k$ and $c_{k+1} = \frac{b_k+c_k}{2}$.

Suppose $b_k \leq a_n \leq \frac{b_k+c_k}{2}$ for **finitely** many $n \in \{n_k + 1, n_k + 2, n_k + 3, \ldots\}$. Then $\frac{b_k+c_k}{2} < a_n \leq c_k$ for **infinitely** many $n \in \{n_k + 1, n_k + 2, n_k + 3, \ldots\}$. In this case, we let n_{k+1} to be the smallest integer in $\{n_k + 1, n_k + 2, n_k + 3, \ldots\}$ such that $\frac{b_k+c_k}{2} < a_{n_{k+1}} \leq c_k$. Let $b_k = \frac{b_k+c_k}{2}$ and $c_{k+1} = c_k$.

By the mathematical induction, we have constructed three sequences $\langle b_k \rangle_{k=1}^{\infty}$, $\langle c_k \rangle_{k=1}^{\infty}$, and the strictly increasing sequence of natural numbers $\langle n_k \rangle_{k=1}^{\infty}$ such that $b_k \leq a_{n_k} \leq c_k$ for all $k = 1, 2, 3, \ldots$.

From construction, $\langle b_k \rangle_{k=1}^{\infty}$ is an increasing sequence bounded above by c_1. So $\lim_{k \to \infty} b_k$ exists. Let $\lim_{k \to \infty} b_k = \beta$. Also, $\langle c_k \rangle_{k=1}^{\infty}$ is a decreasing sequence bounded below by b_1. So $\lim_{k \to \infty} c_k$ exists. Let $\lim_{k \to \infty} c_k = \gamma$. Moreover, $\frac{|c_k - b_k|}{2} = \frac{|c_1 - b_1|}{2^k}$ for all $k = 1, 2, 3, \ldots$. Hence, $|\gamma - \beta| = \lim_{k \to \infty} |b_k - c_k| = \lim_{k \to \infty} \frac{|c_1 - b_1|}{2^k} = |c_1 - b_1| \cdot \lim_{k \to \infty} \frac{1}{2^k} = 0$. Hence, $\beta = \lim_{k \to \infty} b_k = \lim_{k \to \infty} c_k = \gamma$. But $b_k \leq a_{n_k} \leq c_k$ for all $k = 1, 2, 3, \ldots$. By the sandwich theorem, we have $\lim_{k \to \infty} b_k = \lim_{k \to \infty} a_{n_k} = \lim_{k \to \infty} c_k$. Therefore, $\langle a_{n_k} \rangle_{k=1}^{\infty}$ is a convergent subsequence of $\langle a_n \rangle_{n=1}^{\infty}$.

Example 3.7.2 Let $\langle a_n \rangle_{n=1}^{\infty}$ be the sequence of integers in the proof of Theorem 2.5.1. We proved that $\langle a_n \rangle_{n=1}^{\infty}$ is a bounded divergent sequence in Example 3.3.1(4). So it has a convergent subsequence $\langle a_{n_k} \rangle_{k=1}^{\infty}$. Since a_n is one of the integers among 0, 1, 2, ..., 9, there must be an integer M such that a_{n_k} is constant for all integers $k \geq M$.

Now, we give a definition of Cauchy sequence.

Definition 3.7.2 A sequence $\langle a_n \rangle_{n=1}^{\infty}$ is called a <u>Cauchy sequence</u> if for every $\varepsilon > 0$, there exists a positive integer M such that $|a_n - a_m| < \varepsilon$ for all integers $m, n \geq M$.

Baron Augustine-Louis Cauchy (1789–1857) was a famous French mathematician, engineer, and physicist. Besides his contribution to the definitions of convergence of a sequence and a Cauchy sequence, Cauchy is probably most famous for his contribution to the complex analysis.

Theorem 3.7.2(2) below is the reason why we introduce the concept of a Cauchy sequence. However, it is difficult to see why it is important from the theorem. Its applications will be given in Sect. 3.9, and hopefully, that will convince you of the importance of Cauchy sequence.

Problem 3.7.3

(1) Prove that a convergent sequence is a Cauchy sequence.
(2) Suppose $\lim_{n \to \infty} a_n = \infty$ (see Definition 3.1.2). If $\langle a_{n_k} \rangle_{k=1}^{\infty}$ is a subsequence of $\langle a_n \rangle_{n=1}^{\infty}$, prove that $\lim_{k \to \infty} a_{n_k} = \infty$.

Theorem 3.7.2

(1) A Cauchy sequence is bounded.
(2) A Cauchy sequence is convergent. (This is the converse of Problem 3.7.3(1).)

Proof of (1): This proof is similar to the proof of Theorem 3.1.3. Let $\langle a_n \rangle_{n=1}^{\infty}$ be a Cauchy sequence. Then there exists a positive integer M such that $|a_n - a_m| < 1$ for all integers $n, m \geq M$. In particular, we have $|a_n - a_M| < 1$ for all integers $n \geq M$. Hence, $-|a_M| - 1 \leq a_M - 1 < a_n < a_M + 1 \leq |a_M| + 1$. Hence, $|a_n| \leq |a_M| + 1$ for all integer $n \geq M$. Let $A = \max\{|a_1|, |a_2|, \ldots, |a_M|\}$. Therefore, $\{|a_n|\}_{n=1}^{\infty}$ is bounded above by $A + |a_M| + 1$. But since $-|a_n| \leq a_n \leq |a_n|$ for all integer n, $\{a_n\}_{n=1}^{\infty}$ is bounded below by $-(A + |a_M| + 1)$, and bounded above by $A + |a_M| + 1$. Hence, a Cauchy sequence is bounded.

Proof of (2): Let $\langle a_n \rangle_{n=1}^{\infty}$ be a Cauchy sequence. Since $\langle a_n \rangle_{n=1}^{\infty}$ is bounded by part (1), it has a convergent subsequence $\langle a_{n_k} \rangle_{k=1}^{\infty}$, and say $\lim_{k \to \infty} a_{n_k} = \alpha$, by Bolzano-Weierstrass Theorem. We will prove that $\lim_{n \to \infty} a_n = \alpha$.

Let $\varepsilon > 0$. Since $\langle a_n \rangle_{n=1}^{\infty}$ is Cauchy, there exists a positive integer M_1 such that $|a_n - a_m| < \frac{\varepsilon}{2}$ for all integers $m, n \geq M_1$. Since $\lim_{k \to \infty} a_{n_k} = \alpha$, there exists a positive integer M_2 such that $|a_{n_k} - \alpha| < \frac{\varepsilon}{2}$ for all integers $k \geq M_2$.

Let $m, k \geq M_1 + M_2$. Then $|a_m - \alpha| \leq |a_m - a_{n_k}| + |a_{n_k} - \alpha|$ by the triangle inequality. Since $k \geq M_1 + M_2 > M_2$, we have $n_k \geq k > M_2$ so that $|a_{n_k} - \alpha| < \frac{\varepsilon}{2}$. Again, we know that $n_k \geq k$ so that $n_k \geq k \geq M_1 + M_2 \geq M_1$. Together with $m \geq M_1 + M_2 \geq M_1$, we have $|a_m - a_{n_k}| < \frac{\varepsilon}{2}$. Hence, we have $|a_m - \alpha| \leq |a_m - a_{n_k}| + |a_{n_k} - \alpha| < \frac{\varepsilon}{2} + \frac{\varepsilon}{2} = \varepsilon$. Thus, we have shown that if $m \geq M_1 + M_2$ is an integer, we have $|a_m - \alpha| < \varepsilon$. Therefore, we have shown $\lim_{n \to \infty} a_n = \alpha$.

3.8   Rational Powers of a Positive Number

Let us start this section by generalizing Theorem 2.5.1.

Theorem 3.8.1 Let $x > 1$ be a real number. Suppose $k \geq 2$ is an integer. Then there exists a number $L > 1$ such that $L^k = x$.

(We denote this number L by $x^{\frac{1}{k}}$. In other words, we **define** the number $x^{\frac{1}{k}}$ to be L for every number $x > 1$. And $x^{\frac{1}{k}} > 1$. By "$x^{\frac{1}{k}}$ exists", we mean that $x^{\frac{1}{k}}$ is a real number and it makes sense to talk about such notation **after** this theorem is proved.)

Proof Let a_0 be the largest integer less than x such that $a_0^k \leq x < (a_0 + 1)^k$. Let a_1 be the integer among $0, 1, 2, \ldots, 9$ such that $\left(a_0 + \frac{a_1}{10}\right)^k \leq x < \left(a_0 + \frac{a_1}{10} + \frac{1}{10}\right)^k$.

Let $n \geq 1$ be an integer such that $a_i \in \{0, 1, 2, \ldots, 9\}$ is defined every integer $0 \leq i \leq n$ so that

$$\left(a_0 + \frac{a_1}{10} + \cdots + \frac{a_k}{10^k}\right)^k \leq x < \left(a_0 + \frac{a_1}{10} + \cdots + \frac{a_k}{10^k} + \frac{1}{10^k}\right)^k$$

Then we define a_{n+1} to be the integer among $0, 1, 2, \ldots, 9$ such that

$$\left(a_0 + \frac{a_1}{10} + \cdots + \frac{a_n}{10^n} + \frac{a_{n+1}}{10^{n+1}}\right)^k \leq x < \left(a_0 + \frac{a_1}{10} + \cdots + \frac{a_n}{10^n} + \frac{a_{n+1}}{10^{n+1}} + \frac{1}{10^{n+1}}\right)^k$$

By mathematical induction, a_n is defined for all integers $n \geq 0$.

For simplicity, let $b_n = a_0 + \frac{a_1}{10} + \cdots + \frac{a_n}{10^n}$ for every $n = 0, 1, 2, \ldots$. Hence, we defined an increasing sequence $\langle b_n \rangle_{n=1}^{\infty}$.

Let $n \geq 1$ be an integer. Then we have.

$b_n = a_0 + \frac{a_1}{10} + \cdots + \frac{a_n}{10^n} \leq a_0 + \frac{9}{10} + \frac{9}{10^2} + \frac{9}{10^3} + \cdots = a_0 + 1$. This shows that $\langle b_m \rangle_{m=1}^{\infty}$ is bounded above by $a_0 + 1$. Hence, we know that $\lim_{m \to \infty} b_m$ exists. Let $\lim_{n \to \infty} b_n = L$. Note that $1 \leq b_0 \leq b_n \leq L < b_n + \frac{1}{10^n} < b_1 + 1$ for each integer $n \geq 1$. Hence, we know that $L \geq 1$. Suppose $L = 1$. Then $a_n = 0$ for all integer $n \geq 1$. Then we must have $x = 1$. This is a contradiction to $x > 1$. This shows that $L > 1$.

Claim 1: We will show that $L^k \geq x$.

Let $\varepsilon > 0$. Then there is an integer n such that $k\left(\frac{1}{10^n}\right) \cdot (b_1 + 2)^{k-1} < \varepsilon$. Then we have

$$x < \left(b_n + \frac{1}{10^n}\right)^k \leq \left(L + \frac{1}{10^n}\right)^k$$

$$= L^k + \binom{k}{1} L^{k-1} \left(\frac{1}{10^n}\right)^1 + \binom{k}{2} L^{k-2} \left(\frac{1}{10^n}\right)^2 + \cdots$$

$$+ \binom{k}{k-1} L^1 \left(\frac{1}{10^n}\right)^{k-1} + \left(\frac{1}{10^n}\right)^k \qquad \text{by the binomial theorem}$$

$$< L^k + \left(\frac{1}{10^n}\right) \cdot \left\{ \binom{k}{1} L^{k-1} + \binom{k}{2} L^{k-2} + \cdots + \binom{k}{k-1} L^1 + 1 \right\}$$

$$= L^k + \left(\frac{1}{10^n}\right) \cdot \left\{ k L^{k-1} + \frac{k(k-1)}{2 \cdot 1} L^{k-2} + \cdots + \frac{k(k-1)\cdots 2}{(k-1)\cdots 2 \cdot 1} L^1 + 1 \right\}$$

$$= L^k + k\left(\frac{1}{10^n}\right) \cdot \left\{ L^{k-1} + \frac{1}{2} \cdot \frac{(k-1)}{1} L^{k-2} + \cdots + \frac{1}{k-1} \cdot \frac{(k-1)\cdots 2}{(k-2)\cdots 2 \cdot 1} L^1 + 1 \right\}$$

$$= L^k + k\left(\frac{1}{10^n}\right) \cdot \left\{ L^{k-1} + \frac{1}{2}\binom{k-1}{1} L^{k-2} + \frac{1}{3}\binom{k-1}{2} L^{k-3} + \cdots + \frac{1}{k-1}\binom{k-1}{k-2} L^1 + 1 \right\}$$

$$< L^k + k\left(\frac{1}{10^n}\right) \cdot \left\{ L^{k-1} + \binom{k-1}{1} L^{k-2} + \binom{k-1}{2} L^{k-3} + \cdots + \binom{k-1}{k-1} L^1 + 1 \right\}$$

$$= L^k + k\left(\frac{1}{10^n}\right) \cdot (L+1)^{k-1} \qquad \text{by the binomial theorem}$$

$$= L^k + k\left(\frac{1}{10^n}\right) \cdot (b_1 + 2)^{k-1}$$

$$< L^k + \varepsilon \quad \text{since } k\left(\frac{1}{10^n}\right) \cdot (b_1 + 2)^{k-1} < \varepsilon$$

Therefore, we have shown that $L^k > x - \varepsilon$ for every $\varepsilon > 0$. This proves that $L^k \geq x$.

Claim 2: We will show that $L^k \leq x$.

Let $\varepsilon > 0$. Then there is a $0 < \delta < 1$ such that $\delta < \frac{\varepsilon}{k(L+1)^{k-1}}$. Since $\lim_{n \to \infty} b_n = L$, there is an integer M such that $b_n \geq L - \delta$ for all integers $n \geq M$. Let $n \geq M$. Then we have

$x > b_n^k \geq (L - \delta)^k$

$$= L^k + \binom{k}{1} L^{k-1}(-\delta)^1 + \binom{k}{2} L^{k-2}(-\delta)^2 + \ldots + \binom{k}{k-1} L^1(-\delta)^{k-1} + (-\delta)^k$$

$$= L^k - k\delta \cdot \left\{ L^{k-1} + \frac{1}{2}\binom{k-1}{1} L^{k-2}(-\delta)^1 + \ldots + \frac{1}{k-1}\binom{k-1}{k-2} L^1(-\delta)^{k-2} + (-\delta)^{k-1} \right\}$$

$$> L^k - k\delta \cdot \left\{ L^{k-1} + \frac{1}{2}\binom{k-1}{1} L^{k-2}\delta^1 + \frac{1}{3}\binom{k-1}{2} L^{k-3}\delta^2 + \ldots + \frac{1}{k-1}\binom{k-1}{k-2} L^1\delta^{k-2} + \delta^{k-1} \right\}$$

$$> L^k - k\delta \cdot \left\{ L^{k-1} + \binom{k-1}{1} L^{k-2} + \binom{k-1}{2} L^{k-3} + \ldots + \binom{k-1}{k-2} L^1 + 1 \right\}$$

$$= L^k - k\delta \cdot (L+1)^{k-1} \qquad \text{since } \delta < 1$$

$$> L^k - \varepsilon \qquad \text{since } \delta < \frac{\varepsilon}{k(L+1)^{k-1}}.$$

Therefore, we have shown that $x > L^k - \varepsilon$ for every $\varepsilon > 0$. This proves that $x \geq L^k$. Therefore, by Claims 1 and 2, we have shown that $L^k = x$.

Definition 3.8.1

(1) The number $a^{\frac{1}{k}}$ is defined for every number $a > 1$ by Theorem 3.8.1.

(2) Let $0 < a < 1$, and let $n \geq 2$ be an integer. Then we define $a^{\frac{1}{n}} = \frac{1}{\left(\frac{1}{a}\right)^{\frac{1}{n}}}$. This makes

sense since $\frac{1}{a} > 1$ and $\left(\frac{1}{a}\right)^{\frac{1}{n}}$ exists by Theorem 3.8.1. Hence, for every $a > 0$, and for every integer $n \geq 2$, the number $a^{\frac{1}{n}} = \sqrt[n]{a}$ exists and it is a real number. (Note that $1^{\frac{1}{n}} = 1$.)

(3) Let $a > 0$. If p is a rational number, then $p = \frac{m}{n}$ for some integer m and a positive integer n. We define a^p to be $\left(a^{\frac{1}{n}}\right)^m$, i.e., $a^p = \left(a^{\frac{1}{n}}\right)^m$.

(4) Let $a > 0$, and let p, q be rational numbers. Then we define that

$$\left(a^p\right)^q = \left(a^q\right)^p = a^{pq} \text{ and } a^p a^q = a^{p+q}$$

(5) Let $a, b > 0$, and let p be a rational number. Then $a^p b^p = (ab)^p$.

(6) If $a < 0$, then $a^{\frac{1}{n}} = -|a|^{\frac{1}{n}}$ is defined only when n is an **odd integer**.

Remark 3.8.1

(1) Let $a > 0$, and let n, m be non-zero integers. Then $\left(\frac{1}{a}\right)^{\frac{1}{n}} = \frac{1}{a^{\frac{1}{n}}}$ and $a^{\frac{m}{n}} = \left(a^{\frac{1}{n}}\right)^m$.

(2) Let $a > 0$. Then a^r is defined when r is a rational number. However, a^r is **not** defined when r is an irrational number, **yet**. One of the goals in the remaining chapters is to understand exponential numbers of the form a^r better. Note that $0^0 = 1$ and $0^m = 0$, where m is a positive integer. We do not define 0^r when r is a negative number.

Theorem 3.8.2 Let $a > 1$. Let $1 \le n < m$ be integers. We have the followings:

(1) $a^n > 1$
(2) $0 < a^{-n} < 1$
(3) $1 < a^n < a^m$
(4) $0 < a^{-m} < a^{-n} < 1$

Problem 3.8.1 Prove Theorem 3.8.2.

(The Mathematical induction may be useful to prove (1).)

Theorem 3.8.3 Let $1 < a < b$. Let n be a positive integer. Let $0 < p < q$ be rational numbers. Then we have the followings:

(1) $a^{\frac{1}{n}} > 1$. Hence, $a^p > 1$.
(2) $a^{\frac{1}{n}} < b^{\frac{1}{n}}$. Hence, $a^p < b^p$.
(3) $a^{\frac{1}{n+1}} < a^{\frac{1}{n}}$. Hence, $a^p < a^q$.

Proof of (1): We have $a^{\frac{1}{n}} > 1$ by Theorem 3.8.1. Hence, by Theorem 3.8.2(1), we have that $a^p > 1$.

Proof of (2): Note that $\frac{b}{a} > 1$ so that $\frac{b^{\frac{1}{n}}}{a^{\frac{1}{n}}} = \left(\frac{b}{a}\right)^{1/n} > 1$. Hence, $a^{\frac{1}{n}} < b^{\frac{1}{n}}$. We have $a^p < b^p$ by Theorem 3.8.2(3).

Proof of (3): Since $a^{\frac{1}{n+1}} > 1$, we have $a^{\frac{1}{n+1}} = a^{1-\frac{n}{n+1}} = \frac{a}{a^{\frac{n}{n+1}}} > 1$. Hence, $a^{\frac{n}{n}} < a^{\frac{n}{n+1}}$ so that $a^{\frac{1}{n}} = \left(a^{\frac{n}{n}}\right)^{\frac{1}{n}} < \left(a^{\frac{n}{n+1}}\right)^{\frac{1}{n}} = a^{\frac{1}{n+1}}$. We have $a^p < a^q$ by Theorem 3.8.2(3).

Lemma 3.8.1 Let $a > 1$. Then $\lim_{n\to\infty} a^{\frac{1}{n}} = 1$.

Proof Since $\left(a^{\frac{1}{n}}\right)_{n=1}^{\infty}$ is decreasing by Theorem 3.8.3(2) and bounded below by 1, it converges. Let n be a positive integer. Then

$$\left(a^{\frac{1}{n}} - 1\right)\left(a^{\frac{n-1}{n}} + a^{\frac{n-2}{n}} + \cdots + 1\right) = \left(a^{\frac{1}{n}}\right)^n - 1 = a - 1. \text{ So}$$

$$0 < \left(a^{\frac{1}{n}} - 1\right) = \frac{a-1}{\left(a^{\frac{n-1}{n}} + a^{\frac{n-2}{n}} + \cdots + 1\right)} < \frac{a-1}{n} \text{ since } a^{\frac{n-1}{n}} + a^{\frac{n-2}{n}} + \cdots + 1 > n. \text{ Since}$$

$\lim_{n\to\infty} \frac{a-1}{n} = 0$, $\lim_{n\to\infty} \left(a^{\frac{1}{n}} - 1\right) = 0$ by the sandwich theorem. Therefore,

$$\lim_{n\to\infty} a^{\frac{1}{n}} = \lim_{n\to\infty}\left[\left(a^{\frac{1}{n}} - 1\right) + 1\right] = \lim_{n\to\infty}\left(a^{\frac{1}{n}} - 1\right) + 1 = 0 + 1 = 1.$$

Theorem 3.8.4 Let $a > 0$. Let $\langle b_n \rangle$ be a sequence of rational numbers converging to 0. Then $\lim_{n \to \infty} a^{b_n} = 1$.

Proof Let $a > 0$. Since $\lim_{n \to \infty} 1^{b_n} = 1$, we assume that $a \neq 1$.

Suppose $a > 1$.

(Claim 1) Suppose $\langle b_n \rangle$ is a sequence of **positive** rational numbers converging to 0. Then we will show that $\lim_{n \to \infty} a^{b_n} = 1$.

Let $\varepsilon > 0$. There exists an integer $M > 0$ such that $a^{\frac{1}{m}} - 1 = \left| a^{\frac{1}{m}} - 1 \right| < \varepsilon$ for all $m \geq M$ by Lemma 3.8.1. Since $\langle b_n \rangle$ is a sequence of positive rational numbers converging to 0, there is an integer $N > 0$ such that $|b_n - 0| = |b_n| = b_n < \frac{1}{M}$ for every $n \geq N$. Let $n \geq N$. Then $\left| a^{b_n} - 1 \right| = a^{b_n} - 1 < a^{\frac{1}{M}} - 1 < \varepsilon$. This proves $\lim_{n \to \infty} a^{b_n} = 1$.

(Claim 2) Suppose $\langle b_n \rangle$ is a sequence of rational numbers converging to 0. (We are **not** assuming b_n to be positive.) Then, we will show that $\lim_{n \to \infty} a^{b_n} = 1$.

For every integer $n > 0$, we have $-|b_n| \leq b_n \leq |b_n|$. Since $a^{-|b_n|} \leq 1 \leq a^{|b_n|}$, and since either $a^{b_n} = a^{|b_n|}$ or $a^{b_n} = a^{-|b_n|}$, we have $a^{-|b_n|} \leq a^{b_n} \leq a^{|b_n|}$.

By Claim 1, we have $\lim_{n \to \infty} a^{|b_n|} = 1$. So we have $\lim_{n \to \infty} a^{-|b_n|} = \lim_{n \to \infty} \frac{1}{a^{|b_n|}} = \frac{1}{\lim_{n \to \infty} a^{|b_n|}} = 1$. Hence, by the sandwich theorem, we have $\lim_{n \to \infty} a^{b_n} = 1$.

Now, suppose $0 < a < 1$ and $\langle b_n \rangle$ is a sequence of rational numbers converging to 0. Let $c = \frac{1}{a}$. Then $c > 1$ and $\lim_{n \to \infty} c^{b_n} = 1$ by Claim 2. Hence,

$$\lim_{n \to \infty} a^{b_n} = \lim_{n \to \infty} \left(\frac{1}{c} \right)^{b_n} = \lim_{n \to \infty} \frac{1}{c^{b_n}} = \frac{1}{\lim_{n \to \infty} c^{b_n}} = \frac{1}{1} = 1$$

Therefore, $\lim_{n \to \infty} a^{b_n} = 1$ for any $a > 0$ and for any sequence b_n of rational numbers converging to 0.

3.9   Irrational Powers of a Positive Number

In this section, we will consider numbers like $2^{\sqrt{2}}$. We need one lemma.

Lemma 3.9.1 Let $a > 1$. For every $\varepsilon > 0$, there exists a $\delta > 0$ such that $|a^r - 1| < \varepsilon$ whenever r is a rational number such that $|r| < \delta$.

Proof Let $\varepsilon > 0$. Since $\lim_{n \to \infty} a^{\frac{1}{n}} = 1 = \lim_{n \to \infty} a^{-\frac{1}{n}}$ by Theorem 3.8.4, there exists a positive integer M such that $\left| a^{\frac{1}{n}} - 1 \right| < \varepsilon$ and $\left| a^{-\frac{1}{n}} - 1 \right| < \varepsilon$ whenever $n \geq M$. In particular, we have $\left| a^{\frac{1}{M}} - 1 \right| < \varepsilon$ and $\left| a^{-\frac{1}{M}} - 1 \right| < \varepsilon$. Let r be a rational number such that $-\frac{1}{M} < r < \frac{1}{M}$. If $0 \le r < \frac{1}{M}$, then, by Theorem 3.8.3, we have $a^{-\frac{1}{M}} < 1 \le a^r < a^{\frac{1}{M}}$. If $-\frac{1}{M} < r < 0$, then $\frac{1}{M} > -r > 0$ so that $a^{-r} < a^{\frac{1}{M}}$ or $a^{-\frac{1}{M}} < a^r < 1 < a^{\frac{1}{M}}$. At any rate, we have $a^{-\frac{1}{M}} < a^r < a^{\frac{1}{M}}$. Hence, $a^{-\frac{1}{M}} - 1 < a^r - 1 < a^{\frac{1}{M}} - 1$ so that $|a^r - 1| < \max\left\{\left|a^{\frac{1}{M}} - 1\right|, \left|a^{-\frac{1}{M}} - 1\right|\right\} < \varepsilon$. Therefore, if we choose $\delta = \frac{1}{M}$, then we have $|a^r - 1| < \varepsilon$ whenever r is a rational number such that $|r| < \delta$. This completes the proof.

What are Cauchy sequences good for? One example is that we need Cauchy sequences and Theorem 3.7.2 to prove the next theorem in order to **define** a^r when r is an **irrational** number.

Theorem 3.9.1

(1) Let $\langle r_n \rangle_{n=1}^{\infty}$ be a convergent sequence of **rational** positive numbers. Let $a > 1$. Then the sequence $\langle a^{r_n} \rangle_{n=1}^{\infty}$ converges.

(2) Let $\langle r_n \rangle_{n=1}^{\infty}$ and $\langle p_n \rangle_{n=1}^{\infty}$ be convergent sequences of positive rational numbers converging to the same number. Let $a > 1$. Then $\lim_{n \to \infty} a^{r_n} = \lim_{n \to \infty} a^{p_n}$.

Proof of (1): By Theorem 3.7.2, it suffices to show that $\langle a^{r_n} \rangle_{n=1}^{\infty}$ is a Cauchy sequence.

(Here, showing $\langle a^{r_n} \rangle_{n=1}^{\infty}$ being a **Cauchy sequence** is the **only** way to prove its convergence. See Remark 3.9.1(2) below.)

Let $\varepsilon > 0$. Since $\langle r_n \rangle_{n=1}^{\infty}$ is a convergent sequence, $\{|r_n|\}_{n=1}^{\infty}$ is bounded above by a rational number $q \ge 1$. Then $a^{r_n} \le a^{|r_n|} < a^q$ for all n by Theorem 3.8.3. By Lemma 3.9.1, there exists a $\delta > 0$ such that $\left|a^{r_n - r_m} - 1\right| < \frac{\varepsilon}{a^q}$ whenever $|r_n - r_m| < \delta$. Since $\langle r_n \rangle_{n=1}^{\infty}$ is Cauchy, there is an integer $N > 0$ such that $|r_n - r_m| < \delta$ whenever $n, m \ge N$. Let $n, m \ge N$. Then,

$$\left|a^{r_n} - a^{r_m}\right| = \left|a^{r_m}\right| \cdot \left|a^{r_n - r_m} - 1\right|$$
$$\le a^{|r_m|} \cdot \left|a^{r_n - r_m} - 1\right|$$
$$\le a^q \cdot \left|a^{r_n - r_m} - 1\right|$$
$$< a^q \cdot \frac{\varepsilon}{a^q} = \varepsilon$$

Therefore, $\langle a^{r_n} \rangle_{n=1}^{\infty}$ converges since it is a Cauchy sequence.

Proof of the part (2): Since $\lim_{n \to \infty} r_n = \lim_{n \to \infty} p_n$, we have $\lim_{n \to \infty} (r_n - p_n) = 0$. Hence,

$$\lim_{n\to\infty} \left(a^{r_n} - a^{p_n}\right) = \lim_{n\to\infty} a^{p_n} \cdot \left(a^{r_n - p_n} - 1\right)$$
$$= \left(\lim_{n\to\infty} a^{p_n}\right) \cdot \left[\lim_{n\to\infty} \left(a^{r_n - p_n} - 1\right)\right]$$
$$= \left(\lim_{n\to\infty} a^{p_n}\right) \cdot (1 - 1) = 0$$

since $\lim_{n\to\infty} a^{r_n - p_n} = 1$ by Theorem 3.8.4, and since $\lim_{n\to\infty} a^{p_n}$ exists by part (1).

Therefore, we have $\lim_{n\to\infty} a^{r_n} = \lim_{n\to\infty} a^{p_n}$.

Remark 3.9.1

(1) Lemma 3.9.1 is saying that $\lim_{r\to 0} a^r = 1$ if $a > 1$ in term of the limit of a function that will be defined in Chap. 4. We needed this here in order to control the size of $\left|a^{r_n - r_m} - 1\right|$ in the proof of Theorem 3.9.1(1).

(2) Let $\langle r_n \rangle_{n=1}^{\infty}$ be a convergent sequence of rational numbers such that $\lim_{n\to\infty} r_n = r$ for some $r \in \mathbb{R}$. Let $a > 1$. Please note that we could not prove $\lim_{n\to\infty} a^{r_n} = a^r$ since a^r is **not** defined when r is irrational. As a matter of fact, **because of** Theorem 3.9.1, the next definition makes sense.

Definition 3.9.1

(1) Let $a > 1$. If r is an irrational positive number, we **define** a^r by $a^r = \lim_{n\to\infty} a^{r_n}$, where $\langle r_n \rangle_{n=1}^{\infty}$ is a sequence of **rational** positive numbers that converges to r. If p is a negative number, then $-p > 0$ and we define $a^p = \frac{1}{a^{-p}}$.

(2) If $0 < b < 1$, then $\frac{1}{b} > 1$. We define b^r to be $\frac{1}{\left(\frac{1}{b}\right)^r}$. That is, $b^r = \frac{1}{\left(\frac{1}{b}\right)^r}$.

(3) $1^r = 1$ for any real number r.

Remark 3.9.2 Now, we can finally talk about a number like $2^{\sqrt{2}}$. For example, in the proof of Theorem 2.5.1, we constructed a sequence $\langle b_n \rangle_{n=1}^{\infty}$ that converges to $\sqrt{2}$. Notice that the sequence $\langle b_n \rangle_{n=1}^{\infty}$ is a sequence of rational numbers. Hence, 2^{b_n} makes sense from Sect. 3.8. By Theorem 3.9.1, $\left(2^{b_n}\right)_{n=1}^{\infty}$ is a Cauchy sequence and converges. So we defined $2^{\sqrt{2}}$ by $2^{\sqrt{2}} = \lim_{n\to\infty} 2^{b_n}$.

Theorem 3.9.2 Let $a > 0$. If $\langle r_n \rangle_{n=1}^{\infty}$ is a sequence of **real** numbers that converges to $r \in \mathbb{R}$, then we **have** $\lim_{n\to\infty} a^{r_n} = a^r$.

Proof For each integer $n > 0$, there exists a rational number b_n such that $|r_n - b_n| < \frac{1}{n}$ and $\left|a^{r_n} - a^{b_n}\right| < \frac{1}{n}$ by Definition 3.9.1.

First, we will show that the sequence $\langle b_n \rangle_{n=1}^{\infty}$ converges to r.

Let $\varepsilon > 0$. Since $\langle r_n \rangle_{n=1}^{\infty}$ converges to r, there is an integer $N > 0$ such that $|r - r_n| < \frac{\varepsilon}{2}$ for all integer $n \geq N$. Also, there is an integer $M > 0$ such that $\frac{1}{n} < \frac{\varepsilon}{2}$ for all integer $n \geq N$. Let $n \geq N + M$. Then $|r - b_n| \leq |r - r_n| + |r_n - b_n| < \frac{\varepsilon}{2} + \frac{1}{n} < \frac{\varepsilon}{2} + \frac{\varepsilon}{2} = \varepsilon$. Hence, the sequence $\langle b_n \rangle_{n=1}^{\infty}$ converges to r.

By Theorem 3.9.1(2), $\lim_{n \to \infty} a^{b_n} = a^r$.

Secondly, we will prove that $\lim_{n \to \infty} a^{r_n} = a^r$.

Let $\varepsilon > 0$. Since $\lim_{n \to \infty} a^{b_n} = a^r$, there is an integer $N > 0$ such that $\left| a^r - a^{b_n} \right| < \frac{\varepsilon}{2}$ for all integer $n \geq N$. Also, there is an integer $M > 0$ such that $\frac{1}{n} < \frac{\varepsilon}{2}$ for all integer $n \geq N$. Let $n \geq N + M$. Then

$$\left| a^r - a^{r_n} \right| \leq \left| a^r - a^{b_n} \right| + \left| a^{b_n} - a^{r_n} \right|$$
$$= \left| a^r - a^{b_n} \right| + \left| a^{r_n} - a^{b_n} \right| < \frac{\varepsilon}{2} + \frac{1}{n} < \frac{\varepsilon}{2} + \frac{\varepsilon}{2} = \varepsilon.$$

Therefore, this shows that $\lim_{n \to \infty} a^{r_n} = a^r$.

Definition 3.9.2

(1) Let $a > 0$, and let $p, q \in \mathbb{R}$. Then we define that

$$\left(a^p \right)^q = \left(a^q \right)^p = a^{pq} \quad \text{and} \quad a^p a^q = a^{p+q}.$$

(2) Let $a, b > 0$, and let $p \in \mathbb{R}$. Then $a^p b^p = (ab)^p$.

Theorem 3.9.3 Let $a > 1$. If $0 < \alpha < \beta$, then $1 < a^\alpha < a^\beta$.

Proof This proof is a consequence of Definition 3.8.1 and Theorem 3.8.2. Let $\langle r_n \rangle_{n=1}^{\infty}$ and $\langle s_n \rangle_{n=1}^{\infty}$ be sequences of rational numbers converging to α and β, respectively. Let r, s be rational numbers such that $0 < \alpha < r < s < \beta$. Without loss of generality, assume that $r_n < r < s < s_n$ for all integers $n > 0$. Then we have

$$a^\alpha = \lim_{n \to \infty} a^{r_n} \leq a^r \quad \text{and} \quad a^\beta = \lim_{n \to \infty} a^{s_n} \geq a^s \text{ by Theorem 3.8.2.}$$

Since $r < s$, we also have $a^r < a^s$ by Theorem 3.8.2. Hence, $a^\alpha < a^\beta$. We leave to the readers to show that $1 < a^\alpha$.

Remark 3.9.3: What is a real number?

One short answer to this question is the following; a real number is the limit of a Cauchy sequence. Because of the construction of real numbers from rational numbers, a real number can be thought of as the limits of Cauchy sequences of numbers defined by rational numbers. In a way, because of Definition 3.9.1, we have finally completed the construction of real numbers.

3.10 The Number e

The definition of the number e is given to be $\lim_{m\to\infty} \left(1 + \frac{1}{m}\right)^m = e$ in most textbooks. We take a slightly different approach to this by taking the definition of e as in Definition 3.10.1 below. This is because the proof of the existence of the limit $\lim_{m\to\infty} \left(1 + \frac{1}{m}\right)^m$ is not easy. So, we first establish that $\lim_{m\to\infty} \left(1 + \frac{1}{m}\right)^m$ exists, and then we will prove that $\lim_{m\to\infty} \left(1 + \frac{1}{m}\right)^m = e$ using Definition 3.10.1.

This section is based on *Differential and Integral Calculus, Second Edition*, Volume 1, 43–44, by R. Courant, Interscience Publisher, 1954 [6].

Problem 3.10.1 Show that $\sum_{n=0}^{\infty} \frac{1}{n!} = 1 + \sum_{n=1}^{\infty} \frac{1}{n!}$ converges.

Definition 3.10.1 Let $e = \sum_{n=0}^{\infty} \frac{1}{n!} = 1 + \sum_{n=1}^{\infty} \frac{1}{n!}$.

One of the most difficult things in a calculus course is to prove that the derivative of e^x is e^x, $\frac{d}{dx} e^x = e^x$, from Definition 3.10.1. Because it is difficult, most calculus books give a lighter approach to this by simply saying "it is", or give a **questionable** definition of e by defining e to be the number such that $\lim_{h\to 0} \frac{e^h - 1}{h} = 1$. It is not obvious the existence of a number e such that $\lim_{h\to 0} \frac{e^h - 1}{h} = 1$. How are we to approximate e from this definition? However, by letting $n = 3$, $e = \sum_{n=0}^{\infty} \frac{1}{n!}$ gives $\sum_{n=0}^{3} \frac{1}{n!} = 1 + 1 + \frac{1}{2} + \frac{1}{6} = 2 + \frac{2}{3} \approx 2.67$, a good approximation of e. **One of the goals** for the remainder of this textbook is to prove $\frac{d}{dx} e^x = e^x$ based on Definition 3.10.1, we will prove this in Theorem 7.4.4 even though we use $\frac{d}{dx} e^x = e^x$ starting in Chap. 5 before Chap. 7.

Problem 3.10.2 Show that $\sum_{n=1}^{m} n! < (m+1)!$ for each positive integer m. This shows that the sequence $\langle n! \rangle_{n=1}^{\infty}$ diverges very rapidly.

Definition 3.10.2 Let $S_m = \sum_{n=0}^{m} \frac{1}{n!}$ for every positive integer m. So S_m is the m-th partial sum of $\sum_{n=0}^{\infty} \frac{1}{n!}$, and $\lim_{m\to\infty} S_m = e$. Note that $e > \sum_{n=0}^{3} \frac{1}{n!} = 1 + 1 + \frac{1}{2} + \frac{1}{6} = 2 + \frac{2}{3} > 2.5$.

Definition 3.10.3 Let $T_m = \left(1 + \frac{1}{m}\right)^m$ for every positive integer m.

We will show that $\lim\limits_{m\to\infty} T_m$ exists and $\lim\limits_{m\to\infty} T_m = e$.

Lemma 3.10.1 We have $0 < e - S_m < \frac{2}{(m+1)!}$ for each positive integer m.
(This shows that the convergence of the m-th partial sums S_m to e is vary fast!)

Proof

$$e - S_m = \sum_{n=m+1}^{\infty} \frac{1}{n!} = \frac{1}{(m+1)!} \cdot \left(1 + \frac{1}{(m+2)} + \frac{1}{(m+2)(m+3)} + \cdots\right)$$

$$< \frac{1}{(m+1)!} \cdot \left(1 + \frac{1}{2} + \frac{1}{2^2} + \cdots\right) = \frac{2}{(m+1)!} \quad \text{since} \quad 1 + \frac{1}{2} + \frac{1}{2^2} + \cdots = 2.$$

Theorem 3.10.1 $\lim\limits_{m\to\infty} T_m = e$. That is $\lim\limits_{m\to\infty} \left(1 + \frac{1}{m}\right)^m = e$.

Proof Let $m > k > 3$ be integers. Then

$$T_m = \left(1 + \frac{1}{m}\right)^m = 1 + \binom{m}{1}\frac{1}{m^1} + \binom{m}{2}\frac{1}{m^2} + \cdots + \binom{m}{k}\frac{1}{m^k} + \cdots + \binom{m}{m}\frac{1}{m^m} \quad \text{by binomial theorem}$$

$$= 1 + m\frac{1}{m^1} + \frac{m(m-1)}{2!} \cdot \frac{1}{m^2} + \cdots + \frac{m(m-1)\ldots(m-k+1)}{k!} \cdot \frac{1}{m^k} + \cdots + \frac{m(m-1)\ldots 1}{m!} \cdot \frac{1}{m^m}$$

$$= 1 + 1 + \frac{1}{2!}\left(1 - \frac{1}{m}\right) + \frac{1}{3!}\left(1 - \frac{1}{m}\right)\left(1 - \frac{2}{m}\right)$$

$$+ \cdots + \frac{1}{m!}\left(1 - \frac{1}{m}\right)\left(1 - \frac{2}{m}\right)\cdots\left(1 - \frac{m-1}{m}\right) \quad \text{(i)}$$

$$\geq 1 + 1 + \frac{1}{2!}\left(1 - \frac{1}{m}\right) + \frac{1}{3!}\left(1 - \frac{1}{m}\right)\left(1 - \frac{2}{m}\right)$$

$$+ \cdots + \frac{1}{k!}\left(1 - \frac{1}{m}\right)\left(1 - \frac{2}{m}\right)\cdots\left(1 - \frac{k-1}{m}\right) \quad \text{since } m > k \quad \text{(ii)}$$

$$\geq 1 + 1 + \frac{1}{2!}\left(1 - \frac{1}{k}\right) + \frac{1}{3!}\left(1 - \frac{1}{k}\right)\left(1 - \frac{2}{k}\right)$$

$$+ \cdots + \frac{1}{k!}\left(1 - \frac{1}{k}\right)\left(1 - \frac{2}{k}\right)\cdots\left(1 - \frac{k-1}{k}\right) \quad \text{since } \frac{1}{m} < \frac{1}{k},$$

$$= 1 + \binom{k}{1}\frac{1}{k^1} + \binom{k}{2}\frac{1}{k^2} + \cdots + \binom{k}{k}\frac{1}{k^k} = \left(1 + \frac{1}{k}\right)^k$$

$$= T_k.$$

This shows that the sequence $\langle T_m \rangle_{m=3}^{\infty}$ is an **increasing sequence**.
From the line (i) above, we can see that
$T_m < \sum_{n=0}^{m} \frac{1}{n!} = S_m$. Since $\lim\limits_{m\to\infty} S_m = e$, $\lim\limits_{m\to\infty} T_m \leq e$.
From the line (ii), we have

$$T_m \geq 1 + 1 + \frac{1}{2!}\left(1 - \frac{1}{m}\right) + \frac{1}{3!}\left(1 - \frac{1}{m}\right)\left(1 - \frac{2}{m}\right)$$
$$+ \cdots + \frac{1}{k!}\left(1 - \frac{1}{m}\right)\left(1 - \frac{2}{m}\right)\cdots\left(1 - \frac{k-1}{m}\right).$$

This shows that

$$\lim_{m \to \infty} T_m \geq \lim_{m \to \infty}\left\{1 + 1 + \frac{1}{2!}\left(1 - \frac{1}{m}\right) + \cdots + \frac{1}{k!}\left(1 - \frac{1}{m}\right)\left(1 - \frac{2}{m}\right)\cdots\left(1 - \frac{k-1}{m}\right)\right\}$$
$$= 1 + 1 + \frac{1}{2!} + \cdots + \frac{1}{k!} = S_k.$$

In other words, we have shown that $\lim_{m \to \infty} T_m \geq S_k$ for every integer $k > 3$. Since $\lim_{m \to \infty} S_m = e$, we have shown that $\lim_{m \to \infty} T_m \geq e$.

Therefore, we have $\lim_{m \to \infty} T_m = e$.

The next theorem is for your information.

Theorem 3.10.2 The number e is an irrational number.

Proof Suppose $e = \frac{q}{p}$ for some positive integers p and q. Then $(p!)e = q(p-1)!$. But $(p!)e = (p!)\sum_{n=0}^{\infty} \frac{1}{n!} = (p!)\sum_{n=0}^{p} \frac{1}{n!} + (p!)\sum_{n=p+1}^{\infty} \frac{1}{n!}$.

So $(p!)\sum_{n=0}^{p} \frac{1}{n!} + (p!)\sum_{n=p+1}^{\infty} \frac{1}{n!} = q(p-1)!$.

Hence, $(p!)\sum_{n=p+1}^{\infty} \frac{1}{n!} = q(p-1)! - (p!)\sum_{n=0}^{p} \frac{1}{n!}$.

This shows that $(p!)\sum_{n=p+1}^{\infty} \frac{1}{n!}$ is an integer. This implies that $\left[q(p-1)!\right] - (p!)\sum_{n=0}^{p} \frac{1}{n!} = (p!)\sum_{n=p+1}^{\infty} \frac{1}{n!}$ is an integer. But $(p!)\sum_{n=p+1}^{\infty} \frac{1}{n!} = \frac{1}{p+1} + \sum_{n=p+1}^{\infty} \frac{1}{(p+1)^n} = \frac{1}{p+1} \cdot \frac{1}{1 - \frac{1}{p+1}} = \frac{1}{p}$. This is a contradiction to $(p!)\sum_{n=p+1}^{\infty} \frac{1}{n!}$ being an integer. Therefore, e is an irrational number.

3.11   Metric and Topological Spaces

The word "space" is used frequently in mathematics to study sets with some special properties, as in a vector space, which may be familiar to many of the readers. A vector space can be a study of the three-dimensional space $\mathbb{R}^3$, or any other finite dimensional space $\mathbb{R}^n$ for some integer $n \geq 1$. What we are studying in this textbook can be partially thought of as a study of the vector space $\mathbb{R}^1$. A vector space can be infinite dimensional vector space as in a Banach space. All of these vector spaces have the notion of "distances" or "metrics" together with the addition of two elements and multiplication of their elements by real or complex numbers.

The space with only the notion of "distance" or "metric" without the additions of two elements nor the multiplications by real or complex numbers is simply called a metric space. So all vector spaces are metric spaces.

An abstract space defined by "open sets" in place of metric is called a topological space. All metric spaces are topological spaces.

By generalizing the notion of the absolute value $|p - a|$, where $p, a \in \mathbb{R}$, to the *distance (metric) between p and a*, where $p, a \in \mathbb{R}^3$, we obtain a metric space $\mathbb{R}^3$, for example. Hence, the set X mentioned in the next definition can be thought of as $\mathbb{R}, \mathbb{R}^2, \mathbb{R}^3$, or any nonempty subsets of these.

Definition 3.11.1 Let X be a non-empty set. A function $d : X \times X \to \mathbb{R}$ is said to be a metric on X if the following conditions are satisfied:

(1) $d(x, y) = d(y, x)$ for all $x, y \in X$,
(2) $d(x, y) = 0$ if $x = y \in X$, and $d(x, y) > 0$ if $x \neq y \in X$, and
(3) $d(x, y) \leq d(x, z) + d(z, y)$ for all $x, y, z \in X$.

(The inequality (3) is called the triangle inequality.)

The set X with the metric d is said to be a metric space and we say (X, d) is a metric space.

Example 3.11.1

(1) For every $x, y \in \mathbb{R}$, let $d(x, y) = |x - y|$. Then $(\mathbb{R}, d)$ is a metric space. This metric d is said to be the usual metric on $\mathbb{R}$.

If we replace $\mathbb{R}$ by its nonempty subset X of $\mathbb{R}$, we can still define $d(x, y) = |x - y|$ for all $x, y \in X$. And (X, d) is a metric space. The space (X, d) is said to be a subspace of $(\mathbb{R}, d)$. This (X, d) for some $X \subset \mathbb{R}$ is exactly the space we are studying, and what we will continue to study in this textbook. Note that $\{x \in \mathbb{R} : d(x, 0) < 1\} = (-1, 1)$, and it is an **open** interval. By generalizing this open interval, we can obtain **open** sets to form a topological space.

(2) For every $(x, y), (a, b) \in \mathbb{R}^2$, let $d((x, y), (a, b)) = \sqrt{(x - a)^2 + (y - b)^2}$. Then $(\mathbb{R}^2, d)$ is a metric space. This metric d is said to be the usual metric on $\mathbb{R}^2$. The set $\{(x, y) : d((x, y), (0, 0)) < 1\}$ is the interior of the unit **open** circle centered at the origin.

(3) For every $(x, y, z), (a, b, c) \in \mathbb{R}^3$, let

$$d((x, y, z), (a, b, c)) = \sqrt{(x - a)^2 + (y - b)^2 + (z - c)^2}$$

Then $(\mathbb{R}^3, d)$ is a metric space. This metric d is said to be the <u>usual metric</u> on $\mathbb{R}^3$.

Proving the triangle inequality for this metric d is not easy. But let us take this for granted. The set $\{(x, y, z) : d((x, y, z), (0, 0, 0)) < 1\}$ is the interior of the unit **open** sphere centered at the origin.

(4) For every $x, y \in \mathbb{R}$, let $d(x, y) = 1$ if $x \neq y$ and $(x, y) = 0$ if $x = y$. Then this is a metric on $\mathbb{R}$. Note that $\{x : d(x, 0) < 1\} = \{0\}$.

(5) For every $(x, y), (a, b) \in \mathbb{R}^2$, let $d((x, y), (a, b)) = max\{|x - a|, |y - b|\}$. Then $(\mathbb{R}^2, d)$ is a metric space. Then $\{(x, y) : d((x, y), (0, 0)) < 1\} = (-1, 1) \times (-1, 1)$. It is an open square.

(6) For every $(x, y), (a, b) \in \mathbb{R}^2$, let $d((x, y), (a, b)) = |x - a| + |y - b|$. Then $(\mathbb{R}^2, d)$ is a metric space. Then the set $\{(x, y) : d((x, y), (0, 0)) < 1\}$ is an open diamond square of side lengths $\sqrt{2}$.

Definition 3.11.2 Let (X, d) be a metric space. A <u>sequence</u> in X is a function $a : \mathbb{N} \to X$ from the set of natural numbers $\mathbb{N}$ into the set X. Usually, we write a_n in place of $a(n)$, and $\langle a_n \rangle_{n=1}^{\infty}$ in place of $a : \mathbb{N} \to X$.

Let $p \in X$. A sequence a_n in X <u>converges</u> to p if, for every $\varepsilon > 0$, there exists a positive integer M such that $d(a_n, p) < \varepsilon$ whenever $n \geq M$.

A sequence $\langle a_n \rangle$ in X is said to be a <u>Cauchy sequence</u> p if, for every $\varepsilon > 0$, there exists a positive integer M such that $d(a_n, a_m) < \in$ whenever $n, m \geq M$.

If all Cauchy sequence in a metric space converges, then the metric space is said to be <u>complete</u>. The set $\mathbb{R}$, closed intervals in $\mathbb{R}$, and the Cantor set (all with the usual metric) are examples of complete metric spaces. A <u>Banach space</u> is a complete vector space.

Definition 3.11.2 is a natural extension of Definition 3.1.1. And the amazing thing about the metric space is that many theorems on sequences in this chapter have related theorems in metric spaces. And that is the reason why the study of this introduction to analysis is important. It is the base for Topology and topological spaces. Advanced topics like vector spaces, the complex numbers, the Hilbert spaces, the Banach spaces are all examples of metric spaces with additional properties.

Definition 3.11.3 Let (X, d) be a metric space. Let $p \in X$ and $\varepsilon > 0$. The ε-<u>neighborhood</u> is denoted by $N_\varepsilon(p)$, and defined by $N_\varepsilon(p) = \{x \in X : d(p, x) < \varepsilon\}$.

Let (X, d) be a metric space. A subset U of X is <u>open</u> if for every $x \in U$, there is an $\varepsilon > 0$ such that $N_\varepsilon(p) \subset U$. Hence, X itself is an open set. Also, by definition, the empty set $\emptyset$ is an open set. A subset F of X is <u>closed</u> if $X - F$ is open. Hence, the empty set $\emptyset$ and X are closed sets.

Let A be a nonempty subset of X. Then a point $p \in X$ is a <u>limit point</u> (a <u>cluster point,</u> or an <u>accumulation point</u>) of A if $N_\varepsilon(p) \cap (A - \{p\}) \neq \emptyset$ for every $\varepsilon > 0$. In other words, if there is a sequence a_n in A converging to a point $p \in X$, then p is a limit point of A. Let A'

be the set of all limit points of A. The set A' is called the derived set of A. The closure of A is denoted by $\overline{A}$, and defined by $\overline{A} = A \cup A'$. It can be proven that a subset F of X is closed if, and only if $F = \overline{F}$. A point in A that is not a limit point of A is called an isolated point of A. Hence, a point $p \in A$ is an isolated point of A if $N_\varepsilon(p) \cap (A - \{p\}) = \emptyset$ for some $\varepsilon > 0$.

Remark 3.11.1 Let τ be the set of all open subsets in X. Then τ has the following properties:

(1) $X, \emptyset \in \tau$;

(2) if $U_k \in \tau$ for all $k \in K$, where K is some index set, then $\bigcup_{k \in K} U_k \in \tau$, and

(3) if $U_1, U_2, \ldots, U_n \in \tau$ for some integer $n > 0$, then $\bigcap_{k=1}^{n} U_k \in \tau$.

From this observation, let τ be a set of subsets in X with the following properties:

(1) $X, \emptyset \in \tau$;

(2) if $U_k \in \tau$ for all $k \in K$, where K is some index set, then $\bigcup_{k \in K} U_k \in \tau$, and

(3) if $U_1, U_2, \ldots, U_n \in \tau$ for some integer $n > 0$, then $\bigcap_{k=1}^{n} U_k \in \tau$.

Then we say τ is a topology on X, (X, τ) is said to be a topological space, and elements in τ are said to be open sets in X.

We state the next theorem without a proof.

Theorem 3.11.1 Let (X, d) be a metric space. Let A be a nonempty subset of X.

(1) A sequence $\langle a_n \rangle$ of points in X converges to a point $p \in X$ if for every $\varepsilon > 0$, there is an integer M such that $a_n \in N_\varepsilon(p)$ whenever $n \geq m$.

(2) A sequence $\langle a_n \rangle$ of points in X converges to a point $p \in X$ if for every open set U containing p, there is an integer M such that $a_n \in U$ whenever $n \geq m$.

(3) (**Definition** A point $p \in X$ is a limit point of A if there exists a sequence $\langle a_n \rangle$ of points in $A - \{p\}$ converges to p.) A point $p \in \overline{A}$ if, and only if there exists a sequence $\langle a_n \rangle$ of A converging to p.

Example 3.11.2 Suppose $\mathbb{R}$ has the usual metric.

(1) A finite subset A of $\mathbb{R}$ has no limit points. That is, all points in the finite set A are isolated points. So the set $A = \{0, 1\}$ has no limit points, and $\overline{A} = A$.

(2) The set $A = \{\frac{1}{n} : n = 1, 2, 3, \ldots\}$ has one limit point 0. Hence, $\overline{A} = A \cup \{0\}$.

(3) If $A = \{\frac{1}{n} + \frac{1}{m} : n = 1, 2, 3, \ldots; m = 1, 2, 3, \ldots\}$, then the set of all limit points of A is the set $A' = \{\frac{1}{n} : n = 1, 2, 3, \ldots\} \cup \{0\}$.

(4) If $A = (0, 1)$, the open interval between 0 and 1, then 0 and 1 are limit point of A. Every point in A is also a limit point of A. Hence, $\overline{A} = A \cup \{0, 1\} = [0, 1]$. So $[0, 1]$ is not only a closed interval, but it is also a closed set.

(5) The set of all limit points of the interval $(0, 1) \cap Q$ is $[0,1]$.

(6) The set of all limit points of Q is $\mathbb{R}$. The closure of Q, the set of all rational numbers, is $\mathbb{R}$.

(7) The set of all limit points of the Cantor set is the Cantor set.

Remark 3.11.2 The study of topological/metric spaces with a moderate amount of the set theory is called the <u>point set topology</u>. Theorems 3.10.1–3.10.2 are the beginning of the point set topology. Point set topology that concentrates on a "compact and connected" metric space is called the <u>continuum theory</u>. The study of nice subsets in $\mathbb{R}^n$ (like surfaces in $\mathbb{R}^3$) with the usual metric and algebraic structure is called the <u>algebraic topology</u>.

References

1. W. Dunham, *Euler's Miracle*, Euleriana **1**(2), Article 5 (2021)
2. A. Kheyfits, Problem 623. College Math. J. **29** (1998)
3. C. Kicey, S. Goel, A Series for lnk. Am. Math. Mon. **105**(6), 552–554 (1998)
4. H. Katsuura, A new infinite series representation of in k. American Math. Monthly **122**(4), 376 (2015)
5. S. Willard, *General Topology*. Addison-Wesley Publishing Co. (1970)
6. R. Courant, *Differential and Integral Calculus, Second Edition*, Volume 1, 43–44, Interscience Publisher (1954)

Continuous Functions

4

4.1   Definition of a Continuous Function

Intuitively, we learned in a calculus class that if the graph of a function on the plane is connected, then the function is continuous. Not only is this idea not workable for proving theorems it is **actually wrong**. When the domain of a function is not connected, then the graph of a continuous function cannot be connected (see Problem 4.1.1). On the other hand, there is a discontinuous function whose graph is connected (see Example 4.1.4). Let us start by introducing the definition of a continuous function.

Definition 4.1.1 Let X, Y be nonempty subsets of $\mathbb{R}$. A function $f : X \to Y$ is said to be <u>continuous</u> at $p \in X$ if for every $\varepsilon > 0$, there exist a $\delta > 0$ such that $|f(x) - f(p)| < \varepsilon$ whenever $x \in X$ and $|x - p| < \delta$. If f is continuous at **every** point of X, then we say simply f is <u>continuous</u>.

(Some mathematicians say that "continuous at $x = p$" in place of "continuous at $p \in X$". This is unfortunate since I do not see any point in saying "$x =$".)

Remark 4.1.1

(1) The use of "for every $\varepsilon > 0$" is very similar to the definition used in the convergence of a sequence. In Definition 3.1.2, we said that the integer M is depending on the choice of ε, and M can be thought as a function of ε. In this definition of the continuity of a function, this number δ is depending on the choice of ε, and δ can be thought as a function of ε.

© The Author(s), under exclusive license to Springer Nature Switzerland AG 2025   109
H. Katsuura, *Introduction to Analysis*, Synthesis Lectures on Mathematics & Statistics,
https://doi.org/10.1007/978-3-031-67954-4_4

(2) In most calculus textbooks, a function $f : X \to Y$ is continuous at $p \in X$ if $\lim\limits_{x \to p} f(x) = f(p)$, by definition. But in this textbook, **this is** Theorem 4.2.3 in the next section, and **it is not our definition of continuity**.

(3) The continuity of a function is defined only at points in the domain of the function. Hence, all functions are **discontinuous** at any point **outside** of its domain by default. For this, please see the next Example 4.1.1.

Example 4.1.1

(1) In calculus, many authors say that the function $f(x) = \frac{1}{x}$ is discontinuous at 0 since "$\lim\limits_{x \to 0} f(x)$ does not exists". Let us analyze this more carefully. The domain of this function is not defined explicitly. So the function f is usually thought to be $f : (-\infty, 0) \cup (0, \infty) \to \mathbb{R}$ defined by $f(x) = \frac{1}{x}$ for every $x \in (-\infty, 0) \cup (0, \infty)$. And the set $(-\infty, 0) \cup (0, \infty)$ is said to be the <u>implied domain</u> of f. So, the function f is not continuous at 0 because 0 is **not even in the domain** of the function f, and not because "$\lim\limits_{x \to 0} f(x)$ does not exist". Even talking about the continuity of f at 0 seems silly. As you will see in Theorem 4.1.3, this function f **is continuous** everywhere on its **domain**.

Unfortunately, many examples of discontinuous functions in an elementary calculus book are of this type, saying that a function is discontinuous at a point outside of its implied domain.

(2) Let $f : (0, \infty) \to \mathbb{R}$ defined by $f(x) = x$ for every $x \in (0, \infty)$. Then f is not continuous at 0 since f is not even defined at 0.

Example 4.1.2 A function $f : \mathbb{R} \to \mathbb{R}$ is defined by $f(x) = 2x - 6$ for every $x \in \mathbb{R}$. We prove that f is continuous at -2.

Preparation
Let $\varepsilon > 0$. Then we want to find a $\delta > 0$ such that if $|x - (-2)| = |x + 2| < \delta$, then $|f(x) - f(-2)| = |(2x - 6) - (2(-2) - 6)| = |2x + 10| < \varepsilon$.

This is correct. But this is not a good construction because of the simplifications $|x - (-2)| = |x + 2|$ and $|(2x - 6) - (2(-2) - 6)| = |2x + 10|$. It is better not to do the usual multiplications because these muddy the use of the inequality $|x - (-2)| < \delta$. Keep in mind what the subject is as follow:

We want to find a $\delta > 0$ such that if $|x - (-2)| < \delta$. Here, $|x - (-2)|$ is the subject, so we do not want to change it to $|x + 2|$. Then

$$|f(x) - f(-2)| = |(2x - 6) - (2(-2) - 6)| = |2\{x - (-2)\}| = 2|x - (-2)| < \varepsilon.$$

In short, we want to find a $\delta > 0$ such that if $|x - (-2)| < \delta$, then $2|x - (-2)| < \varepsilon$.

Now, we can write the proof.

Proof Let $\varepsilon > 0$. (Yes, you have to write this. See Remark 3.1.2.) Let $x \in \mathbb{R}$ such that $|x - (-2)| < \frac{\varepsilon}{2}$. (Here, we are choosing δ to be $\frac{\varepsilon}{2}$ without renaming $\frac{\varepsilon}{2}$ by δ. Please read the comment in the next example.) Then we have

$$|f(x) - f(-2)| = |(2x - 6) - (2(-2) - 6)|$$
$$= |2\{x - (-2)\}| = 2|x - (-2)| < 2 \cdot \frac{\varepsilon}{2} = \varepsilon$$

Therefore, f is continuous at -2.

The above example leads us to the next example.

Example 4.1.3 A function $f : \mathbb{R} \to \mathbb{R}$ is defined by $f(x) = 2x - 6$ for every $x \in \mathbb{R}$. We prove that f is continuous.

Preparation
Let $p \in \mathbb{R}$ and let $\varepsilon > 0$. Then we want $|f(x) - f(p)| = |(2x - 6) - (2p - 6)| = 2|x - p| < \varepsilon$. This gives us the following proof.

Proof Let $p \in \mathbb{R}$ and let $\varepsilon > 0$. Let $x \in \mathbb{R}$ such that $|x - p| < \frac{\varepsilon}{2}$. Then $|f(x) - f(p)| = |(2x - 6) - (2p - 6)| = 2|x - p| < 2 \cdot \frac{\varepsilon}{2} = \varepsilon$. Hence, f is continuous at $p \in \mathbb{R}$. Therefore, f is continuous.

(A **comment** on this proof: We did not change $\frac{\varepsilon}{2}$ to δ. Since it is easy to write $\frac{\varepsilon}{2}$, there is no need to change it to δ. Changing $\frac{\varepsilon}{2}$ to δ may be seen as a kind act to show the readers how δ in the definition of continuity was chosen. But it also can be interpreted as a lack of understanding of the definition of continuity on the part of the author of the proof. Also, it is an insult to rename a perfectly good specific number $\frac{\varepsilon}{2}$ the more general δ. Please see Example 3.1.3.)

As simple as it may look, let us start by the following problem:

Problem 4.1.1

(1) Let $f : \{0, 1\} \to \mathbb{R}$ be a function defined by $f(0) = 2$ and $f(1) = 1$. Prove that f is continuous.
(2) Let $X = \{\frac{1}{n} : n \in \mathbb{N}\}$. Let $f : X \to \mathbb{R}$ be a function. (Yes, let f be a function defined anyway you like.) Prove that f is continuous.
(3) Let $X = (0, 1) \cup (2, 3]$. Let $c \in \mathbb{R}$. Let $f : X \to \mathbb{R}$ be a function defined by $f(x) = c$ for every $x \in X$. Prove that f is continuous.

(4) Let C be the Cantor set. Let $f : C \to \mathbb{R}$ be a function defined by $f(x) = \frac{1}{3}$ if $x \le \frac{1}{3}$ and $f(x) = \frac{2}{3}$ if $x \ge \frac{2}{3}$ for all $x \in C$. Prove that f is continuous.

(Note that the use of the Cantor set in (4) is to make the problem look difficult, but it is not all that different from the part (3). Also note that all the graphs of the functions in (1)–(4) are **not** connected.)

A solution to (1): Let $p \in \{0, 1\}$. We prove that f is continuous at p. Let $\varepsilon > 0$. Let $x \in \{0, 1\}$ such that $|x - p| < \frac{1}{2}$. Then $x = p$ and $|f(x) - f(p)| = |f(p) - f(p)| = 0 < \varepsilon$. Therefore, f is continuous at p.

A solution to (2): Let $p \in X$. Then $p = \frac{1}{n}$ for some $n \in \mathbb{N}$. We prove that f is continuous at p. Let $\varepsilon > 0$. Let $\delta = \frac{1}{n} - \frac{1}{n+1} > 0$. Let $x \in X$ such that $|x - p| < \delta$. Then $-\left(\frac{1}{n} - \frac{1}{n+1}\right) < x - \frac{1}{n} < \frac{1}{n} - \frac{1}{n+1}$. This implies that

(*) $\frac{1}{n+1} < x < \frac{2}{n} - \frac{1}{n+1} = \frac{n+2}{n(n+1)}$.

Note that $\frac{1}{n} < \frac{1}{n} \cdot \frac{n+2}{n+1} = \frac{n+2}{n(n+1)}$.

Case 1: If $n = 1 = p$, then $x > \frac{1}{2}$ from (*). Since $x \in X$, we have that $x = p$.

Case 2: Suppose $n > 1$. Then $\frac{1}{n-1} - \frac{n+2}{n(n+1)} = \frac{2}{n(n+1)(n-1)} > 0$ so that $\frac{n+2}{n(n+1)} < \frac{1}{n-1}$.
Hence, we have that $\frac{1}{n+1} < x < \frac{1}{n-1}$ from (*). Since $x \in X$, we must have $x = \frac{1}{n} = p$.
In either case, we have $|f(x) - f(p)| = |f(p) - f(p)| = 0 < \varepsilon$. Therefore, f is continuous at p.

Problem 4.1.2 A function $f : \mathbb{R} \to \mathbb{R}$ is defined by $f(x) = \pi x$ for every $x \in \mathbb{R}$.

(1) Prove that f is continuous at π.
(2) Prove that f is continuous.

Problem 4.1.3

(1) A function $f : \mathbb{R} \to \mathbb{R}$ is defined by $f(x) = x^2$ for every $x \in \mathbb{R}$. Prove that f is continuous at -2.
(2) A function $f : \mathbb{R} \to \mathbb{R}$ is defined by $f(x) = \frac{2}{3}x^2 + 7x$ for every $x \in \mathbb{R}$. Prove that f is continuous.

Problem 4.1.4

(1) A function $f : \mathbb{R} \to \mathbb{R}$ is defined by $f(x) = x^3$ for every $x \in \mathbb{R}$. Prove that f is continuous at -2.
(2) A function $f : \mathbb{R} \to \mathbb{R}$ is defined by $f(x) = \frac{2}{3}x^3 + 7x$ for every $x \in \mathbb{R}$. Prove that f is continuous.

Problem 4.1.5 One way to understand the definition of continuity is to negate it.

(1) Let X, Y be nonempty subsets of $\mathbb{R}$. Complete the following sentence:

A function $f : X \to Y$ is said to be <u>not continuous</u> at $p \in X$ if $\cdots$.

(2) Let $f : \mathbb{R} \to \mathbb{R}$ be a function defined by $f(x) = 1$ for every $x \in (-\infty, 0]$ and $f(x) = 5$ for every $x \in (0, \infty)$. Prove that f is not continuous at 0.
(3) Let $f : (-\infty, 0) \cup (0, \infty) \to \mathbb{R}$ be a function defined by $f(x) = 1$ for every $x \in (-\infty, 0)$ and $f(x) = 5$ for every $x \in (0, \infty)$. Is f continuous or discontinuous?

A solution to (1): A function $f : X \to Y$ is said to be <u>not continuous (discontinuous)</u> at $p \in X$ if for every $\delta > 0$, there is an $\varepsilon > 0$ and an $x \in X$ such that $|x - p| < \delta$ and $|f(x) - f(p)| \geq \varepsilon$.
A solution to (2): Let $\delta > 0$. Then $\frac{\delta}{2} \in (0, \infty)$, $\left|0 - \frac{\delta}{2}\right| = \frac{\delta}{2} < \delta$, and $\left|f(0) - f\left(\frac{\delta}{2}\right)\right| = |1 - 5| = 4 > 0$. Therefore, f is not continuous at 0.
(Note: In this proof, the choice for the ε is 4. But we did not rename 4 by saying "Let $\varepsilon = 4$" because 4 would be insulted to be renamed ε.)

Example 4.1.4 Let $f : \left[0, \frac{2}{\pi}\right] \to \mathbb{R}$ be a function defined by $f(x) = \sin \frac{1}{x}$ for every $0 < x \leq \frac{2}{\pi}$ and $f(0) = 0$. We prove that f is not continuous at 0.

Let $\delta > 0$. Then there is an integer $n \geq 1$ such that $\frac{1}{2n\pi + \frac{\pi}{2}} < \delta$. Then $0 < \frac{1}{2n\pi + \frac{\pi}{2}} < \frac{1}{2\pi} < \frac{2}{\pi}$ so that $\frac{1}{2n\pi + \frac{\pi}{2}} \in \left[0, \frac{2}{\pi}\right]$. Moreover, $f\left(\frac{1}{2n\pi + \frac{\pi}{2}}\right) = \sin\left(2n\pi + \frac{\pi}{2}\right) = \sin \frac{\pi}{2} = 1$. Hence, $\left|f\left(\frac{1}{2n\pi + \frac{\pi}{2}}\right) - f(0)\right| = |1 - 0| = 1 > 0$. Therefore, f is not continuous at 0. Even though the function f is not continuous, its graph is **connected** (Fig. 4.1).

Theorem 4.1.1 Let n be a positive integer. A function $f : \mathbb{R} \to \mathbb{R}$ is defined by $f(x) = x^n$ for every $x \in \mathbb{R}$. Then the function f is continuous.

Fig. 4.1 A sketch of the function $y = f(x)$

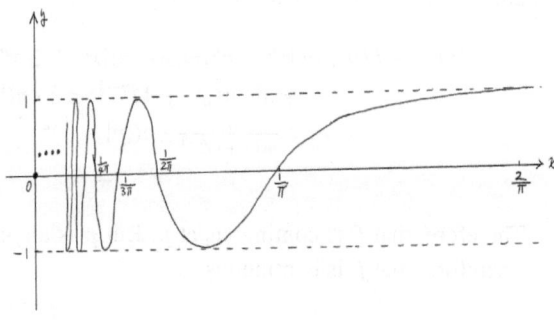

Preparation of the proof: Let $p \in \mathbb{R}$ and let $\varepsilon > 0$. Then we have to find a $\delta > 0$ such that $|f(x) - f(p)| = |x^n - p^n| < \varepsilon$ when $|x - p| < \delta$. What do we know about $x^n - p^n$?

The only thing we can do is to factor. So we factor it to get $|x^n - p^n| = |x - p||x^{n-1} + x^{n-2}p + \cdots + xp^{n-2} + p^{n-1}|$ (See Problem 1.4.2).

What can we do about $|x^{n-1} + x^{n-2}p + \cdots + xp^{n-2} + p^{n-1}|$? Suppose we know what δ is. It is usually a small number.

(*) If $|x - p| < \delta$, then $p - \delta < x < p + \delta$, or $|x| < max\{|p - \delta|, |p + \delta|\} \le |p| + \delta$.

Since $|x^{n-1} + x^{n-2}p + \cdots + xp^{n-2} + p^{n-1}| \le |x|^{n-1} + |x|^{n-2}|p| + \cdots + |x||p|^{n-2} + |p|^{n-1}$, we can replace $|x|$ and $|p|$ by $|p| + \delta$ to obtain $|x^{n-1} + x^{n-2}p + \cdots + xp^{n-2} + p^{n-1}| \le n(|p| + \delta)^{n-1}$. So we have

(**) $|x^n - p^n| = |x - p||x^{n-1} + x^{n-2}p + \cdots + xp^{n-2} + p^{n-1}| < \delta \cdot n(|p| + \delta)^{n-1}$.

We want $\delta \cdot n(|p| + \delta)^{n-1}$ to be less than ε. But δ inside of $(|p| + \delta)^{n-1}$ gets in the way. Is there a way to get rid of this δ? Read (*) carefully. The number δ in (*) is only to give a rough estimate on $|x|$. It really does not matter what δ is, as long as it is a positive number. Would it matter if δ is 10 or 20? Not at all. But for simplicity, let δ in (*) to be 1. Then we can conclude that $|x^{n-1} + x^{n-2}p + \cdots + xp^{n-2} + p^{n-1}| \le n(|p| + 1)^{n-1}$ and by (**), we have $|x^n - p^n| = |x - p||x^{n-1} + x^{n-2}p + \cdots + xp^{n-2} + p^{n-1}| < \delta \cdot n(|p| + 1)^{n-1}$. Can we choose $\delta \cdot n(|p| + 1)^{n-1} < \varepsilon$? Yes, we can. We have to be mindful of δ satisfy both $0 < \delta \le \frac{\varepsilon}{n(|p|+1)^{n-1}}$ and $0 < \delta \le 1$. Do not forget $0 < \delta \le 1$. Without this, we cannot control (*). Now, we can write this formally.

Proof Let $p \in \mathbb{R}$. We will show that f is continuous at p. Let $\varepsilon > 0$. Let δ be a positive number smaller than $\frac{\varepsilon}{n(|p|+1)^{n-1}}$ and 1. That is, $0 < \delta < min\left\{\frac{\varepsilon}{n(|p|+1)^{n-1}}, 1\right\}$. Let $x \in \mathbb{R}$ such that $|x - p| < \delta$.

Claim 1: We will prove that $|x^{n-1} + x^{n-2}p + \cdots + xp^{n-2} + p^{n-1}| \le n(|p| + 1)^{n-1}$.

The inequality $|x - p| < \delta \le 1$ implies that $p - 1 < x < p + 1$. Hence, $|x| < max\{|p - 1|, |p + 1|\} \le |p| + 1$ by the triangular inequality. Moreover,

$$|x^{n-1} + x^{n-2}p + \cdots + xp^{n-2} + p^{n-1}| \le |x|^{n-1} + |x|^{n-2}|p| + \cdots + |x||p|^{n-2} + |p|^{n-1}$$
$$\le n(|p| + 1)^{n-1}.$$

Claim 2: We will show that $|f(x) - f(p)| < \varepsilon$.

$$|f(x) - f(p)| = |x^n - p^n| = |x - p||x^{n-1} + x^{n-2}p + \cdots + xp^{n-2} + p^{n-1}|$$
$$< \delta \cdot n(|p| + 1)^{n-1} < \delta \cdot n(|p| + 1)^{n-1}$$
$$< \frac{\varepsilon}{n(|p|+1)^{n-1}} \cdot n(|p| + 1)^{n-1} \quad \text{by Claim 1,}$$
$$= \varepsilon.$$

Therefore, that f is continuous at p. But p is an arbitrary element of $\mathbb{R}$. Therefore, that f is continuous.

Theorem 4.1.2 Let $n \geq 2$ be an integer. A function $f : (1, \infty) \to \mathbb{R}$ is defined by $f(x) = x^{\frac{1}{n}}$ for every $x \in (1, \infty)$. Then the function f is continuous.

Preparation of the proof: Let $p \in (1, \infty)$ and let $\varepsilon > 0$. Then we have to find a $\delta > 0$ such that $|f(x) - f(p)| = \left| x^{\frac{1}{n}} - p^{\frac{1}{n}} \right| < \varepsilon$ when $|x - p| < \delta$. What do we know about $x^{\frac{1}{n}} - p^{\frac{1}{n}}$? This is similar to the previous proof.

The only thing we can do is to factor. So we factor it to get
$$\left| x^{\frac{n}{n}} - p^{\frac{n}{n}} \right| = \left| x^{\frac{1}{n}} - p^{\frac{1}{n}} \right| \left| x^{\frac{n-1}{n}} + x^{\frac{n-2}{n}} p^{\frac{1}{n}} + \cdots + x^{\frac{1}{n}} p^{\frac{n-2}{n}} + p^{\frac{n-1}{n}} \right| \quad \text{(see Problem 1.4.2)}.$$
Hence,

$$|f(x) - f(p)| = \left| x^{\frac{1}{n}} - p^{\frac{1}{n}} \right| = \frac{|x - p|}{\left| x^{\frac{n-1}{n}} + x^{\frac{n-2}{n}} p^{\frac{1}{n}} + \cdots + x^{\frac{1}{n}} p^{\frac{n-2}{n}} + p^{\frac{n-1}{n}} \right|}$$

What can we do about $\left| x^{\frac{n-1}{n}} + x^{\frac{n-2}{n}} p^{\frac{1}{n}} + \cdots + x^{\frac{1}{n}} p^{\frac{n-2}{n}} + p^{\frac{n-1}{n}} \right|$? Since we want to bound the term $\left| x^{\frac{1}{n}} - p^{\frac{1}{n}} \right|$ by ε, we want to bound $\frac{1}{\left| x^{\frac{n-1}{n}} + x^{\frac{n-2}{n}} p^{\frac{1}{n}} + \cdots + x^{\frac{1}{n}} p^{\frac{n-2}{n}} + p^{\frac{n-1}{n}} \right|}$ from above. But this is equivalent to bound
$$\left| x^{\frac{n-1}{n}} + x^{\frac{n-2}{n}} p^{\frac{1}{n}} + \cdots + x^{\frac{1}{n}} p^{\frac{n-2}{n}} + p^{\frac{n-1}{n}} \right| \text{ from below.}$$

Since $p > 1$ and $x > 1$, we have $\left| x^{\frac{n-1}{n}} + x^{\frac{n-2}{n}} p^{\frac{1}{n}} + \cdots + x^{\frac{1}{n}} p^{\frac{n-2}{n}} + p^{\frac{n-1}{n}} \right| > n \geq 1$.

Hence, $\left| x^{\frac{1}{n}} - p^{\frac{1}{n}} \right| = \frac{|x-p|}{\left| x^{\frac{n-1}{n}} + x^{\frac{n-2}{n}} p^{\frac{1}{n}} + \cdots + x^{\frac{1}{n}} p^{\frac{n-2}{n}} + p^{\frac{n-1}{n}} \right|} < \frac{1}{n} |x - p| \leq |x - p|$

Now, we write this formally.

Proof Let $p \in (1, \infty)$. We will show that f is continuous at p. Let $\varepsilon > 0$. Let $x \in (1, \infty)$ such that $|x - p| < \varepsilon$. Then since $x > 1$ and $p > 1$, we have
$$(1) \quad x^{\frac{n-1}{n}} + x^{\frac{n-2}{n}} p^{\frac{1}{n}} + \cdots + x^{\frac{1}{n}} p^{\frac{n-2}{n}} + p^{\frac{n-1}{n}} > n.$$
Since $|x - p| = \left| x^{\frac{n}{n}} - p^{\frac{n}{n}} \right| = \left| x^{\frac{1}{n}} - p^{\frac{1}{n}} \right| \left| x^{\frac{n-1}{n}} + x^{\frac{n-2}{n}} p^{\frac{1}{n}} + \cdots + x^{\frac{1}{n}} p^{\frac{n-2}{n}} + p^{\frac{n-1}{n}} \right|$, we have

$$|f(x) - f(p)| = \left| x^{\frac{1}{n}} - p^{\frac{1}{n}} \right| = \frac{|x-p|}{\left| x^{\frac{n-1}{n}} + x^{\frac{n-2}{n}} p^{\frac{1}{n}} + \cdots + x^{\frac{1}{n}} p^{\frac{n-2}{n}} + p^{\frac{n-1}{n}} \right|}$$
$$< \frac{1}{n} |x - p|$$
$$\leq |x - p| < \varepsilon$$

Therefore, that f is continuous at p. But p is arbitrary element of $(1, \infty)$. Therefore, that f is continuous.

Remark 4.1.1 Theorems 4.1.1 and 4.1.2 suggest the following question: Let $r \in \mathbb{R}$. A function $f : (0, \infty) \to \mathbb{R}$ is defined by $f(x) = x^r$ for every $x \in (0, \infty)$. Is the function f continuous? YES, but this is a difficult question to answer. We will prove this in Theorem 4.3.12. For now, we practice using Definition 4.1.1.

The next problem is slightly more difficult than the ones above. So I have divided it into four parts. The goal is the part (4). Please complete (1)–(3) before attempting (4). These are to help you to complete (4).

Problem 4.1.6 A function $f : (0, \infty) \to \mathbb{R}$ is defined by $f(x) = \frac{1}{x}$ for every $x \in (0, \infty)$.

(1) Prove that f is continuous at 2.
(2) Prove that f is continuous at $\frac{1}{3}$.
(3) Prove that f is continuous at $\frac{1}{4}$.
(4) Prove that f is continuous.

(As you can see from (1)–(3), if p is a large positive number, it is not difficult to prove that f is continuous at p. However, if p is very close to 0, showing continuity of f at p becomes difficult. You have to control x away from 0 by choosing δ very carefully. Here, we use the number p itself to control the size of $\frac{1}{x}$. A solution is given in the proof the next theorem.)

Theorem 4.1.3 A function $f : \mathbb{R} - \{0\} \to \mathbb{R}$ is defined by $f(x) = \frac{1}{x}$ for every $x \in \mathbb{R} - \{0\}$. Then f is continuous.

Preparation: Let $p > 0$. If $|x - p| < \frac{p}{2}$, then $0 < \frac{p}{2} < x < \frac{3p}{2}$ so that $\frac{1}{x} < \frac{2}{p}$. Hence, $\left| \frac{1}{x} - \frac{1}{p} \right| = \frac{|x-p|}{px} < \frac{2}{p} \cdot \frac{|x-p|}{p} = \frac{2}{p^2}|x - p|$. This allows us to write a proof.

Proof We will prove that f is continuous at $p > 0$. Let $\varepsilon > 0$. Let $0 < \delta < \min\left\{\frac{p}{2}, \frac{p^2\varepsilon}{2}\right\}$. Let $x \in \mathbb{R} - \{0\}$ such that $|x - p| < \delta$. Since $\delta < \frac{p}{2}$, we have $|x - p| < \delta < \frac{p}{2}$. Then $\frac{p}{2} < x < \frac{3p}{2}$ so that $0 < \frac{1}{x} < \frac{2}{p}$. Hence,

$$|f(x) - f(p)| = \left| \frac{1}{x} - \frac{1}{p} \right| = \frac{|x - p|}{px} < \frac{2}{p} \cdot \frac{|x - p|}{p}$$

$$= \frac{2}{p^2}|x - p| < \frac{2}{p^2} \cdot \frac{p^2\varepsilon}{2} = \varepsilon \quad \text{since } |x - p| < \delta < \frac{p^2\varepsilon}{2}.$$

Therefore, f is continuous at $p > 0$. Showing the continuity of f at $p < 0$ is left in the next problem.

Problem 4.1.7 A function $f : \mathbb{R} - \{0\} \to \mathbb{R}$ is defined by $f(x) = \frac{1}{x}$ for every $x \in \mathbb{R} - \{0\}$. Prove that f is continuous at $p < 0$.

Problem 4.1.8 Let $f : \mathbb{R} \to \mathbb{R}$ be a continuous function. Let X be a nonempty subset of $\mathbb{R}$. Let $g : X \to \mathbb{R}$ be a function defined by $g(x) = f(x)$ for every $x \in X$. Prove that g is continuous. (This function g is said to be the *restriction* of f *restricted* to X, and the function g is usually denoted by $f|_X$. See Definition 1.3.2.)

Problem 4.1.9

(1) Let $f : \mathbb{R} \to \mathbb{R}$ be a function defined by $f(x) = \frac{1}{x}$ for every $x \in (-\infty, 0) \cup (0, \infty)$ and $f(0) = -10$. Prove that f is not continuous at 0. Not by contradiction please. Unlike the function f defined in Example 4.1.1(1), this function f is defined at 0.
(2) Let $f : \mathbb{R} \to \mathbb{R}$ be a function defined by $f(x) = 0$ if $x \in \mathbb{R}$ is rational and $f(x) = -1$ if $x \in \mathbb{R}$ is irrational. Prove that f is not continuous at any point. Not by contradiction please.

Recall Theorem 3.6.1: Let $f : C \to [0, 1]$ be a function defined by $f\left(\sum_{n=1}^{\infty} \frac{s_n}{3^n}\right) = \sum_{n=1}^{\infty} \frac{(s_n/2)}{2^n} = \sum_{n=1}^{\infty} \frac{s_n}{2^{n+1}}$ for every $\sum_{n=1}^{\infty} \frac{s_n}{3^n} \in C$. This function f was defined in Theorem 3.6.1, and it was called the <u>Cantor function</u>. See Fig. 3.6.6. for its graph. The Cantor function is onto. So, we prove the next theorem as a sequel to Theorem 3.6.1. Also, this theorem should be surprising to you since the graph of this function is **"totally" disconnected**.

Theorem 4.1.4 The Cantor function $f : C \to [0, 1]$ is continuous.

Proof Let $\varepsilon > 0$. We want to show f is continuous at $s = \sum_{n=1}^{\infty} \frac{s_n}{3^n} \in C$ for some $s_n \in \{0, 2\}$ for every integer $n > 0$. Let $M > 0$ be an integer such that $\frac{1}{2^M} < \varepsilon$. Let $x \in C$ such that $|s - x| < \frac{1}{3^M}$. Since $x \in C$, $x = \sum_{n=1}^{\infty} \frac{x_n}{3^n}$ for some $x_n \in \{0, 2\}$ for all $n = 1, 2, 3, \cdots$.

We want to show that $s_n = x_n$ for all $n = 1, 2, 3, \cdots, M$. Suppose $n > 0$ is the smallest integer such that $s_n \neq x_n$. Then $(s_n = 0 \text{ and } x_n = 2)$ or $(s_n = 2 \text{ and } x_n = 0)$. Hence, we have $|s - x| = \left| \sum_{k=1}^{\infty} \frac{s_k}{3^k} - \sum_{k=1}^{\infty} \frac{x_k}{3^k} \right| = \left| \sum_{k=n}^{\infty} \frac{s_k - x_k}{3^k} \right| \geq \frac{|s_k - x_k|}{3^k} - \sum_{k=n}^{\infty} \frac{|s_k - x_k|}{3^k} = \frac{2}{3^n} - \sum_{k=n}^{\infty} \frac{2}{3^k} = \frac{2}{3^n} - \frac{2}{3^{n+1}} \cdot \frac{1}{1 - \frac{1}{3}} = \frac{1}{3^n}$. Since $|s - x| < \frac{1}{3^M}$, we must have $\frac{1}{3^M} \geq \frac{1}{3^n}$, i.e., $n \geq m$. Hence, $s_n = x_n$ for all $n = 1, 2, 3, \cdots, M$.

Now, we have

$$|f(s) - f(x)| = \left| f\left(\sum_{n=1}^{\infty} \frac{s_n}{3^n} \right) - f\left(\sum_{n=1}^{\infty} \frac{x_n}{3^n} \right) \right|$$

$$= \left| \sum_{n=1}^{\infty} \frac{s_n}{2^{n+1}} - \sum_{n=1}^{\infty} \frac{x_n}{2^{n+1}} \right| = \left| \sum_{n=M+1}^{\infty} \frac{s_n}{2^{n+1}} - \sum_{n=M+1}^{\infty} \frac{x_n}{2^{n+1}} \right| = \left| \sum_{n=M+1}^{\infty} \frac{s_n - x_n}{2^{n+1}} \right|$$

$$\leq \sum_{n=M+1}^{\infty} \frac{|s_n - x_n|}{2^{n+1}} \leq \sum_{n=M+1}^{\infty} \frac{2}{2^{n+1}} = \sum_{n=M+1}^{\infty} \frac{1}{2^n}$$

$$= \frac{1}{2^{M+1}} \cdot \sum_{n=0}^{\infty} \frac{1}{2^n} = \frac{1}{2^{M+1}} \cdot 2 = \frac{1}{2^M} < \varepsilon$$

since $M > 0$ was chosen to be an integer such that $\frac{1}{2^M} < \varepsilon$.

This proves that f is continuous.

Remark 4.1.3

(1) We will prove that the Cantor function $f : C \to [0, 1]$ is nowhere differentiable in the next chapter.

(2) Let $A_1 = \left(\frac{1}{3}, \frac{2}{3} \right)$. Let $B_1 = [0, 1] - A_1 = \left[0, \frac{1}{3} \right] \cup \left[\frac{2}{3}, 1 \right]$.

Let $A_2 = \left(\frac{1}{3^2}, \frac{2}{3^2} \right) \cup \left(\frac{7}{3^2}, \frac{8}{3^2} \right)$,

Let $B_2 = B_1 - A_2 = \left[0, \frac{1}{3^2} \right] \cup \left[\frac{2}{3^2}, \frac{3}{3^2} \right] \cup \left[\frac{6}{3^2}, \frac{7}{3^2} \right] \cup \left[\frac{8}{3^2}, \frac{9}{3^2} \right]$.

Note that A_2 is the union of two middle one third open intervals of closed interval of B_1. Let n be a positive integer such that B_n is defined as a union of 2^n disjoint closed intervals. Then we let A_{n+1} is the union of 2^n open intervals that are exactly the middle one third of the closed intervals in B_n. Then we define $B_{n+1} = B_n - A_{n+1}$.

(See the definition of the set U_n in Example 3.6.7. Then $A_n = U_n$ for every integer $n > 0$.)

By the mathematical induction, B_n is defined for all positive integer n such that $B_{n+1} \subset B_n$. Then $C = \bigcap_{n=1}^{\infty} B_n = [0, 1] - \bigcup_{n=1}^{\infty} A_n$.

Let $f : C \to [0, 1]$ be the Cantor function. If (a, b) is one of the open intervals of A_n for some n, then $a, b \in C$ and $f(a) = f(b)$.

Let us define $F : [0, 1] \to [0, 1]$ by the following:

If $x \in C$, then $F(x) = f(x)$.

If $x \notin C$, then $x \in A_n$ for some n. Then $x \in \left(\frac{i}{3^n}, \frac{i+1}{3^n} \right)$ for some open interval $\left(\frac{i}{3^n}, \frac{i+1}{3^n} \right)$ of A_n. We define $F(x) = f\left(\frac{i}{3^n} \right) = f\left(\frac{i+1}{3^n} \right)$.

Then the graph of the function $y = F(x)$ resembles an infinite staircase. So, some people call this function $F : [0, 1] \to [0, 1]$ the devil's staircase function or Cantor's staircase function. It is **continuous** and its graph is **connected** (Fig. 4.2).

Fig. 4.2 A sketch of a graph
of the devil's staircase function

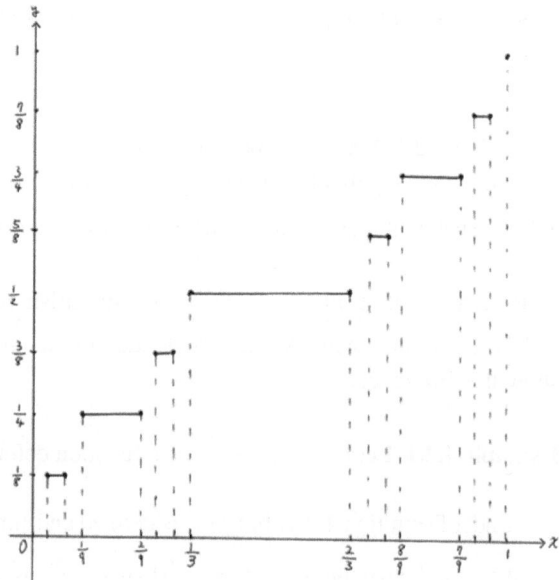

4.2   Limits and Continuity of Functions

We start by defining "limits" of a function in this section. As you can see in Definition
4.2.1, the limit of a function is very similar to the definition of continuity, and also it
is similar to the definition of the limit of a sequence. As a matter of fact, the purpose
of considering limits of functions is to apply what we learned on limits of sequences in
Chap. 3 to the continuity of functions.

Let $X \neq \emptyset$ be a subset of $\mathbb{R}$. Recall that $p \in \overline{X}$ if there exists a sequence $\langle a_n \rangle$ in X
converging to p, and $\overline{X}$ is said to be the <u>closure</u> of the set X. We often read $\overline{X}$ as the "X
closure". If $X = \overline{X}$, then X is a closed set. A point $p \in \overline{X}$ is said to be a <u>limit point</u> of X
if there exists there exists a sequence $\langle a_n \rangle$ in $X - \{p\}$ converging to p. So $\overline{X}$ is the union
of X and the set of all limit points of X. (A set with finite elements from $\mathbb{R}$ is a closed
set. Please see Example 3.11.2 for more examples.)

Definition 4.2.1 Let $f : X \to \mathbb{R}$ be a function. Let $p \in \overline{X}$. Let $L \in \mathbb{R}$. We say $\lim_{x \to p} f(x) = L$,
or $\lim_{x \to p, x \in X} f(x) = L$, if for every $\varepsilon > 0$, there exists a $\delta > 0$ such that $|f(x) - L| < \varepsilon$
whenever $x \in X$ and $|x - p| < \delta$.

We read $\lim_{x \to p} f(x) = L$ as "$f(x)$ converges to L as x approaches to p" or "the <u>limit of the</u>
<u>function</u> $f(x)$ is L as x approaches to p".

Some textbook authors adopt the following definition in place of Definition 4.2.1 of $\lim_{x \to p} f(x) = L$.

Definition $\widehat{4.2.1}$ Let $X \neq \emptyset$ be a subset of $\mathbb{R}$, and let $f : X \to \mathbb{R}$ be a function. Let $p \in \overline{X}$. Let $L \in \mathbb{R}$. We say that $\underset{x \to p}{\mathrm{Lim}} f(x) = L$, or $\underset{x \to p, x \in X}{\mathrm{Lim}} f(x) = L$, if for every $\varepsilon > 0$, there exists a $\delta > 0$ such that $|f(x) - L| < \varepsilon$ whenever $x \in X$ and $0 < |x - p| < \delta$.

Here, we write $\underset{x \to p}{\mathrm{Lim}} f(x)$ in order to **distinguish** $\lim_{x \to p} f(x)$ in Definition 4.2.1.

The limits of functions under these definitions **are** deferent. The next two examples show the difference.

Example 4.2.1 Let $f : \{0, 1\} \to \mathbb{R}$ be a function defined by $f(0) = 2$ and $f(1) = 1$.

(1) Under Definition 4.2.1, $\lim_{x \to 0} f(x)$ is defined and $\lim_{x \to 0} f(x) = 2$. To see this, let $\varepsilon > 0$. Let $x \in \{0, 1\}$ such that $|x - 0| < \frac{1}{2}$. Then $x = 0$ and $|f(x) - 2| = |f(0) - 2| = |2 - 2| = 0 < \varepsilon$. Therefore, $\lim_{x \to 0} f(x) = 2$ under Definition 4.2.1.

(2) Since there is no $x \in \{0, 1\}$ such that $0 < |x - 0| < \frac{1}{2}$, the limit $\underset{x \to 0}{\mathrm{Lim}} f(x)$ does **not** exist under Definition $\widehat{4.2.1}$.

Example 4.2.2 First, let us **negate** Definition 4.2.1. Let $\emptyset \neq X \subset \mathbb{R}$, and let $f : X \to \mathbb{R}$ be a function. Let $p \in \overline{X}$ and $L \in \mathbb{R}$. Then $\lim_{x \to p} f(x) \neq L$ if there is a number $\varepsilon > 0$ such that for every $\delta > 0$, there is an $x \in X$ such that $|x - p| < \delta$ and $|f(x) - L| \geq \varepsilon$. If $\lim_{x \to p} f(x) \neq L$ for every $L \in \mathbb{R}$, we say that $\lim_{x \to p} f(x)$ **does not exist**.

(1) Now, let $f : \mathbb{R} \to \mathbb{R}$ be a function defined by $f(x) = 2$ if $x \neq 0$, and $f(0) = 0$. Under Definition 4.2.1, $\lim_{x \to 0} f(x)$ does not exists. To see this, let $L \in \mathbb{R}$. We will show that $\lim_{x \to 0} f(x) \neq L$. Let $\delta > 0$.

Case 1: Suppose $L = 2$. Let $\delta > 0$. Let $x = 0$. Then $x \in \mathbb{R}$ and $|x - 0| = 0 < \delta$. Moreover, $|f(x) - L| = |0 - 2| = 2 > 0$. Therefore, $\lim_{x \to 0} f(x) \neq L$.

Case 2: Suppose $L \neq 2$. Let $\delta > 0$. Let $x = \frac{\delta}{2}$. Then $x \in \mathbb{R}$ and $|x - 0| = x = \frac{\delta}{2} < \delta$. But $|f(x) - L| = |2 - L| > 0$ since $L \neq 2$. Therefore, $\lim_{x \to 0} f(x) \neq L$.

Therefore, $\lim_{x \to 0} f(x)$ does **not** exists.

(2) Again, let $f : \mathbb{R} \to \mathbb{R}$ be a function defined by $f(x) = 2$ if $x \neq 0$, and $f(0) = 0$. Under
Definition $\widehat{4.2.1}$, $\operatorname*{Lim}_{x \to 0} f(x) = 2$. To see this, let $\varepsilon > 0$. Let $\delta > 0$. Let $x \in \mathbb{R}$ such that
$0 < |x - 0| < \delta$. Then $|f(x) - 2| = |2 - 2| = 0 < \varepsilon$. Therefore, $\operatorname*{Lim}_{x \to 0} f(x) = 2$.

(3) Let $f : \mathbb{R} - \{0\} \to \mathbb{R}$ be a function defined by $f(x) = 2$ for every $x \in \mathbb{R} - \{0\}$. Then
$\lim_{x \to 0} f(x) = 2 = \operatorname*{Lim}_{x \to 0} f(x)$.

Remark 4.2.1 We use Definition 4.2.1 **for the limit convergence of a function.** We do
not use Definition $\widehat{4.2.1}$. In either definition of limits, please pay attention to the condition
"**whenever** $x \in X$". This condition "whenever $x \in X$" is not emphasized enough in many
textbooks. Because the point p, in the closure $\overline{X}$, may not be a point inside X, where X is
the domain of f, I think some textbooks add "$0 <$" in "$0 < |x - p| < \delta$". Since we have
"whenever $x \in X$" in Definition 4.2.1, there is no need of saying "$0 <$" in "$0 < |x - p| < \delta$".

One additional important reason why we use Definition 4.2.1 will be given in Remark
4.2.2.

Problem 4.2.1

(1) Let $f : \mathbb{R} - \{0\} \to \mathbb{R}$ be a function defined by $f(x) = -2$ for every $x \in \mathbb{R} - \{0\}$. Prove
that $\lim_{x \to 0} f(x) = -2$.

(2) Let $f : \mathbb{R} \to \mathbb{R}$ be a function defined by $f(x) = 1$ if x is rational, and $f(x) = -1$ if x is
irrational. Use Definition 4.2.1 to prove that $\lim_{x \to p} f(x)$ does not exist for any $p \in \mathbb{R}$. Not
by contradiction, please. Compare this to Problem 4.1.8(2).

(3) Let $f : \mathbb{R} \to \mathbb{R}$ be a function defined by $f(x) = 3x^2 - 5x - 3$ for every $x \in \mathbb{R}$. Prove
that $\lim_{x \to 1} f(x) = -5$

**The next theorem is the bridge between the continuity of a function and limits of
sequences.**

Theorem 4.2.1 Let $\emptyset \neq X \subset \mathbb{R}$, and let $f : X \to \mathbb{R}$ be a function. Let $p \in \overline{X}$ and $L \in \mathbb{R}$.

(1) If $\lim_{x \to p} f(x) = L$, then for every sequence $\langle a_n \rangle$ in X converging to p, we have $\lim_{n \to \infty} f(a_n) = L$. In other words, the sequence $\langle f(a_n) \rangle_{n=1}^{\infty}$ converges to L.

(2) Suppose, for every sequence $\langle a_n \rangle$ in X converging to p, we have $\lim_{n \to \infty} f(a_n) = L$. Then
$\lim_{x \to p} f(x) = L$.

Proof of (1) is left to the readers.

Proof of (2): Suppose, for every sequence $\langle a_n \rangle$ in X converging to p, we have $\lim_{n \to \infty} f(a_n) = L$. On the contrary, suppose $\lim_{x \to p} f(x) \neq L$. Then there exists an $\varepsilon > 0$ such that, for every $\delta > 0$, there exists $x \in X$ and $|x - p| < \delta$ such that $|f(x) - L| \geq \varepsilon$. So for every positive integer n, there exists $a_n \in X$ and $|a_n - p| < \frac{1}{n}$ such that $|f(a_n) - L| \geq \varepsilon$. Since $|a_n - p| < \frac{1}{n}$ for every integer n, $\langle a_n \rangle$ is a sequence in X converging to p such that $\lim_{n \to \infty} f(a_n) \neq L$. This is a contradiction to our assumption $\lim_{n \to \infty} f(a_n) = L$. Therefore, we must have $\lim_{x \to p} f(x) = L$.

Problem 4.2.2 Let $\emptyset \neq X \subset \mathbb{R}$, and let $f : X \to \mathbb{R}$ be a function. Let $p \in \overline{X}$. If $\lim_{x \to p} f(x) = L$, then for every sequence $\langle a_n \rangle$ in X converging to p, prove that $\lim_{n \to \infty} f(a_n) = L$. This is to prove Theorem 4.2.1(1).

Definition 4.2.3 Let $\emptyset \neq X \subset \mathbb{R}$, and let $f : X \to \mathbb{R}$ and $g : X \to \mathbb{R}$ be functions. Then $f + g : X \to \mathbb{R}$ and $f \cdot g : X \to \mathbb{R}$ are functions defined by $(f + g)(x) = f(x) + g(x)$ and $(f \cdot g)(x) = f(x) \cdot g(x)$ for every $x \in X$. Let $\hat{X} = \{x \in X : g(x) \neq 0\}$. Then the function $f/g = \frac{f}{g} : \hat{X} \to \mathbb{R}$ is defined by $(f/g)(x) = f(x)/g(x)$ for every $x \in \hat{X}$.

Let $n \geq 0$ is an integer, and if $b_k \in \mathbb{R}$ for each $k = 0, 1, 2, \ldots, n$ such that $b_n \neq 0$, then a function $f : X \to \mathbb{R}$ defined by $f(x) = b_0 + b_1 x + b_2 x^2 + \cdots + b_n x^n$ for every $x \in X$, then f is said to be a <u>polynomial</u> of <u>degree</u> n.

If f and g are polynomials, then f/g is called a <u>rational function</u>.

Theorem 4.2.2 Let $\emptyset \neq X \subset \mathbb{R}$, and let $f, g : X \to \mathbb{R}$ be functions. Let $p \in \overline{X}$. Suppose $\lim_{x \to p} f(x)$ and $\lim_{x \to p} g(x)$ exist. Then the following are true.

(1) $\lim_{x \to p} (f + g)(x) = \lim_{x \to p} f(x) + \lim_{x \to p} g(x)$,

(2) $\lim_{x \to p} (f \cdot g)(x) = \lim_{x \to p} f(x) \cdot \lim_{x \to p} g(x)$,

(3) if $g(x) \neq 0$ for every $x \in X$ and $\lim_{x \to p} g(x) \neq 0$, then $\lim_{x \to p} \left(\frac{f}{g} \right)(x) = \dfrac{\lim_{x \to p} f(x)}{\lim_{x \to p} g(x)}$.

Proof of (1) is by Theorem 3.1.4 and Theorem 4.2.1. Let $\langle a_n \rangle$ be a sequence in X converging to p. Since $\lim_{x \to p} f(x)$ and $\lim_{x \to p} g(x)$ exist, we have $\lim_{x \to p} f(x) = \lim_{n \to \infty} f(a_n)$ and $\lim_{x \to p} g(x) = \lim_{n \to \infty} g(a_n)$. Hence,

$$\lim_{x \to p} f(x) + \lim_{x \to p} g(x) = \lim_{n \to \infty} f(a_n) + \lim_{n \to \infty} g(a_n) \quad \text{by Theorem 4.2.1}$$

$$= \lim_{n \to \infty} \{f(a_n) + g(a_n)\} \quad \text{by Theorem 3.1.4}$$

$$= \lim_{n \to \infty} (f + g)(a_n) \quad \text{by the definition of the sum of two functions}$$

$$= \lim_{x \to p} (f + g)(x) \quad \text{by Theorem 4.2.1.}$$

Since the proofs of (2) and (3) are also by Theorem 3.1.4 and Theorem 4.2.1 and similar to the above proof, we leave the proofs of (2) and (3) to the reader.

The next theorem is one of the reasons why the limit of a function is important.

Theorem 4.2.3 Let $\emptyset \neq X \subset \mathbb{R}$, and let $f : X \to \mathbb{R}$ be a function. Let $p \in X$. (Note that we did not write $p \in \overline{X}$.) Then the following statements are equivalent.

(1) The function f is continuous at p.
(2) If $\langle a_n \rangle$ is a sequence in X converging to p, we have $\lim_{n \to \infty} f(a_n) = f(p)$.
(3) $\lim_{x \to p, x \in X} f(x) = f(p)$.

Proof of (2) and (3) being equivalent: Theorem 4.2.1 proves the equivalence of (2) and (3).

Proof of (1) implies (3). Suppose the function f is continuous at p. Then for every $\varepsilon > 0$, there exists a $\delta > 0$ such that $|f(x) - f(p)| < \varepsilon$ whenever $x \in X$ and $|x - p| < \delta$. But this exactly $\lim_{x \to p} f(x) = f(p)$ by Definition 4.2.1.

Proof of (3) implies (1): Suppose $\lim_{x \to p} f(x) = f(p)$. Then for every $\varepsilon > 0$, there exists a $\delta > 0$ such that $|f(x) - f(p)| < \varepsilon$ whenever $x \in X$ and $|x - p| < \delta$. Hence, f is continuous at p.

Remark 4.2.2

(1) In Example 4.2.1 (2), we saw that if $f : \{0, 1\} \to \mathbb{R}$ be a function defined by $f(0) = 2$ and $f(1) = 1$, then $\widehat{\lim_{x \to 0}} f(x)$ does not exist under Definition $\widehat{4.2.1}$, while f is a continuous function as we saw in Problem 4.1.1(1). This shows that under **Definition** $\widehat{4.2.1}$ for limits, the **statement** "a function $f : X \to \mathbb{R}$ is continuous at $p \in X$ if and only if $\widehat{\lim_{x \to p}} f(x) = f(p)$" **is not true!** But "a function $f : X \to \mathbb{R}$ is continuous at $p \in X$ if and only if $\lim_{x \to p} f(x) = f(p)$" **is true.** Please be careful.

(2) Some textbooks use Definition $\widehat{4.2.1}$ to define the continuity of a function $f : X \to \mathbb{R}$ at $p \in X$ as follow: "If $\widehat{\lim_{x \to p}} f(x) = f(p)$, then f is continuous at $p \in X$". If this is the case, the function $f : \{0, 1\} \to \mathbb{R}$ defined by $f(0) = 2$ and $f(1) = 1$ is no longer a continuous function. Of course, we do not use this definition.

(3) If $f : (0, \infty) \to \mathbb{R}$ defined by $f(x) = x$ for every $x \in (0, \infty)$, then $\lim_{x \to 0} f(x) = 0$. But f is not continuous at 0 since $0 \notin (0, \infty)$. See Example 4.1.1(2).

Remark 4.2.3 Please note that any function is **discontinuous** at points outside of the domain of the function by default. Please see Example 4.1.1.

Theorem 4.2.4 Let $\emptyset \neq X \subset \mathbb{R}$, and let $f : X \to \mathbb{R}$ and $g : X \to \mathbb{R}$ be functions continuous at $p \in X$.

(1) The function $f + g : X \to \mathbb{R}$ defined by $(f + g)(x) = f(x) + g(x)$ for every $x \in X$ is continuous at p.
(2) The function $f \cdot g : X \to \mathbb{R}$ defined by $(f \cdot g)(x) = f(x) \cdot g(x)$ for every $x \in X$ is continuous at p.
(3) Suppose $g(x) \neq 0$ for every $x \in X$. The function $f / g : X \to \mathbb{R}$ defined by $(f / g)(x) = f(x) / g(x)$ for every $x \in X$ is continuous at p.

Proof is by Theorems 4.2.2 and 4.2.3.

Example 4.2.2 Let $f : (-\infty, 0) \cup (0, \infty) \to \mathbb{R}$ be a function defined by $f(x) = x$ for every $x \in (-\infty, 0) \cup (0, \infty)$. Then $f(x) \neq 0$ for every $x \in (-\infty, 0) \cup (0, \infty)$ and f is continuous. Hence, $\frac{1}{f} : (-\infty, 0) \cup (0, \infty) \to \mathbb{R}$ defined by $\frac{1}{f}(x) = \frac{1}{f(x)} = \frac{1}{x}$ for every $x \in (-\infty, 0) \cup (0, \infty)$ is continuous by Theorem 4.2.3 (see Theorem 4.1.3 for comparison).

Theorem 4.2.5

(1) All polynomial functions are continuous.
(2) All rational functions are continuous (at points where the denominator does not vanish). In other words, rational functions are continuous on their domains.

Proof of (1) and (2) are by Theorem 4.2.4 and Theorem 4.1.1.

Theorem 4.2.6 Let X, Y, Z be nonempty subsets of $\mathbb{R}$, and let $f : X \to Y, g : Y \to Z$ be functions. Suppose the function f is continuous at $p \in X$, and suppose the function g is continuous at $f(p)$. Let $F : X \to Z$ be a function defined by $F(x) = g(f(x))$ for every $x \in X$. Then F is continuous at p.

(**Definition**: We write $F : X \to Z$ by $g \circ f : X \to Z$, and $g \circ f$ is said to be the composition of f with g. Hence, the composition of two continuous functions is continuous.)

Proof We have $\lim_{x \to p} f(x) = f(p)$ and $\lim_{y \to f(p)} g(y) = g(f(p))$. So we have $\lim_{x \to p} F(x) = \lim_{x \to p} g(f(x)) = \lim_{y \to f(p)} g(y)$, by substituting $y = f(x)$,
$= g(f(p))$.

By Theorem 4.2.3, F is continuous at p.

Example 4.2.3 Let $r \in \mathbb{Q} \cap (0, \infty)$. Then $r = \frac{p}{q}$ for some positive integers p and q. So let $F :$ $(0, \infty) \to \mathbb{R}$ be a function defined by $f(x) = x^r$ for every $x \in (0, \infty)$. By Theorem 4.1.1, the function $f : \mathbb{R} \to \mathbb{R}$ defined by $f(x) = x^p$ for every $x \in \mathbb{R}$ is continuous. By Theorem 4.1.2, the function $g : (0, \infty) \to \mathbb{R}$ defined by $g(x) = x^{\frac{1}{q}}$ for every $x \in (0, \infty)$ is continuous. Then $F(x) = x^r = \left(x^{\frac{1}{q}} \right)^p = f_p(g_q(x))$ for every $x \in (0, \infty)$. By Theorem 4.2.6, the function F is continuous. So if $r \in \mathbb{Q} \cap (-\infty, 0)$ by letting $r = -s$, we have $x^r = \frac{1}{x^s}$. This shows that if $r \in \mathbb{Q}$, then the function $F : (0, \infty) \to \mathbb{R}$ defined by $F(x) = x^r$ for every $x \in (0, \infty)$ is continuous. This brings us to the natural **question**: If r is an **irrational** number, **is** the function $G : (0, \infty) \to \mathbb{R}$ defined by $G(x) = x^r$ for every $x \in (0, \infty)$ continuous? We will answer this in Theorem 4.3.12 as a consequence of the next theorem.

Theorem 4.2.7 (*Continuity of an exponential function*) Let $a > 0$ and $c \in \mathbb{R}$. Let $f : \mathbb{R} \to \mathbb{R}$ be a function defined by $f(x) = a^{cx}$ for every $x \in \mathbb{R}$. Then f is continuous.

Proof Let $r \in \mathbb{R}$. Let $\langle r_n \rangle_{n=1}^{\infty}$ is a sequence of real numbers that converges to r. Then

$$\lim_{x \to r} f(x) = \lim_{n \to \infty} f(r_n) \quad \text{by Theorem 4.2.1}$$

$$= \lim_{n \to \infty} a^{cr_n} = a^{cr} \quad \text{by \textbf{Theorem 3.9.2}}$$

$$= f(r)$$

Therefore, the function f is continuous.

Remark 4.2.4 The above proof seems short and simple. However, we really had to work hard in Chap. 3 to obtain Theorem 3.9.2.

Axiom: Trigonometric functions $\cos : \mathbb{R} \to [-1, 1]$ and $\sin : \mathbb{R} \to [-1, 1]$ are continuous.

Remark 4.2.5 The continuity of sine and cosine functions is something that should be proved. But we take a shortcut and **assume** the continuity of **sine** and **cosine** functions. Remember that all angles are measured in radian. In case you missed it, I gave a brief review of trigonometry in Remark 1.3.1 and Example 1.3.7.

Theorem 4.2.8 Prove that $\lim_{x \to 0} \frac{\sin x}{x} = 1$.

Fig. 4.3 A figure used to prove $\lim\limits_{x\to 0} \frac{\sin x}{x} = 1$

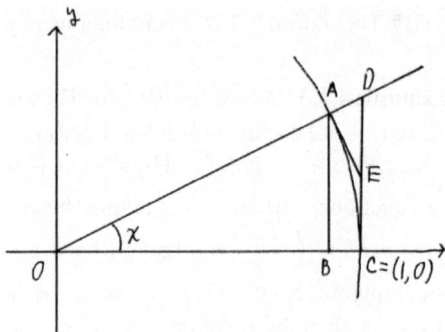

Proof Let $x > 0$. In Fig. 4.3, the portion $\overset{\frown}{AC}$ is a part of the circle of radius 1, the segments AB and DC are perpendicular to the x-axis, and EA is perpendicular to the line OA. So we have the following inequality:

$$|AB| < \overset{\frown}{AC} < |AE| + |EC|.$$

Also, from the right triangle EAD, we have $|AE| < |DE|$ so that $|AE| + |EC| < |CD|$. Hence, we have $|AB| < \overset{\frown}{AC} < |CD|$

However, $\overset{\frown}{AC} = x$. That is, x is the arc length of the portion $\overset{\frown}{AC}$ of the circle of radius 1 by the **definition of radian measurement** of the angle $\angle AOC$ as in Fig. 4.3. And we also have $|AB| = \sin x$ and $|CD| = \tan x = \frac{\sin x}{\cos x}$. Hence, $|AB| < \overset{\frown}{AC} < |CD|$ becomes $\sin x < x < \frac{\sin x}{\cos x}$.

Dividing both sides by $\sin x > 0$, we have $1 < \frac{x}{\sin x} < \frac{1}{\cos x}$. The reciprocal gives us $\cos x < \frac{\sin x}{x} < 1$. Since $\lim\limits_{x\to 0} \cos x = \cos 0 = 1$, we have $\lim\limits_{x\to 0, x>0} \frac{\sin x}{x} = 1$.

By substituting $x = -t$, we have

$$\lim_{x\to 0, x<0} \frac{\sin x}{x} = \lim_{t\to 0, t>0} \frac{\sin(-t)}{-t} = \lim_{t\to 0, t>0} \frac{-\sin t}{-t} = \lim_{t\to 0, t>0} \frac{\sin t}{t} = 1.$$

Therefore, $\lim\limits_{x\to 0} \frac{\sin x}{x} = 1$.

Remark 4.2.6 The above proof is a standard proof of $\lim\limits_{x\to 0} \frac{\sin x}{x} = 1$ in most elementary calculus textbooks. However, some textbooks use the inequality

(area of $\triangle AOB$) < (area of the part of the circle AOC) < (area of $\triangle DOC$).

In this proof, one needs to know that the area of a circle of radius r is πr^2 to obtain (area of the part of the circle AOC) = $\frac{x}{2}$.

Next is a modified Archimedes method to obtain the area of a circle in order to demonstrate that the use of $\lim\limits_{x\to 0} \frac{\sin x}{x} = 1$ is essential. It is an example of the method

Fig. 4.4 A part of a regular
n-gon inscribed in a circle

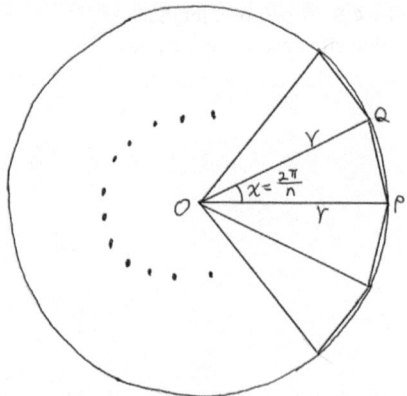

of exhaustion. Hence, the proof of $\lim\limits_{x \to 0} \frac{\sin x}{x} = 1$ using the area of AOC is a "circular" argument.

Theorem 4.2.9 The area $\mathcal{A}$ of a circle of radius r is πr^2.

Proof Let $n \geq 1$ be an integer.

(Case 1) We first consider a regular n-gon **inscribed** in a circle of radius $r > 0$ as in Fig. 4.4. Let P and Q be two adjacent vertices of the n-gon, and let O be the center. Then the angle $\angle POQ = \frac{2\pi}{n}$. Let $x = \frac{2\pi}{n}$. Then the altitude of $\triangle POQ$ from O is given by $r \cos \frac{x}{2}$ and the base $|PQ| = 2r \sin \frac{x}{2}$. So, we have

$$(\text{area of} \triangle POQ) = \frac{1}{2}\left(r \cos \frac{x}{2}\right)\left(2r \sin \frac{x}{2}\right) = r^2 \cos \frac{x}{2} \cdot \sin \frac{x}{2} = \frac{1}{2}r^2 \sin x.$$

Since $x = \frac{2\pi}{n}$, we have $n = \frac{2\pi}{x}$ so that the

$(\text{area of the } n\text{-gon}) = n \cdot \frac{1}{2}r^2 \sin x = \left(\frac{2\pi}{x}\right) \cdot \frac{1}{2}r^2 \sin x = \pi r^2 \frac{\sin x}{x}$.

By taking the limit $n \to \infty$, the area of the n-gon approximates the area $\mathcal{A}$ of the circle of radius r from inside. But if $n \to \infty$, then we have $x = \frac{2\pi}{n} \to 0$, and

$\lim\limits_{n \to \infty} (\text{area of the} n - \text{gon}) = \lim\limits_{x \to 0} \pi r^2 \frac{\sin x}{x} = \pi r^2 \lim\limits_{x \to 0} \frac{\sin x}{x} = \pi r^2$. Hence, $\mathcal{A} \geq \pi r^2$.

(Case 2) Now we consider a regular n-gon that **circumscribes** a circle of radius $r > 0$ as in Fig. 4.5. Let S and T be two adjacent vertices of the circumscribing n-gon, and let O be the center. Then the angle $\angle SOT = \frac{2\pi}{n}$. Let $y = \frac{\pi}{n}$. Then the altitude of $\triangle SOT$ from O is r and the base $|ST| = 2r \tan y$. So, we have

$$(\text{area of } \triangle SOT) = \frac{1}{2}r(2r \tan y) = r^2 \frac{\sin y}{\cos y}.$$

Since $y = \frac{\pi}{n}$, we have $n = \frac{\pi}{y}$ so that the

Fig. 4.5 A part of a regular
n-gon circumscribing a circle

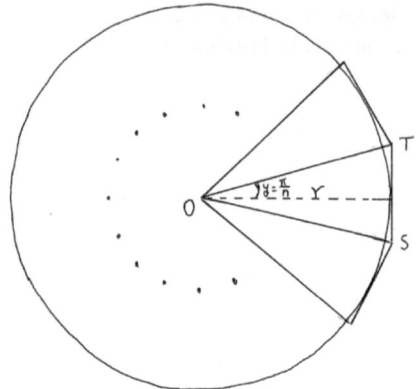

$$(\text{area of the } n\text{-gon}) = n \cdot r^2 \frac{\sin y}{\cos y} = \left(\frac{\pi}{y}\right) \cdot r^2 \frac{\sin y}{\cos y} = \pi r^2 \cos y \cdot \frac{\sin y}{y}.$$

Since $y \to 0$ as $n \to \infty$, and since $\lim\limits_{y \to 0} \cos y = 1$, we have

$$\lim_{n \to \infty} (\text{area of the } n - \text{gon}) = \pi r^2 \lim_{y \to 0} \cos y \cdot \frac{\sin y}{y} = \pi r^2. \text{ Hence, } \mathcal{A} \le \pi r^2.$$

From the above two cases, we have, $\mathcal{A} = \pi r^2$.

Example 4.2.4 Let $\tan : \left(-\frac{\pi}{2}, \frac{\pi}{2}\right) \to \mathbb{R}$ be defined by $\tan x = \frac{\sin x}{\cos x}$ for every $x \in \left(-\frac{\pi}{2}, \frac{\pi}{2}\right)$.
We prove that this function is continuous.

By Example 4.2.1, the function $f : (-\infty, 0) \cup (0, \infty) \to \mathbb{R}$ defined by $f(x) = \frac{1}{x}$ for every
$x \in (-\infty, 0) \cup (0\infty)$ is continuous. So $f(\cos x) = \frac{1}{\cos x}$ is continuous at any $x \in \left(-\frac{\pi}{2}, \frac{\pi}{2}\right)$
by Theorem 4.2.4. By Theorem 4.2.2(3), $\tan x = \frac{\sin x}{\cos x} = \sin x \cdot f(\cos x)$ is continuous at any
point $x \in \left(-\frac{\pi}{2}, \frac{\pi}{2}\right)$.

Definition 4.2.2 Let $\emptyset \ne X \subset \mathbb{R}$. Let $f : X \to \mathbb{R}$ be a function. Let $p \in \overline{X}$. If L is a number
such that, for every $A > 0$, there exists a $\delta > 0$ such that $f(x) > A$ whenever $x \in X$ and
$|x - p| < \delta$, then we write $\lim\limits_{x \to p} f(x) = \infty$. The value $f(x)$ is said to approach infinity as x
approaches p.

If $\lim\limits_{x \to p} \left[-f(x)\right] = \infty$, then we say $\lim\limits_{x \to p} f(x) = -\infty$.
Let $f : \mathbb{R} \to \mathbb{R}$ be a function. Let $L \in \mathbb{R}$. We say that $\lim\limits_{x \to \infty} f(x) = L$ if for every $\varepsilon > 0$,
there exists a number $M > 0$ such that $|f(x) - L| < \varepsilon$ whenever $x \in \mathbb{R}$ and $x > M$.
If $\lim\limits_{x \to \infty} f(-x) = L$, then we write $\lim\limits_{x \to -\infty} f(x) = L$.

All limits of the types $\lim\limits_{x \to p} f(x)$ are defined by using Definitions 4.2.1 or 4.2.2.

Remark 4.2.7 Recall that a sequence $\langle a_n \rangle_{n=1}^{\infty}$ is a function $a : \mathbb{N} \to \mathbb{R}$. Suppose a sequence $\langle a_n \rangle_{n=1}^{\infty}$ converges to L. Then the definition of $\lim_{n \to \infty} a_n = L$ is exactly the limit of the function $\lim_{n \to \infty} a(n) = L$.

Problem 4.2.3 Let $\emptyset \neq X \subset \mathbb{R}$, and let $f : X \to \mathbb{R}$ be a function. Let $p \in \overline{X}$.

(1) If $\lim_{x \to p} f(x) = \infty$, then for every sequence $\langle a_n \rangle$ in X converging to p, prove that $\lim_{n \to \infty} f(a_n) = \infty$.

(2) Suppose, for every sequence $\langle a_n \rangle$ in X converging to p, we have $\lim_{n \to \infty} f(a_n) = \infty$. Prove that $\lim_{x \to p} f(x) = \infty$.

Problem 4.2.4 Let $f : (0, \infty) \to \mathbb{R}$ be a function defined by $f(x) = \frac{1}{x}$ for every $x \in (0, \infty)$.

(1) Prove that $\lim_{x \to 0} f(x) = \infty$. (Because the domain of the function f is $(0, \infty)$, $\lim_{x \to 0} f(x)$ means the one-sided limit $\lim_{x \to 0, x > 0} f(x)$.)

(2) Prove that $\lim_{x \to \infty} f(x) = 0$.

Theorem 4.2.10 (The sandwich theorem) Let $\emptyset \neq X \subset \mathbb{R}$, and let $f, g, h : X \to \mathbb{R}$ be functions. Let $p \in \overline{X}$ or $p = \infty$ when $X = (0, \infty)$. Suppose $f(x) \leq g(x) \leq h(x)$ for all $x \in X$.

Then $\lim_{x \to p} f(x) \leq \lim_{x \to p} g(x) \leq \lim_{x \to p} h(x)$.

Proof This is an application of the sandwich theorem, Theorem 3.2.2, and we leave the proof to the readers.

4.3 Theorems on Continuous Functions

Some of the most important theorems on continuous functions are theorems in this section. These theorems were probably thought of as trivial by most mathematicians before 1800. Also, these theorems were discussed in entry-level calculus classes without proofs, so not knowing the details of these proofs did not cause much difficulties in applications. But because of a better understanding of real numbers and limits in the nineteenth century, the need to prove these theorems emerged. We do not know who was the first to prove these theorems although they are very important theorems.

Theorem 4.3.1 Let $a < b$. Suppose $f : [a, b] \to \mathbb{R}$ is continuous. Then the set $f([a, b])$ is bounded.

Proof We will only show that $f([a, b])$ is bounded above. Showing that $f([a, b])$ is bounded below is left to the readers.

Suppose $f([a, b])$ is not bounded above. Then there is a sequence $\langle y_n \rangle_{n=1}^{\infty}$ in $f([a, b])$ such that $y_n \geq n$ for each positive integer n. So $\lim_{n \to \infty} y_n = \infty$.

Let $x_n \in f^{-1}(y_n)$ for each positive integer n. Then $\langle x_n \rangle_{n=1}^{\infty}$ is a sequence in $[a, b]$. Since $[a, b]$ is bounded, the sequence $\langle x_n \rangle_{n=1}^{\infty}$ has a convergent subsequence, say $\langle x_{n_k} \rangle_{k=1}^{\infty}$, by the Bolzano-Weierstrass theorem. Let $\lim_{k \to \infty} x_{n_k} = c \in [a, b]$. By the continuity of f, we have $\lim_{k \to \infty} f(x_{n_k}) = f(c)$. But this implies that $\lim_{k \to \infty} f(x_{n_k}) = \lim_{k \to \infty} y_{n_k} = f(c)$. This is a contradiction to $\lim_{k \to \infty} y_{n_k} = \lim_{n \to \infty} y_n = \infty$ (see Problem 3.7.2).

Therefore, the set $f([a, b])$ is bounded above.

Theorem 4.3.2 (The Extreme Value Theorem) Let $a < b$. Suppose $f : [a, b] \to \mathbb{R}$ is continuous.

(1) There exists $c \in [a, b]$ such that $f(c) = \max f([a, b])$.
(2) There exists $d \in [a, b]$ such that $f(d) = \min f([a, b])$.

Proof of (1): It suffices to show that there exists $c \in [a, b]$ such that $f(c) = \sup f([a, b])$. This proof is similar to the proof of the previous theorem. By Theorem 4.3.1, $\sup f([a, b])$ exists. For every positive integer n, there exists $y_n \in f([a, b])$ such that $y_n \geq \sup f([a, b]) - \frac{1}{n}$ since $\sup f([a, b]) - \frac{1}{n}$ is not an upper bound of $f([a, b])$. Hence, $\left(\sup f([a, b]) - \frac{1}{n} \right) \leq y_n \leq \sup f([a, b])$ for every n. Since $\lim_{n \to \infty} \left(\sup f([a, b]) - \frac{1}{n} \right) = \sup f([a, b])$, we have $\lim_{n \to \infty} y_n = \sup f([a, b])$. Let $x_n \in f^{-1}(y_n)$ for each positive integer n. Then $\langle x_n \rangle_{n=1}^{\infty}$ is a sequence in $[a, b]$. So by Bolzano-Weierstrass theorem, $\langle x_n \rangle_{n=1}^{\infty}$ has a convergent subsequence, say $\langle x_{n_k} \rangle_{k=1}^{\infty}$ converging to $c \in [a, b]$. By the continuity of f, we have $\lim_{k \to \infty} f(x_{n_k}) = f(c)$. But this implies that $\lim_{k \to \infty} f(x_{n_k}) = \lim_{k \to \infty} y_{n_k} = f(c)$. Since $\lim_{n \to \infty} y_n = \sup f([a, b])$, we must have $f(c) = \sup f([a, b])$. Since $c \in [a, b]$, we have $f(c) = \max f([a, b])$.

Proof of (2) is left to the reader.

Remark 4.3.1 Please make sure why it suffices to show that there exists $c \in [a, b]$ such that $f(c) = \sup f([a, b])$ in the above proof. Make sure that you understand the distinction between the notations $\sup f([a, b])$ and $\max f([a, b])$. Also, please make sure why $\lim_{k \to \infty} y_{n_k} = f(c)$ in the above proof.

As you saw in the above proof and in the proof of Theorem 3.7.2 earlier, Bolzano-Weierstrass theorem is an abstract yet very important theorem used almost exclusively to prove other important theorems. This theorem led to the concept called "compactness" in topology. See Definition 4.5.2.

Theorem 4.3.3 (The Intermediate Value Theorem) Let $a < b$. Suppose $f : [a, b] \to \mathbb{R}$ is a continuous function such that $f(a) \neq f(b)$. Let p be a number between $f(a)$ and $f(b)$. Then there exists a $c \in [a, b]$ such that $f(c) = p$.

Proof Assume that $f(a) < p < f(b)$. Since $f^{-1}((-\infty, p]) \subset [a, b]$, $\sup f^{-1}((-\infty, p])$ exists. Let $\alpha = \sup f^{-1}((-\infty, p])$ for simplicity. Since $f^{-1}((-\infty, p]) \subset [a, b]$, we know that $\alpha \in [a, b]$ since $\alpha \in [a, b]$ is closed. For every positive integer n, there exists $x_n \in f^{-1}((-\infty, p])$ such that $x_n \geq \alpha - \frac{1}{n}$ since $\alpha - \frac{1}{n}$ is not an upper bound of $f^{-1}((-\infty, p])$. Hence, $\alpha \geq x_n \geq \alpha - \frac{1}{n}$ for every n. Since $\lim_{n\to\infty} (\alpha - \frac{1}{n}) = \alpha$, we have $\lim_{n\to\infty} x_n = \alpha \in [a, b]$. By the continuity of f and since $f(x_n) \leq p$, we have $\lim_{n\to\infty} f(x_n) = f(\alpha) \leq p$.

We want to show that $f(\alpha) = p$. On the contrary, suppose $f(\alpha) < p$. Then $p - f(\alpha) > 0$. By the continuity of f, there is a $\delta > 0$ such that

$|f(x) - f(\alpha)| < p - f(\alpha)$ whenever $x \in [a, b]$ and $|x - \alpha| < \delta$.

Since $f(\alpha) < p < f(b)$, we must have $\alpha < b$. Hence, there exists $x \in [a, b]$ such that $\alpha < x < \alpha + \delta$. Since $|x - \alpha| < \delta$, we have $|f(x) - f(\alpha)| < p - f(\alpha)$. In particular, we have $f(x) - f(\alpha) < p - f(\alpha)$. so that $f(x) < p$. Thus $x \in f^{-1}((-\infty, p])$. This is a contradiction to $x > \alpha = \sup f^{-1}((-\infty, p])$. Therefore, we must have $f(\alpha) = p$.

This proves the theorem when $f(a) < p < f(b)$. Similarly, we can prove the theorem when $f(b) < p < f(a)$.

Corollary 4.3.4 Let $a < b$. Suppose $f : [a, b] \to \mathbb{R}$ is continuous. Let p be a number such that $\min f([a, b]) \leq p \leq \max f([a, b])$. Then there exists a $c \in [a, b]$ such that $f(c) = p$.

In short, if $f : [a, b] \to \mathbb{R}$ is continuous, then $f([a, b])$ is the closed interval $\big[\min f([a, b]), \max f([a, b])\big]$, i.e., $f([a, b]) = \big[\min f([a, b]), \max f([a, b])\big]$.

Proof Let $\min f([a, b]) \leq p \leq \max f([a, b])$. By Theorem 4.3.2, there exists $u, v \in [a, b]$ such that $f(u) = \min f([a, b])$ and $f(v) = \max f([a, b])$. If $p = \min f([a, b])$ or $p = \max f([a, b])$, then $f(u) = p$ or $f(v) = p$. So assume $u < v$. Then by applying Theorem 4.3.3 to $f|_{[u,v]}$, there is a point $c \in (u, v) \subset [a, b]$ such that $f(c) = p$. The case $u > v$ is similar. This proves this corollary.

A short statement of Corollary 4.3.4 is the next corollary.

Corollary 4.3.5 Let $a < b$. Suppose $f : [a, b] \to \mathbb{R}$ is continuous. Then $f([a, b])$ is a closed interval or a set with one element when f is a constant function.

Definition 4.3.1 Let $a < b$. Let $[a, b] \subset X \subset \mathbb{R}$. Let $f : X \to \mathbb{R}$ be a function. The function f is said to be strictly increasing on $[a, b]$ if $a < x_1 < x_2 < b$ implies $f(x_1) < f(x_2)$. The function f is said to be strictly decreasing if $a < x_1 < x_2 < b$ implies $f(x_1) > f(x_2)$.

The next theorem seems trivial. But I think the proof is surprisingly difficult.

Theorem 4.3.6 Let $a < b$, and $c < d$. Let $f : [a, b] \rightarrow [c, d]$ be a 1–1, onto, and continuous function. If $f(a) < f(b)$, then $f(a) = c, f(b) = d$, and f is strictly increasing. If $f(a) > f(b)$, then $f(a) = d, f(b) = c$, and f is strictly decreasing.

Proof We will prove that if $f(a) < f(b)$, then $f(a) = c, f(b) = d$, and f is strictly increasing. Assume $f(a) < f(b)$.

Claim 1: We prove that $f(a) = c$ and $f(b) = d$.

Suppose $f(a) > c$. Since f is onto, there is $a < s \leq b$ such that $f(s) = c$. Since $f(a) < f(b)$, we must have $s < b$ and $c < f(a) < f(b)$. By the intermediate value theorem, there is $s < t < b$ such that $f(t) = f(a)$. Since $a < t$, $f(t) = f(a)$ is a contradiction to f being 1–1. This shows that $f(a) = c$. Similarly, we can show that $f(b) = d$.

Claim 2: Suppose $a < x < b$. We prove that $f(a) < f(x)$.

Since $a \neq x$, $f(a) \neq f(x)$. Suppose $f(x) < f(a)$. Since $f(a) < f(b)$, we have $f(a) \in [f(x), f(b)]$. By the intermediate value theorem, there is $c \in [x, b]$ such that $f(c) = f(a)$. Since $a < c$, this is a contradiction to f being 1–1. Hence, $f(a) < f(x)$.

Claim 3: Suppose $a < x < b$. We prove that $f(x) < f(b)$.

Since $b \neq x$, $f(b) \neq f(x)$. Suppose $f(x) > f(b)$. Since $f(a) < f(b)$, we have $f(b) \in [f(a), f(x)]$. By the intermediate value theorem, there is $c \in [a, x]$ such that $f(c) = f(b)$. Since $c < b$, this is a contradiction to f being 1–1. Hence, $f(x) < f(b)$.

Claim 4: Suppose $a < x_1 < x_2 < b$. We prove that $f(x_1) < f(x_2)$. By Claims 2 and 3, we have $f(a) < f(x_1) < f(b)$. Now, replace a by x_1 in Claims 2 and 3. Then we have $f(x_1) < f(x_2)$.

If $f(a) > f(b)$, we can prove that f is strictly decreasing in a similar way.

Problem 4.3.1 Construct a function $f : [0, 1] \rightarrow [0, 1]$ that is 1–1, onto, not strictly increasing, and not strictly decreasing. By Theorem 4.3.6, this function f cannot be continuous.

Problem 4.3.2 Suppose $f(\{-1\} \cup (0, 1]) \rightarrow [0, 1]$ defined by $f(-1) = 0$ and $f(x) = x$ for every $x \in (0, 1]$.

(1) Prove that the function f is continuous at -1.
(2) Prove that f is 1–1, and onto. This shows that $f^{-1} : [0, 1] \rightarrow \{-1\} \cup (0, 1]$ is the inverse **function** of f.
(3) Prove that $f^{-1} : [0, 1] \rightarrow \{-1\} \cup (0, 1]$ is **not** continuous at 0. (Again, without contradiction.)

In the next theorem, please note that the domain of the function f being a **finite closed interval** $[a, b]$ is very **important** as demonstrated in Problem 4.3.2.

Theorem 4.3.7 (Continuity of an inverse function) Let $a < b$ and $c < d$. Let $f : [a, b] \to [c, d]$ be a 1–1, onto, and continuous function. Then $f^{-1} : [c, d] \to [a, b]$ is continuous.

Proof Since f is 1–1, either $f(a) < f(b)$ or $f(a) > f(b)$. Suppose $f(a) < f(b)$. By Theorem 4.3.6, we have f is increasing, and $f(a) = c, f(b) = d$. Hence, $f^{-1}(c) = a$ and $f^{-1}(d) = b$.

Let $p \in [c, d]$. We will prove that f^{-1} is continuous at p. Let $\varepsilon > 0$. We need to consider three cases; $c < p < d, p = c$, or $p = d$.

Case 1: Suppose $c < p < d$. Then $a < f^{-1}(p) < b$. Then there is an ε' such that $\varepsilon > \varepsilon' > 0$ and $a < f^{-1}(p) - \varepsilon' < f^{-1}(p) < f^{-1}(p) + \varepsilon' < b$. Since f is strictly increasing, we have

$$c = f(a) < f\left(f^{-1}(p) - \varepsilon'\right) < f\left(f^{-1}(p)\right) = p < f\left(f^{-1}(p) + \varepsilon'\right) < f(b) = d.$$

Let $\delta > 0$ such that $f\left(f^{-1}(p) - \varepsilon'\right) < p - \delta < p < p + \delta < f\left(f^{-1}(p) + \varepsilon'\right)$.

Let $|p - y| < \delta$. Then $p - \delta < y < p + \delta$ and $p \in [c, d]$. By the intermediate value theorem, there are $s, x, t \in [a, b]$ such that $f(s) = p - \delta, f(x) = y, f(t) = p + \delta$. Since f is increasing and since $p - \delta < y < p + \delta$, we must have $s < x < t$. This shows that

(1) $f^{-1}(p - \delta) = s < f^{-1}(y) = x < t = f^{-1}(p + \delta)$.

Since $f\left(f^{-1}(p) - \varepsilon'\right) < p - \delta$, we must have

(2) $f^{-1}(p) - \varepsilon' < f^{-1}(p - \delta)$.

Since $p + \delta < f\left(f^{-1}(p) + \varepsilon'\right)$, we must have

(3) $f^{-1}(p + \delta) < f^{-1}(p) + \varepsilon'$.

From inequalities (1)–(3), we have

$$f^{-1}(p) - \varepsilon' < f^{-1}(y) < f^{-1}(p) + \varepsilon'.$$

Hence, $\left|f^{-1}(p) - f^{-1}(y)\right| < \varepsilon' < \varepsilon$.
This shows that f^{-1} is continuous at p.

Case 2: Suppose $p = c$. We show that f^{-1} is continuous at c. Then $a = f^{-1}(c) < b$. Then there is an ε' such that $\varepsilon > \varepsilon' > 0$ and $a = f^{-1}(c) < f^{-1}(c) + \varepsilon' < b$. Since f is strictly increasing, we have

$$c = f(a) < f\left(f^{-1}(c) + \varepsilon'\right) < f(b) = d.$$

Let $\delta > 0$ such that $c < c + \delta < f\left(f^{-1}(c) + \varepsilon'\right)$.

Let $|c - y| < \delta$. Then $c \le y < c + \delta$. By the intermediate value theorem, there are $x, t \in [a, b]$ such that $f(x) = y, f(t) = c + \delta$. Since f is increasing and since $y < c + \delta$, we must have $x < t$. This shows that

(4) $f^{-1}(y) = x < t = f^{-1}(c + \delta).$

Since $c + \delta < f\left(f^{-1}(c) + \varepsilon'\right)$, we must have

(5) $f^{-1}(c + \delta) < f^{-1}(c) + \varepsilon'.$

From inequalities (4) and (5), we have

$a = f^{-1}(c) \le f^{-1}(y) < f^{-1}(c) + \varepsilon'.$

Hence, $\left|f^{-1}(c) - f^{-1}(y)\right| < \varepsilon' < \varepsilon.$

This shows that f^{-1} is continuous at c.

Case 3: Suppose $p = d$. Showing f^{-1} is continuous at d is similar to Case 2.

Therefore, $f^{-1} : [c, d] \to [a, b]$ is continuous. The case $f(a) > f(b)$ is left to the readers.

Corollary 4.3.8 The restricted trigonometric functions $\cos : [0, \pi] \to [-1, 1]$ and $\sin : \left[-\frac{\pi}{2}, \frac{\pi}{2}\right] \to [-1, 1]$ are 1–1, onto, and continuous. So, their inverses, $\cos^{-1} : [0, \pi] \to [-1, 1]$ and $\sin^{-1} : \left[-\frac{\pi}{2}, \frac{\pi}{2}\right] \to [-1, 1]$, are continuous.

Example 4.3.1 Let $f : (-1, 1) \to \mathbb{R}$ be a function defined by $f(x) = \frac{x}{1-x^2}$ for all $x \in (-1, 1)$. Then f is a 1–1 and onto function by Example 1.3.6(7). Hence, $f^{-1} : \mathbb{R} \to (-1, 1)$ exists. In order to see the continuity of f^{-1}, let $p \in \mathbb{R}$. Since $\lim_{x \to 1} f(x) = \infty$ and $\lim_{x \to -1} f(x) = -\infty$, there is a number $0 < R < 1$ such that $p \in \left(-\frac{R}{1-R^2}, \frac{R}{1-R^2}\right)$. By Theorem 4.3.7, the continuity of $f|_{[-R,R]} : [-R, R] \to \left[-\frac{R}{1-R^2}, \frac{R}{1-R^2}\right]$ implies that the continuity of $f|_{[-R,R]}^{-1} : \left[-\frac{R}{1-R^2}, \frac{R}{1-R^2}\right] \to [-R, R]$. I leave to the readers to prove that this implies the continuity of f^{-1} at p.

Alternately, Example 1.3.6(7) implies that

$$f^{-1}(y) = \begin{cases} \frac{-1+\sqrt{1+4y^2}}{2y} & \text{if } y \ne 0 \\ 0 & \text{if } y = 0 \end{cases} \quad \text{for all } y \in \mathbb{R}.$$

Hence, f^{-1} is continuous when $y \neq 0$. We can conclude that f^{-1} is continuous if $\lim_{y \to 0} \frac{-1+\sqrt{1+4y^2}}{2y} = 0$ is shown. Please **show** that $\lim_{y \to 0} \frac{-1+\sqrt{1+4y^2}}{2y} = 0$.

Problem 4.3.3 Let $g : (0, \infty) \to \mathbb{R}$ be a function. If $g|_{\left[\frac{1}{e^n}, e^n\right]}$ is continuous for every integer $n \geq 1$, prove that g is continuous.

Theorem 4.3.9 Let $f : \mathbb{R} \to (0, \infty)$ be a function defined by $f(x) = e^x$ for every $x \in \mathbb{R}$. Then the function f is 1–1, onto, and continuous. Moreover, $f^{-1} : (0, \infty) \to \mathbb{R}$ is also continuous.

Proof The function f is continuous by Theorem 4.2.7. Let $p \in (0, \infty)$. Then there is an integer $n > 0$ such that $\frac{1}{2^n} < p < 2^n$. Since $e > 2$, we have $e^{-n} = \frac{1}{e^n} < \frac{1}{2^n} < p < 2^n < e^n$. Hence, $p \in \left[\frac{1}{e^n}, e^n\right] \subset f([-n, n]) \subset f(\mathbb{R})$. Since f is continuous, there exists $c \in [-n, n]$ such that $f(c) = p$ by the intermediate value theorem. This proves that f is onto.

By Theorem 3.9.3, f is increasing. This shows that f is 1–1.

Therefore, $f|_{[-n,n]} : [-n, n] \to \left[\frac{1}{e^n}, e^n\right]$ is 1–1 and onto and it is continuous by Problem 4.3.3. By Theorem 4.3.7, $f|_{[-n,n]}^{-1}$ is continuous at p. This proves that f^{-1} is continuous at $p \in (0, \infty)$. That is, the inverse function f^{-1} is continuous.

Definition 4.3.2 Let $f : \mathbb{R} \to (0, \infty)$ be a function defined by $f(x) = e^x$ for every $x \in \mathbb{R}$. Then the <u>natural logarithmic function</u> is $f^{-1} : (0, \infty) \to \mathbb{R}$, and denoted by $f^{-1}(y) = \ln y$ for all $y \in (0, \infty)$. We read $\ln y$ "log y" or "natural log y". If $a > 1$, we define $\log_a y = \frac{\ln y}{\ln a}$ for all $y \in (0, \infty)$. We read $\log_a y$ "log base a of y".

Theorem 4.3.10

(1) The natural logarithmic function $\ln : (0, \infty) \to \mathbb{R}$ is a continuous function.
(2) $\ln 1 = 0$
(3) $\ln e = 1$.
(4) If $a, b \in (0, \infty)$, then $\ln ab = \ln a + \ln b$.
(5) If $a \in (0, \infty)$ and $x \in \mathbb{R}$, then $\ln a^x = x \cdot \ln a$.

Proof of (1)–(3): The part (1) is a repeat of Theorem 4.3.9. The parts (2) and (3) follow from $e^0 = 1 = e^{\ln 1}$ and $e^{\ln e} = e = e^1$, and the exponential function e^x being 1–1.

Proof of (4): $e^{\ln a + \ln b} = e^{\ln a} e^{\ln b} = ab = e^{\ln(ab)}$. Since the exponential function e^x is 1–1, we have $\ln a + \ln b = \ln ab$.

Proof of (5): $e^{\ln a^x} = a^x = \left(e^{\ln a}\right)^x = e^{x \cdot \ln a}$. Since the exponential function e^x is 1–1, $\ln a^x = x \cdot \ln a$.

Theorem 4.3.11 Let $a > 0$. Let $x \in \mathbb{R}$. Then $a^x = e^{x \cdot \ln a}$.

Proof Suppose $a^x = e^y$ for some $y \in \mathbb{R}$. Then

$$x \cdot \ln a = \ln a^x = \ln e^y = y \cdot \ln e = y.$$

Hence, $y = x \ln a$. Therefore, $a^x = e^y = e^{x \cdot \ln a}$. ∎

Remark 4.3.2 Theorem 4.3.11 is a very useful theorem in taking derivatives, limits, and integration of exponential functions.

Problem 4.3.4 (See Example 4.2.1) (1) Let $f : (\mathbb{R} - \{0\}) \to (\mathbb{R} - \{0\})$ be a function defined by $f(x) = \frac{1}{x}$ for every $x \in (\mathbb{R} - \{0\})$. Show that f^{-1} is continuous.

(2) Let $\tan : \left(-\frac{\pi}{2}, \frac{\pi}{2}\right) \to \mathbb{R}$ be defined by $\tan x = \frac{\sin x}{\cos x}$ for every $x \in \left(-\frac{\pi}{2}, \frac{\pi}{2}\right)$. Prove that $\tan^{-1}$ is continuous.

We can finally **answer the question we posed in** Example 4.2.3

Theorem 4.3.12 (Continuity of a power function) If $r \neq 0$ is a real number, and if $F : (0, \infty) \to \mathbb{R}$ is a function defined by $F(x) = x^r$ for every $x \in (0, \infty)$, then the function F is continuous.

Proof By Theorem 4.2.7, the function $f : \mathbb{R} \to (0, \infty)$ defined by $f(x) = e^{rx}$ for every $x \in \mathbb{R}$ is continuous. Let $g : (0, \infty) \to \mathbb{R}$ be a function defined by $g(x) = \ln x$ for every $x \in (0, \infty)$. It is continuous by Theorem 4.3.10. Let $x \in (0, \infty)$. Then by Theorem 4.3.11, we have

$$x^r = e^{r \cdot \ln x} \text{ so that } F(x) = x^r = e^{r \cdot \ln x} = e^{r \cdot g(x)} = f(g(x)).$$

Hence, F is a composition of two continuous functions. Therefore, F is continuous. ∎

Remark 4.3.3 In a way, one of the main purposes of the construction of real numbers in Chap. 2 and the study of sequences and Cauchy sequences in Chap. 3 can be summarized by Theorem 4.2.7 and Theorem 4.3.12.

Remark 4.3.4 For your information, let $f : (0, \infty) \times \mathbb{R} \to \mathbb{R}$ be a function of two variables defined by $f(x, y) = x^y$ for every $(x, y) \in (0, \infty) \times \mathbb{R}$. Then Theorems 4.2.7 and 4.4.12 imply that this function f of two variables is also continuous from the metric space $(0, \infty) \times \mathbb{R}$ with the usual metric into $\mathbb{R}$ with the usual metric (See Definition 4.5.1).

4.4   The Exponential Function e^x

This section is a generalization of Sect. 3.10.

Problem 4.4.1 Let $x \in \mathbb{R}$. Show that $\sum_{n=0}^{\infty} \frac{x^n}{n!} = 1 + \sum_{n=1}^{\infty} \frac{x^n}{n!}$ converges absolutely using two tests.

(1) Use the geometric series and the comparison test. (Compare this to Problem 3.8.1)
(2) Use the ratio test.

Definition 4.4.1 Let $e : \mathbb{R} \to \mathbb{R}$ be a function defined by $e(x) = \sum_{n=0}^{\infty} \frac{x^n}{n!} = 1 + \sum_{n=1}^{\infty} \frac{x^n}{n!}$ for every $x \in \mathbb{R}$. Because of Problem 4.4.1, this function is defined. Also, note that $e(1) = e$ by the definition of e, and $e(0) = 1$.

Definition 4.4.2 For every positive integer m, let $S_m : \mathbb{R} \to \mathbb{R}$ and $T_m : \mathbb{R} \to \mathbb{R}$ be functions defined by $S_m(x) = \sum_{n=0}^{m} \frac{x^n}{n!}$ and $T_m(x) = \left(1 + \frac{x}{m}\right)^m$ for every $x \in \mathbb{R}$.

Problem 4.4.2 Let $x \in \mathbb{R}$. Show that

$$T_m(x) = 1 + x + \frac{x^2}{2!}\left(1 - \frac{1}{m}\right) + \frac{x^3}{3!}\left(1 - \frac{1}{m}\right)\left(1 - \frac{2}{m}\right) + \cdots$$
$$+ \frac{x^m}{m!}\left(1 - \frac{1}{m}\right)\left(1 - \frac{2}{m}\right)\cdots\left(1 - \frac{m-1}{m}\right)$$

for every positive integer m. Use the binomial theorem.

Lemma 4.4.1 The sequence $\langle T_m(x)\rangle_{m=1}^{\infty}$ is increasing for all $x \in [0, \infty)$.
(Compare this to the proof of Theorem 3.10.1.)

Proof Let $0 \le k \le m$ be integers. Let $x \in [0, \infty)$. Then

$$T_m(x) = 1 + x + \frac{x^2}{2!}\left(1 - \frac{1}{m}\right) + \frac{x^3}{3!}\left(1 - \frac{1}{m}\right)\left(1 - \frac{2}{m}\right) + \cdots$$
$$+ \frac{x^m}{m!}\left(1 - \frac{1}{m}\right)\left(1 - \frac{2}{m}\right)\cdots\left(1 - \frac{m-1}{m}\right) \quad \text{by Problem 4.4.2}$$
$$\ge 1 + x + \frac{x^2}{2!}\left(1 - \frac{1}{m}\right) + \frac{x^3}{3!}\left(1 - \frac{1}{m}\right)\left(1 - \frac{2}{m}\right) + \cdots$$
$$+ \frac{x^k}{k!}\left(1 - \frac{1}{m}\right)\left(1 - \frac{2}{m}\right)\cdots\left(1 - \frac{k-1}{m}\right) \quad \text{since } k \le m$$

$$\geq 1 + x + \frac{x^2}{2!}\left(1 - \frac{1}{k}\right) + \frac{x^3}{3!}\left(1 - \frac{1}{k}\right)\left(1 - \frac{2}{k}\right) + \cdots$$

$$+ \frac{x^k}{k!}\left(1 - \frac{1}{k}\right)\left(1 - \frac{2}{k}\right)\cdots\left(1 - \frac{k-1}{k}\right) \quad \text{again, since } k \leq m$$

$$= T_k(x).$$

Problem 4.4.3 Let $x \in [0, \infty)$. Show that $\langle T_m(x)\rangle_{m=1}^{\infty}$ converges and $\lim\limits_{m \to \infty} T_m(x) \leq e(x)$. (Compare this to the proof of Theorem 3.10.1.)

Problem 4.4.4 Let $x \in [0, \infty)$. Show that $\lim\limits_{m \to \infty} T_m(x) \geq S_k(x)$ for every positive integer k. (Compare this to the proof of Theorem 3.10.1.)

Recall from Sect. 3.10 that $\lim\limits_{m \to \infty}\left(1 + \frac{1}{m}\right)^m = e$ when m is an integer. The next lemma seems clear from this, but the proof is not trivial.

Lemma 4.4.2 $\lim\limits_{x \to \infty}\left(1 + \frac{1}{x}\right)^x = e$. (Here, we are assuming $x \in \mathbb{R}$.) More formally, let $f : (1, \infty) \to \mathbb{R}$ be a function defined by $f(x) = \left(1 + \frac{1}{x}\right)^x$ for every $x \in (1, \infty)$. Then $\lim\limits_{x \to \infty} f(x) = e$.

Proof If $x \in (1, \infty)$, let $\lfloor x \rfloor$ be the largest integer less than or equal to x. Let $x \in (1, \infty)$. Since $\lfloor x \rfloor \leq x < \lfloor x \rfloor + 1$, we have

$$\left(1 + \frac{1}{x}\right)^x \leq \left(1 + \frac{1}{\lfloor x \rfloor}\right)^x \leq \left(1 + \frac{1}{\lfloor x \rfloor}\right)^{\lfloor x \rfloor + 1}, \quad \text{and} \quad \left(1 + \frac{1}{x}\right)^x \geq \left(1 + \frac{1}{\lfloor x \rfloor + 1}\right)^x \geq$$
$$\left(1 + \frac{1}{\lfloor x \rfloor + 1}\right)^{\lfloor x \rfloor}.$$

Hence, we have

$$\left(1 + \frac{1}{\lfloor x \rfloor + 1}\right)^{\lfloor x \rfloor} \leq \left(1 + \frac{1}{x}\right)^x \leq \left(1 + \frac{1}{\lfloor x \rfloor}\right)^{\lfloor x \rfloor + 1} \quad \text{for every } x \in (1, \infty).$$

Now,

$$\lim_{x \to \infty}\left(1 + \frac{1}{\lfloor x \rfloor}\right)^{\lfloor x \rfloor + 1} = \lim_{n \to \infty}\left(1 + \frac{1}{n}\right)^{n+1}$$

$$= \lim_{n \to \infty}\left(1 + \frac{1}{n}\right)^n \lim_{n \to \infty}\left(1 + \frac{1}{n}\right) \quad \text{by Theorems 3.1.4(2) and 3.10.1.}$$

$$= e \cdot 1 = e.$$

$$\lim_{x \to \infty}\left(1 + \frac{1}{\lfloor x \rfloor + 1}\right)^{\lfloor x \rfloor} = \lim_{n \to \infty}\left(1 + \frac{1}{n+1}\right)^n$$

$$= \frac{\lim_{n\to\infty}\left(1+\frac{1}{n+1}\right)^{n+1}}{\lim_{n\to\infty}\left(1+\frac{1}{n+1}\right)} \quad \text{by Theorems 3.1.4(4) and 3.10.1.}$$

$$= \frac{e}{1} = e.$$

By the sandwich theorem, Theorem 4.2.10, we have $\lim_{x\to\infty}\left(1+\frac{1}{x}\right)^{x} = e$.

Theorem 4.4.1 For every $x \in \mathbb{R}$, we have $e(x) = e^x$, i.e., $e^x = \sum_{n=0}^{\infty}\frac{x^n}{n!}$.

Proof Let $x \in [0, \infty)$. From Problems 4.4.3 and 4.4.4, we have $\lim_{m\to\infty} T_m(x) = e(x)$. Let m be a positive integer. Let $\frac{1}{k} = \frac{x}{m}$ or $m = kx$. Here, k may not be an integer. Let n be the largest integer such that $n \le k$. Then

$$\lim_{m\to\infty} T_m(x) = \lim_{m\to\infty}\left(1+\frac{x}{m}\right)^m = \lim_{k\to\infty}\left(1+\frac{x}{kx}\right)^{kx}$$

$$= \lim_{k\to\infty}\left[\left(1+\frac{1}{k}\right)^k\right]^x = \left[\lim_{k\to\infty}\left(1+\frac{1}{k}\right)^k\right]^x.$$

This last equality is by Theorem 4.3.12 (not by Theorem 4.2.7). Hence, we have

$$\lim_{m\to\infty} T_m(x) = \left[\lim_{k\to\infty}\left(1+\frac{1}{k}\right)^k\right]^x = e^x \text{ since } \lim_{k\to\infty}\left(1+\frac{1}{k}\right)^k = e.$$

So we have shown that $e(x) = e^x$ for every $x \in [0, \infty)$. Since $e^{-x} = \frac{1}{e^x}$, we have $e(x) = e^x = \sum_{n=0}^{\infty}\frac{x^n}{n!} = 1 + \sum_{n=1}^{\infty}\frac{x^n}{n!}$ for every $x \in \mathbb{R}$.

Remark 4.4.1 Essentially, we have obtained $e^x = \sum_{n=0}^{\infty}\frac{x^n}{n!}$ without derivatives. See Sect. 7.5 for the Taylor's series of e^x.

4.5   Continuous Functions in Metric Spaces

The purpose of this section is to motivate readers to study metric and topological spaces. So, **all the theorems in this section are stated without proofs.**

Definition 4.5.1 Let (X, d) and $(Y, \hat{d})$ be two nonempty metric spaces. A function f : $X \to Y$ is said to be <u>continuous</u> at $p \in X$ if for every $\varepsilon > 0$, there exist a $\delta > 0$ such that $\hat{d}(f(x), f(p)) < \varepsilon$ whenever $x \in X$ and $d(x, p) < \delta$. If f is continuous at every point of X, then we say simply f is <u>continuous</u>.

  Recall Notations: For every $\delta > 0$ and for every $p \in X$, let

$$N_\delta(p) = \{x \in X : d(p, x) < \delta\}.$$

Then $N_\varepsilon(f(p)) = \{y \in Y : \hat{d}(f(p), f(x)) < \varepsilon\}$.
With these notations we can rewrite continuity as in the next theorem.

Remark 4.5.1 Definition 4.5.1 is a direct extension of Definition 4.1.1. And it is a standard definition of continuity. As I mentioned in Remark 4.2.2, some authors define a function f to be continuous at p if $\lim\limits_{x \to p} f(x) = f(p)$. But this is **not** standard.

Theorem 4.5.1 Let (X, d) and $(Y, \hat{d})$ be nonempty metric spaces. Let $f : X \to Y$ be a function, and let $p \in X$. Then, the following statements are equivalent:

(1) The function f is continuous at p.
(2) For every $\varepsilon > 0$, there exist a $\delta > 0$ such that $f(x) \in N_\varepsilon(f(p))$ whenever $x \in N_\delta(p)$.
     (Note that $N_\delta(p)$ is a subset in X, and $N_\varepsilon(f(p))$ is a subset in Y.)
(3) For every $\varepsilon > 0$, there exist a $\delta > 0$ such that $N_\delta(p) \subset f^{-1}(N_\varepsilon(f(p)))$.

Recall Definitions: Let (X, d) be a metric space. A subset U of X is said to be <u>open</u> if for every $p \in U$, there exists an $\varepsilon > 0$ such that $N_\varepsilon(p) \subset U$. A subset F is <u>closed</u> in X if $X - F$ is open.

  Theorem 4.5.1 is about continuity at a point. The next theorem is global.

Theorem 4.5.2 Let (X, d) and $(Y, \hat{d})$ be metric spaces. Let $f : X \to Y$ be a function. The following statements are equivalent:

(1) The function f is continuous.
(2) For every $p \in X$ and for every $\varepsilon > 0$, there exist a $\delta > 0$ such that $N_\delta(p) \subset f^{-1}(N_\varepsilon(f(p)))$.
(3) For every open subset V of $Y, f^{-1}(V)$. is open in X.
(4) For every closed subset F of $Y, f^{-1}(F)$ is closed in X.

Remark 4.5.2 Please note the simplicity of the statements (3) and (4) in Theorem 4.5.2. This simplicity may have been the motivation to start a new branch of mathematics called

point set topology. By knowing the collection of all open sets (which we call the topology of a set), and by using (3) as the definition of a continuous function between topological spaces, we can analyze many properties (continuous functions, compactness and connectedness, for example) of topological spaces without the use of distance functions (metrics).

Definition 4.5.2 Let (X, d) be a metric space. A subset Y of X with the metric d, i.e., (Y, d), is said to be a <u>subspace</u> of X. A subspace Y is said to be <u>compact</u> if all sequences in Y have convergent subsequences.

Example 4.5.1 Let d be the usual metric on $\mathbb{R}$.

(1) The set $[0,1]$ is closed in $\mathbb{R}$.
(2) Let n be a positive integer. A subset of $\mathbb{R}^n$ is a compact if, and only if, it is closed and bounded. (This is called Heine-Borel Theorem. Compare this to Bolzano-Weierstrass theorem.)

As an extension of Example 4.5.1, we have the following.

Example 4.5.2 Let d be the usual metric on $\mathbb{R}^2$.

(1) The set $[0, 1]^2$ is closed in $\mathbb{R}^2$.
(2) The set $[0, 1]^2$ is a compact subset of $\mathbb{R}^2$.

Remark 4.5.3 For your information:

(1) Let $n \in \{1, 2, 3\}$. If $(\mathbb{R}^n, d)$ is a metric space with d being the usual metric, then a subset of $\mathbb{R}^n$ is compact if, and only if it is closed and bounded.
(2) Let (X, d) and $(Y, \hat{d})$ be metric spaces. Let $f : X \to Y$ be a continuous function. Let $A \subset X$ be compact. Then $f(A)$ is a compact subset of Y. This is to say that a continuous image of a compact space is compact.
(3) We did not give the definition of a connected space. But we have a theorem very similar to (2) that states as follows: A continuous image of a connected space is connected.
(4) As we have seen, many fundamental mathematical theorems in this book were discovered around 1900. Let me add my favorite discovery of that time. Let d be the usual metric on $\mathbb{R}$, and let $\hat{d}$ be the usual metric on $\mathbb{R}^2$ defined in Example 3.10.1. Sometimes, a continuous function from $([0, 1], d)$ into $(\mathbb{R}^2, \hat{d})$ is called a <u>curve</u>. In 1890, an Italian mathematician Giuseppe Peano discovered a continuous function from $([0, 1], d)$ **onto** $([0, 1]^2, \hat{d})$. A continuous function from $([0, 1], d)$ **onto** $([0, 1]^2, \hat{d})$ is called a **space-filling curve**.
(5) It is also interesting to note that any compact metric space is a continuous image of the **Cantor set**. From this, Hahn and Mazurkievic independently proved that a continuous

image of the unit interval [0,1] is a *locally connected*, compact, and *connected* metric space (Hahn- Mazurkievic Theorem). I did not explain what we mean by "locally connected" and "connected" here. You may study these in a topology class.

4.6   Function Continuous Only at the Irrationals

This section can be seen as an excursion to an unexpected example related to the number theory and it is written for readers interested in this type of example. Skipping this section will not interfere in studying the latter part of this book.

Many authors of real analysis textbooks present the following Theorem 4.6.1 as an exercise. Drawing on the references cited at the end of this section, for example, see Problem 4.16 on page 97 of [1], Problem 4 on page 109 of [3], Problem 17.14 on page 95 of [5], and Problem 18 on page 100 of [6]. Some of the authors give the solutions to this problem, and they are difficult. So we give an alternate proof different from any of the above to this problem in this section. It requires some theorems from number theory.

Theorem 4.6.1 Let $f : \mathbb{R} \to \mathbb{R}$ be a function defined by the following rule:

(1) $f(0) = 1$,
(2) if $x \in \mathbb{R}$ is an irrational number, then $f(x) = 0$,
(3) if $x \in \mathbb{R}$ is a non-zero rational number, and if $x = \frac{p}{q}$ for some integers p and $q > 0$ in reduced form, then we let $f(x) = \frac{1}{q}$.

We will show that this function f is continuous at each irrational number, and discontinuous at each rational number.

I knew this theorem for a long time, but I could not prove it in a way satisfactory to me until recently. Showing discontinuity of f at each rational point is not too difficult. However, the proof of the continuity of f at an irrational number is difficult. Actually, showing the continuity of f at any irrational number seemed formidable. I needed a theorem that states that if rational numbers get close to a fixed irrational number, the denominator of the rational numbers as a fraction of two integers must get uniformly large. Bressoud shows how to solve this difficulty on pages 95–96 in [2] by making a "critical observation". But this "critical observation" was not clear to me. So, I consulted Professor Schettler who specializes in number theory. He is a colleague of mine at SJSU. Schettler suggested using what is called "convergents" of the continued fraction expansion of an irrational number. Using convergents, we can make the "critical observation" of Bressoud in [2] into an explicit computation of δ for a given ε in the definition of continuity. We give a brief summary of the convergents and give a proof to the above theorem.

For a details, I highly recommend Chap. 5 of *Irrational Numbers* by Ivan Niven [4]. This book is accessible to undergraduate students and Chap. 5 can be read independently without reading the prior chapters.

Let ξ (reads Xi) be an irrational number. Let $[\xi]$ be the greatest integer part of ξ. We define a sequence of irrational numbers $\langle \xi_n \rangle_{n=0}^{\infty}$ and another sequence $\langle a_n \rangle_{n=0}^{\infty}$ of integers as follow:

$$\xi_0 = \xi \qquad a_0 = [\xi_0]$$
$$\xi_1 = \tfrac{1}{\xi_0 - a_0} \qquad a_1 = [\xi_1]$$
$$\xi_2 = \tfrac{1}{\xi_1 - a_1} \qquad a_2 = [\xi_2]$$

$$\cdots$$

$$\xi_i = \tfrac{1}{\xi_{i-1} - a_{i-1}} \qquad a_i = [\xi_i]$$

Let $h_{-2} = 0$, $h_{-1} = 1$, and $h_i = a_i h_{i-1} + h_{i-2}$ for all integers $i \geq 0$. Also, let $k_{-2} = 1$, $k_{-1} = 0$, and $k_i = a_i k_{i-1} + k_{i-2}$ for all $i \geq 0$.

Definition 4.6.1 Let $z_i = \tfrac{h_i}{k_i}$ for all $i \geq 0$. Then $z_i = \tfrac{h_i}{k_i}$ are called <u>convergents</u> of ξ.

(Note: "Convergent" as a noun is not a word in a standard dictionary, so I think this word was made up for this purpose. It is used in [4].)

We quote the following lemma from Niven's book.

Lemma 4.6.1 Let ξ be an irrational number and $z_i = \tfrac{h_i}{k_i}$ the convergent of ξ for each $i \geq 0$. Then we have the following:

(1) (1) $1 = k_0 < k_1 < k_2 < \cdots$,
(2) (2) $z_0 < z_2 < z_4 < \cdots \xi < \cdots < z_5 < z_3 < z_1$,
(3) $|\xi - z_{i+1}| = \left| \xi - \tfrac{h_{i+1}}{k_{i+1}} \right| < \left| \xi - \tfrac{h_i}{k_i} \right| = |\xi - z_i|$ for all $i \geq 0$,
(4) $\lim_{i \to \infty} z_i = \lim_{i \to \infty} \tfrac{h_i}{k_i} = \xi$
(5) If a and $b > 0$ are integers such that $\left| \xi - \tfrac{a}{b} \right| < \left| \xi - \tfrac{h_{i+1}}{k_{i+1}} \right|$ for some $i \geq 0$, then $b > k_i$.

Example 4.6.1 It is interesting to note that $\left| \sqrt{2} - \tfrac{89}{63} \right| < \left| \sqrt{2} - \tfrac{141}{100} \right|$, yet $63 < 100$. This does not contradict Lemma 4.6.1(5) because $\tfrac{141}{100}$ is not a convergent of $\sqrt{2}$. If $\xi = \sqrt{2}$ in Theorem 4.4.2, then $a_0 = 1$ and $a_n = 2$ for all $n \geq 1$. So some of the convergents of $\sqrt{2}$ are $\tfrac{1}{1}, \tfrac{3}{2}, \tfrac{7}{5}, \tfrac{17}{12}, \tfrac{41}{29}, \tfrac{99}{70}, \tfrac{239}{169}, \tfrac{577}{408}, \cdots$.

Now we are ready to prove Theorem 4.6.1.

Proof of Theorem 4.6.1 Let ξ be an irrational number. We will show that f is continuous at ξ. For all $i \geq 0$, let $z_i = \frac{h_i}{k_i}$ be convergents of ξ. Let $\varepsilon > 0$. Then there exists an integer $n \geq 1$ such that $\frac{1}{k_n} < \varepsilon$ by Lemma 4.6.1 part (1). Let $\delta = \left| \xi - \frac{h_n}{k_n} \right| > 0$. Let x be a number such that $|\xi - x| < \delta$. If x is irrational, then $|f(\xi) - f(x)| = 0 < \varepsilon$. Suppose $x = \frac{a}{b}$ for some integers a and $b > 0$ in the reduced form. Then $|f(\xi) - f(x)| = \frac{1}{b} < \frac{1}{k_n} < \varepsilon$ by part (5) of Lemma 4.6.1. This proves that f is continuous at each irrational number.

Let $q = \frac{a}{b}$ be a rational number for some a and $b > 0$. (If $q = 0$, we let $a = 0$ and $b = 1$ in order to be consistent with the definition of f.) We will show that f is not continuous at q. Let $\delta > 0$. There exists a positive integer n such that $\frac{\sqrt{2}}{n} < \delta$. Then $\left| q - \left(q - \frac{\sqrt{2}}{n} \right) \right| < \delta$ and $q - \frac{\sqrt{2}}{n}$ is an irrational number. Hence, we have $\left| f(q) - f\left(q - \frac{\sqrt{2}}{n} \right) \right| = \frac{1}{b} > 0$. Therefore, f is not continuous at q.
This proves Theorem 4.6.1.

References

1. T.M. Apostol, *Mathematical Analysis*, Second Edition, Addison-Wesley Publishing Co. (1974)
2. D. Bressoud, *A Radical Approach to Real Analysis*, The Mathematical Association of America (1994)
3. R. Courant, F. John, *Introduction to Calculus and Analysis Vol. 1*, Interscience Publisher (1965)
4. I. Niven, *Irrational Numbers*, The Mathematical Association of America (1956)
5. K.A. Ross, *Elementary Analysis: The Theory of Calculus*, Springer (1980)
6. W. Rudin, *Principles of Mathematical Analysis Third Edition*, McGraw-Hill, Inc. (1976)

Differentiation

5.1   Basics of Differentiation

Recall that when we write $\lim_{x \to 0} f(x)$, the variable x has to be from the domain X of the function f. So when we write $\lim_{h \to 0} \frac{f(x+h)-f(x)}{h}$ in the next definition, the expression $\frac{f(x+h)-f(x)}{h}$ is a function of h, and the variable $(x+h)$ has to be in the domain of f, and $h \neq 0$ since the denominator cannot be zero. Hence, $\lim_{h \to 0} \frac{f(x+h)-f(x)}{h}$ means

$$\lim_{h \to 0, x+h \in X, h \neq 0} \frac{f(x+h)-f(x)}{h}.$$

Remark 5.1.1 Let x be an element from the domain X of the function f. By substituting $t = x + h$, we have $\lim_{h \to 0} \frac{f(x+h)-f(x)}{h} = \lim_{t \to x} \frac{f(t)-f(x)}{t-x}$. Here, $\lim_{t \to x} \frac{f(t)-f(x)}{t-x}$ means that t is in the domain X of the function f and $t \neq x$.

Definition 5.1.1

(1) Let X be a subset of $\mathbb{R}$. A point $p \in X$ is an <u>isolated point</u> of X if there is an $\varepsilon > 0$ such that $(p - \varepsilon, p + \varepsilon) \cap X = \emptyset$. Equivalently, a point $p \in X$ is an <u>isolated point</u> of X if $p \in X$ and p is not a limit point of X.

(2) (Fermat) Let X be a subset of $\mathbb{R}$ with no isolated points, and let $f : X \to \mathbb{R}$ be a function. If $x \in X$, and if $\lim_{h \to 0} \frac{f(x+h)-f(x)}{h} = \lim_{t \to x} \frac{f(t)-f(x)}{t-x}$ exists, then we say that f is <u>differentiable</u> at x, and we denote the value of the limit by $f'(x)$ or $\frac{d}{dx}f(x)$. If the function f is differentiable at each point of $Y \subset X$, we treat $f' : Y \to \mathbb{R}$ as a function defined by $f'(x) = \lim_{h \to 0} \frac{f(x+h)-f(x)}{h} = \lim_{t \to x} \frac{f(t)-f(x)}{t-x}$ for every $x \in Y$. In this case, the

H. Katsuura, *Introduction to Analysis*, Synthesis Lectures on Mathematics & Statistics, https://doi.org/10.1007/978-3-031-67954-4_5

function f' is said to be the <u>derivative</u> of f, and f is said to be <u>differentiable</u> on Y or $f|_Y$ is <u>differentiable</u>.

Remark 5.1.2

(1) The sets $(0, 1)$, $[0, 1]$, and $\mathbb{R}$ have no isolated points. Both 0 and 1 are isolated points of the set $X = \{0, 1\}$.

If $X = \left\{\frac{1}{n}\right\}_{n=1}^{\infty} \cup \{0\}$, then all points in $\left\{\frac{1}{n}\right\}_{n=1}^{\infty}$ are all isolated points of X, while 0 is not an isolated point. The point 0 is a limit point of X.

All the points in the Cantor set are limit points, and the Cantor set has no isolated points.

(2) Suppose $p \in X$ is an isolated point of X. Then $(p - \varepsilon, p + \varepsilon) \cap X = \emptyset$ for some $\varepsilon > 0$. In order for a function $f : X \to \mathbb{R}$ to be differentiable at p, the limit $f'(p) = \lim\limits_{h \to 0} \frac{f(p+h)-f(p)}{h}$ has to exist. But $f(p + h)$ is not even defined for every h between $-\varepsilon$ and ε since $(p - \varepsilon, p + \varepsilon) \cap X = \emptyset$. This explains why X cannot have isolated points in Definition 5.1.1.

Except in Sect. 5.3 we are mostly interested in the differentiability of a function defined on an interval and the properties of derivatives except in Sect. 5.3. We will talk about the non-differentiability of the Cantor function in Sect. 5.3.

(3) The definition of the derivative $\lim\limits_{h \to 0} \frac{f(x+h)-f(x)}{h}$ was arrived at by Pierre de Fermat (1607–1656) in an attempt to find the definition of a tangent line to a curve in a plane. And the readers of this textbook know how to find the equation of a tangent line to a given curve at a given point since this was covered in your beginning calculus classes. However, before Fermat's definition of the derivative, only the tangent line to a **circle** was defined (a line perpendicular to the radius of the circle touching the circle at one point). It took about 2000 years since Euclid wrote about that in *The Elements* [1] for someone to make the leap to consider and study tangent lines to a curve. That someone was Fermat. As simple as his definition, his definition of the derivative revolutionized mathematics through differential equations and the fundamental theorem of calculus (Theorem 6.2.7, in the next chapter), revolutionized physics through Newtonian mechanics, revolutionized engineering by making it possible to estimate the structural strength or to estimate power output of steam engines, and brought the industrial revolution to the nineteenth century Europe in my opinion. Fermat was a contemporary of Rene Descartes and Evangelista Torricelli. Fermat is famous for Fermat's Last Theorem in number theory.

Notation 5.1.1
Let X be an interval in $\mathbb{R}$ for the remainder of this section.

Theorem 5.1.1 Let $f : X \to \mathbb{R}$ be a function differentiable at $p \in X$. Then f is continuous at p.

Proof Since f is differentiable at p, we know that $f'(p) = \lim\limits_{t \to p} \frac{f(t)-f(p)}{t-p}$ exists.

(By replacing x by p in $f'(x) = \lim\limits_{t \to x} \frac{f(t)-f(x)}{t-x}$, we have $f'(p) = \lim\limits_{t \to p} \frac{f(t)-f(p)}{t-p}$.)

But then we have

$$\lim_{x \to p} f(x) = \lim_{x \to p} \left[\frac{f(x)-f(p)}{x-p} \cdot (x-p) + f(p) \right]$$

$$= \lim_{x \to p} \frac{f(x)-f(p)}{x-p} \cdot \lim_{x \to p}(x-p) + \lim_{x \to p} f(p) \quad \text{by theorem 4.2.2,}$$

$$= f'(p) \cdot 0 + f(p) = f(p).$$

Therefore, f is continuous at p by Theorem 4.2.3.

The applications of derivatives are covered in calculus courses. So, we will concentrate on basic properties and theorems of derivatives.

Theorem 5.1.2 (*power derivative*) Let n be a positive integer, and let $f : \mathbb{R} \to \mathbb{R}$ be a function defined by $f(x) = x^n$ for every $x \in \mathbb{R}$. Then $f'(x) = nx^{n-1}$ for every $x \in \mathbb{R}$.

Proof Let $x \in \mathbb{R}$. Then

$$\lim_{t \to x} \frac{f(t)-f(x)}{t-x} = \lim_{t \to x} \frac{t^n - x^n}{t-x}$$

$$= \lim_{t \to x} \frac{(t-x)\left(t^{n-1} + t^{n-2}x + t^{n-3}x^2 + \cdots + tx^{n-2} + x^{n-1}\right)}{t-x} \quad \text{by Problem 1.4.2}$$

$$= \lim_{t \to x}\left(t^{n-1} + t^{n-2}x + t^{n-3}x^2 + \cdots + tx^{n-2} + x^{n-1}\right) = nx^{n-1}.$$

Hence, $f'(x) = nx^{n-1}$.

Alternate Proof of Theorem 5.1.2 using the Binomial Theorem:

Let $x \in \mathbb{R}$. Then

$$\lim_{h \to 0} \frac{f(x+h)-f(x)}{h} = \lim_{h \to 0} \frac{(x+h)^n - x^n}{h} = \lim_{h \to 0} \frac{\left(\sum_{k=0}^{n}\binom{n}{k}x^{n-k}h^k\right) - x^n}{h}$$

$$= \lim_{h \to 0} \sum_{k=1}^{n}\binom{n}{k}x^{n-k}h^{k-1}$$

$$= \lim_{h \to 0} \left(nx^{n-1} + \sum_{k=2}^{n} \binom{n}{k} x^{n-k} h^{k-1} \right) = nx^{n-1}.$$

Hence, $f'(x) = nx^{n-1}$.

Theorem 5.1.3

(1) $\sin'x = \cos x$ for every $x \in \mathbb{R}$. Here, $\sin'x = (\sin x)' = \frac{d}{dx} \sin x$.
(2) $\cos'x = -\sin x$ for every $x \in \mathbb{R}$.

Proof of (1): Let $x \in \mathbb{R}$. Then

$$\sin'x = \lim_{h \to 0} \frac{\sin(x+h) - \sin x}{h}$$
$$= \lim_{h \to 0} \frac{\sin x \cos h + \cos x \sin h - \sin x}{h}$$
$$= \lim_{h \to 0} \frac{\sin x(\cos h - 1) + \cos x \sin h}{h}$$
$$= \lim_{h \to 0} \left(\frac{(\cos h - 1)}{h} \sin x + \frac{\sin h}{h} \cos x \right).$$

But $\lim_{h \to 0} \frac{\sin h}{h} = 1$ from Theorem 4.2.8. Hence,

$$\lim_{h \to 0} \frac{(\cos h - 1)}{h} = \lim_{h \to 0} \frac{(\cos h - 1)(\cos h + 1)}{h(\cos h + 1)}$$
$$= \lim_{h \to 0} \frac{-\sin^2 h}{h(\cos h + 1)}$$
$$= \lim_{h \to 0} \left(-\frac{\sin h}{h} \frac{\sin h}{(\cos h + 1)} \right) = 1 \cdot 0 = 0.$$

Therefore,

$$\sin'x = \lim_{h \to 0} \left(\frac{(\cos h - 1)}{h} \sin x + \frac{\sin h}{h} \cos x \right) = 0 \cdot \sin x + 1 \cdot \cos x = \cos x.$$

Proof of (2) is left to the readers.

Theorem 5.1.4 Let $f, g : X \to \mathbb{R}$ be differentiable functions. Let $x \in \mathbb{R}$. Then we have the following:

(1) $(f + g)'(x) = f'(x) + g'(x),$

(2) If c is a constant, then $(cf)'(x) = c \cdot f'(x)$,

(3) the <u>product rule</u>: $(f \cdot g)'(x) = f'(x) \cdot g(x) + f(x) \cdot g'(x)$

(4) If $g'(x) \neq 0$ for every $x \in X$, then $\left(\frac{1}{g}\right)'(x) = -\frac{g'(x)}{g^2(x)}$.

(5) the <u>quotient rule</u>: If $g(x) \neq 0$ for every $x \in X$, then

$$\left(\frac{f}{g}\right)'(x) = \frac{f'(x) \cdot g(x) - f(x) \cdot g'(x)}{g^2(x)}.$$

Proofs of (1) and (2) are left to readers.

Proof of (3):

$$
\begin{aligned}
(f \cdot g)'(x) &= \lim_{h \to 0} \frac{(f \cdot g)(x+h) - (f \cdot g)(x)}{h} \\
&= \lim_{h \to 0} \frac{f(x+h) \cdot g(x+h) - f(x) \cdot g(x)}{h} \\
&= \lim_{h \to 0} \frac{g(x+h) \cdot [f(x+h) - f(x)] + g(x+h) \cdot f(x) - f(x) \cdot g(x)}{h} \\
&= \lim_{h \to 0} \left[\frac{g(x+h) \cdot [f(x+h) - f(x)]}{h} + \frac{f(x) \cdot [g(x+h) - g(x)]}{h} \right] \\
&= \lim_{h \to 0} \frac{g(x+h) \cdot [f(x+h) - f(x)]}{h} + \lim_{h \to 0} \frac{f(x) \cdot [g(x+h) - g(x)]}{h} \\
&= \lim_{h \to 0} \frac{[f(x+h) - f(x)]}{h} \cdot \lim_{h \to 0} g(x+h) + f(x) \cdot \lim_{h \to 0} \frac{[g(x+h) - g(x)]}{h} \\
&= f'(x) \cdot g(x) + f(x) \cdot g'(x)
\end{aligned}
$$

since $\lim_{h \to 0} g(x+h) = g(x)$ by Theorem 5.1.1.

Proof of (4):

$$
\begin{aligned}
\left(\frac{1}{g}\right)'(x) &= \lim_{h \to 0} \frac{\frac{1}{g(x+h)} - \frac{1}{g(x)}}{h} = \lim_{h \to 0} \frac{g(x) - g(x+h)}{h \cdot g(x+h) \cdot g(x)} \\
&= \left[-\lim_{h \to 0} \frac{g(x+h) - g(x)}{h} \right] \cdot \left[\lim_{h \to 0} \frac{1}{g(x+h) \cdot g(x)} \right] \\
&= -\frac{g'(x)}{g^2(x)}.
\end{aligned}
$$

Proof of (5) is obtained by combining (3) and (4).

Corollary 5.1.4-1: Polynomials and rational functions are differentiable. The trigonometric functions of secant, cosecant, tangent, and cotangent are all differentiable. We expect you to know how to take derivatives of these functions.

Next, we find the derivative of the composition of the functions f with g.

Theorem 5.1.5 (*The Chain Rule*) Let $g : X \to \mathbb{R}$ and $f : g(X) \to \mathbb{R}$ be differentiable functions. Let $x \in \mathbb{R}$. Then $(f \circ g)'(x) = f'(g(x)) \cdot g'(x)$.

Proof

$$(f \circ g)'(x) = \lim_{h \to 0} \frac{(f \circ g)(x+h) - (f \circ g)(x)}{h}$$
$$= \lim_{h \to 0} \frac{f(g(x+h)) - f(g(x))}{h}$$
$$= \lim_{h \to 0} \frac{f(g(x+h)) - f(g(x))}{g(x+h) - g(x)} \cdot \frac{g(x+h) - g(x)}{h}$$
$$= \lim_{h \to 0} \frac{f(g(x+h)) - f(g(x))}{g(x+h) - g(x)} \cdot \lim_{h \to 0} \frac{g(x+h) - g(x)}{h}.$$

But $\lim_{h \to 0}\left[g(x+h) - g(x) \right] = 0$ by the continuity of g, so let $g(x+h) - g(x) = H$.
Then we have $g(x+h) = g(x) + H$ and $\lim_{h \to 0} \frac{f(g(x+h)) - f(g(x))}{g(x+h) - g(x)} = \lim_{h \to 0} \frac{f(g(x)+H) - f(g(x))}{H} =$
$f'(g(x))$. Hence, we have

$$\lim_{h \to 0} \frac{f(g(x+h)) - f(g(x))}{g(x+h) - g(x)} \cdot \lim_{h \to 0} \frac{g(x+h) - g(x)}{h} = f'(g(x)) \cdot g'(x).$$

This proves the chain rule.

Example 5.1.1 In Theorem 5.1.3, we left the proof of $\cos'x = -\sin x$ to the readers expecting them to use a similar method to part (1). However, here, we use trig identities $\cos x = \sin\left(x + \frac{\pi}{2}\right), \cos\left(x + \frac{\pi}{2}\right) = -\sin x$, and the chain rule to find the derivative of $\cos x$.

$$\cos'x = \sin'\left(x + \frac{\pi}{2}\right) = \cos\left(x + \frac{\pi}{2}\right) \cdot 1 = -\sin x.$$

Example 5.1.2 Instead of writing **the derivatives of inverse functions** as a theorem, it is more practical to show this as an **application of the chain rule**.

Let $g : [-1, 1] \to \left[-\frac{\pi}{2}, \frac{\pi}{2}\right]$ be a function defined by $g(x) = \sin^{-1}x$ for every $x \in [-1, 1]$. Let $y = \sin^{-1}x$. We want to find $\frac{dy}{dx}$. Since $x = \sin y$, differentiating it with respect to x by thinking of y as a function of x implicitly, we have $1 = \cos y \cdot \frac{dy}{dx}$ by the chain rule. (This is called an **implicit differentiation**.) Hence, solving this equation for $\frac{dy}{dx}$ we have

$$\frac{dy}{dx} = \frac{1}{\cos y} = \frac{1}{\sqrt{1 - \sin^2 y}} = \frac{1}{\sqrt{1 - x^2}}.$$

(Note that $-\frac{\pi}{2} < y\frac{\pi}{2}$ implies that $\cos y = \sqrt{1 - \sin^2 y}$.)

We usually write this as $\frac{d}{dx}\sin^{-1}x = \frac{1}{\sqrt{1-x^2}}$. Since $\frac{1}{\sqrt{1-x^2}}$ is not defined when $x = \pm 1$, g is not differentiable at ± 1. Hence, $g' : (-1, 1) \to \mathbb{R}$. Note that the domain of g' is $(-1,1)$ while the domain of g is $[-1, 1]$.

Since $\lim\limits_{x \to \pm 1} \frac{1}{\sqrt{1-x^2}} = \infty$, the graph of $y = \sin^{-1}x$ has vertical tangent lines $x = \pm 1$.

Problem 5.1.1 Let $g : [-1, 1] \to [0, \pi]$ defined by $g(x) = \cos^{-1}x$ for every $x \in [-1, 1]$. Show that $g'(x) = -\frac{1}{\sqrt{1-x^2}}$.

Remark 5.1.4 For every integer $n > 0$, let $e_n : \mathbb{R} \to \mathbb{R}$ be a function defined by $e_n(x) = \sum\limits_{k=0}^{n} \frac{x^k}{k!}$ for every $x \in \mathbb{R}$. Then we know that $\frac{d}{dx}e_n(x) = \sum\limits_{k=1}^{n} \frac{kx^{k-1}}{k!} = \sum\limits_{k=0}^{n-1} \frac{x^k}{k!} = e_{n-1}(x)$ if $n > 1$. Hence, from Theorem 4.4.1, it is plausible to think that

$$\frac{d}{dx}e^x = \frac{d}{dx}\sum\limits_{k=0}^{\infty} \frac{x^k}{k!} = \frac{d}{dx}\lim\limits_{n \to \infty} e_n(x) = \lim\limits_{n \to \infty}\frac{d}{dx}e_n(x) = \lim\limits_{n \to \infty} e_{n-1}(x) = e^x.$$

However, we do not know that if the interchanging the order of $\frac{d}{dx}$ and $\lim\limits_{n \to \infty}$ is permissible in $\frac{d}{dx}\lim\limits_{n \to \infty} e_n(x) = \lim\limits_{n \to \infty}\frac{d}{dx}e_n(x)$. Hence, we cannot conclude that $\frac{d}{dx}e^x = e^x$. This will be proved in Theorem 7.4.4. Since Theorem 7.4.4 is independent of theorems discussed between here and when Theorem 7.4.4 is proven in Chap. 7, I am quoting this theorem here in this section below. It is too important a theorem to ignore and not to use until we arrive at Sect. 7.4.

As I mentioned in Sect. 3.10, many calculus textbooks define the number e to be a number such that $\lim\limits_{h \to 0} \frac{e^h - 1}{h} = 1$. And therefore, the derivation of $\frac{d}{dx}e^x = e^x$ is simple as in

$$\frac{d}{dx}f(x) = \lim\limits_{h \to 0} \frac{f(x + h) - f(x)}{h}$$
$$= \lim\limits_{h \to 0} \frac{e^{x+h} - e^x}{h} = \lim\limits_{h \to 0} \frac{e^x \cdot (e^h - 1)}{h}$$
$$= e^x \cdot \lim\limits_{h \to 0} \frac{(e^h - 1)}{h} = e^x \cdot 1 = e^x.$$

However, I **cannot** justify the existence of the number e such that $\lim\limits_{h \to 0} \frac{e^h - 1}{h} = 1$ without knowing $\frac{d}{dx} e^x = e^x$ to begin with. But then it is a circular argument, and it does not prove $\frac{d}{dx} e^x = e^x$. So, I did not employ this method.

More traditional calculus textbooks define the natural logarithmic function first by $\ln x = \int\limits_1^x \frac{1}{t} dt$ for all $x > 0$, and let e to be a number $\ln e = 1$. Then those authors define e^x to be the inverse function of $\ln x$. This is mathematically sound. However, this explanation of $\frac{d}{dx} e^x = e^x$ sequentially has to wait until after integration is covered in Chap. 6. So this explanation traditionally offered at this point in the discussion does not offer a plausible reason why $\frac{d}{dx} e^x = e^x$ right now. In the same way, we have to wait to obtain an approximation of e, which first requires the Taylor's series to be introduced in order to derive $e^x = \sum\limits_{k=o}^{\infty} \frac{x^k}{k!}$ at the end of Chap. 7.

That said, even though our explanation of $\frac{d}{dx}\left[\lim\limits_{n \to \infty} e_n(x)\right] = \left[\lim\limits_{n \to \infty} \frac{d}{dx} e_n(x)\right]$ has to wait until Chap. 7, we think this derivation is the most natural way to introduce the number e, the function e^x, and to give a plausible reason why $\frac{d}{dx} e^x = e^x$.

Theorem 7.4.4 (*The derivative of* e^x) $\frac{d}{dx} e^x = e^x$ for every $x \in \mathbb{R}$.

Recall **Definition 4.3.1**: Let $f : \mathbb{R} \to (0, \infty)$ be a function defined by $f(x) = e^x$ for every $x \in \mathbb{R}$. Then the *natural logarithmic function* is $f^{-1} : (0, \infty) \to \mathbb{R}$, and denoted by $f^{-1}(y) = \ln y$ for all $y \in (0, \infty)$.

Theorem 5.1.6

(1) Let $y = \ln x$ for every $x \in (0, \infty)$. Then $\frac{dy}{dx} = \frac{1}{x}$.
(2) Let $y = \ln|x|$ for every $x \in (-\infty, 0) \cup (0, \infty)$. Then $\frac{dy}{dx} = \frac{1}{x}$.

Proof Let $x \in (0, \infty)$. Since $y = \ln|x| = \ln x$, we have $x = e^y$. By the chain rule, we have $1 = e^y \cdot \frac{dy}{dx}$. Hence, $\frac{dy}{dx} = \frac{1}{e^y} = \frac{1}{x}$.

Let $x \in (-\infty, 0)$. Then $y = \ln|x| = \ln(-x)$. Hence $\frac{dy}{dx} = \frac{1}{-x} \cdot (-1) = \frac{1}{x}$.

As a corollary to Theorem 7.4.4, Chain Rule, and Theorem 5.1.6, we have the next important **extension** of Theorem 5.1.2.

Theorem 5.1.7 (*The derivative of power functions*) Let $r \in \mathbb{R}$, and let $f : (0, \infty) \to \mathbb{R}$ be a function defined by $f(x) = x^r$ for every $x \in (0, \infty)$. Then $f'(x) = rx^{r-1}$ for every $x \in (0, \infty)$.

Proof Let $x \in (0, \infty)$. Then

$$f'(x) = \frac{d}{dx}x^r = \frac{d}{dx}e^{r \cdot \ln x} = e^{r \cdot \ln x} \cdot \frac{1}{x} = x^r \cdot \frac{1}{x} = rx^{r-1}.$$

Remark 5.1.5 Suppose n is an **odd positive** integer. Let $y = x^{\frac{1}{n}}$. Then $x = y^n$ so that $1 = n \cdot y^{n-1} \cdot \frac{dy}{dx}$. Hence, $\frac{dy}{dx} = \frac{1}{ny^{n-1}} = \frac{1}{n}x^{-\frac{1}{n}(n-1)} = \frac{1}{n}x^{\frac{1}{n}-1}$ for every $x \in \mathbb{R}$.

Problem 5.1.2

(1) Let $X = \mathbb{R} - \left\{\frac{\pi}{2} \pm k\pi : k = 0, 1, 2, 3, \ldots\right\}$. Then $\tan : X \to \mathbb{R}$ is defined by $\tan x = \frac{\sin x}{\cos x}$ for every $x \in X$. Show that $\frac{d}{dx}\tan x = \sec^2 x$.

(2) Let $f : \left(-\frac{\pi}{2}, \frac{\pi}{2}\right) \to \mathbb{R}$ defined by $f(x) = \tan x$ for every $x \in \left(-\frac{\pi}{2}, \frac{\pi}{2}\right)$. Then f is 1–1 and onto (no need to prove this). So $f^{-1} : \mathbb{R} \to \left(-\frac{\pi}{2}, \frac{\pi}{2}\right)$ exists. Find $\frac{d}{dx}f^{-1}(x) = \frac{d}{dx}\tan^{-1}x$.

(3) Suppose n is a **negative** integer. Let $f : (-\infty, 0) \to \mathbb{R}$ be a function defined by $f(x) = x^n$ for every $x \in (-\infty, 0)$. Prove that $\frac{d}{dx}f = nx^{n-1}$ for every $x \in (-\infty, 0)$.

Example 5.1.3

(1) Let $f : \left[0, \frac{\pi}{2}\right) \cup \left[\pi, \frac{3\pi}{2}\right) \to (-\infty, -1] \cup [1, \infty)$ defined by

$f(x) = \sec x$ for every $x \in \left[0, \frac{\pi}{2}\right) \cup \left[\pi, \frac{3\pi}{2}\right)$. Then f is 1–1 and onto. So the inverse function f^{-1} exists. We will find $\frac{d}{dx}f^{-1}(x)$.

Let $y = f^{-1}(x)$. Hence, $x = \sec y$ so that $1 = \sec y \cdot \tan y \cdot \frac{dy}{dx}$. Thus $\frac{dy}{dx} = \frac{1}{\sec y \cdot \tan y}$. From the identity $\cos^2 y + \sin^2 y = 1$, we have $\tan^2 y + 1 = \sec^2 y$. By noting that $f^{-1} :$ $(-\infty, -1] \cup [1, \infty) \to \left[0, \frac{\pi}{2}\right) \cup \left[\pi, \frac{3\pi}{2}\right)$ and $\tan y \geq 0$ when $y \in \left[0, \frac{\pi}{2}\right) \cup \left[\pi, \frac{3\pi}{2}\right)$, we have $\tan y = \sqrt{\sec^2 y - 1} = \sqrt{x^2 - 1}$. Hence, $\frac{d}{dx}f^{-1}(x) = \frac{1}{\sec y \cdot \tan y} = \frac{1}{x\sqrt{x^2-1}}$ for all $x \in (-\infty, -1) \cup (1, \infty)$.

(2) Let $g : \left[0, \frac{\pi}{2}\right) \cup \left(\frac{\pi}{2}, \pi\right] \to (-\infty, -1] \cup [1, \infty)$ defined by $g(x) = \sec x$ for every

$x \in \left[0, \frac{\pi}{2}\right) \cup \left(\frac{\pi}{2}, \pi\right]$. Then g is 1–1 and onto, and g^{-1} exists. We will find $\frac{d}{dx}g^{-1}(x)$.

Let $y = g^{-1}(x)$. Hence, $x = \sec y$ so that $1 = \sec y \cdot \tan y \cdot \frac{dy}{dx}$. Thus $\frac{dy}{dx} = \frac{1}{\sec y \cdot \tan y}$. We have $\tan^2 y + 1 = \sec^2 y$. By noting that the function g^{-1} is $g^{-1} : (-\infty, -1] \cup [1, \infty) \to$ $\left[0, \frac{\pi}{2}\right) \cup \left(\frac{\pi}{2}, \pi\right]$ and

[$\tan y \geq 0$ when $y \in \left[0, \frac{\pi}{2}\right)$], and [$\tan y \leq 0$ when $y \in \left[\pi, \frac{3\pi}{2}\right)$].

Hence, we have $\tan y = \sqrt{\sec^2 y - 1} = \sqrt{x^2 - 1}$ when $y \in \left[0, \frac{\pi}{2}\right)$, and $\tan y = -\sqrt{\sec^2 y - 1} = -\sqrt{x^2 - 1}$ when $y \in \left[\pi, \frac{3\pi}{2}\right)$. Hence, $\frac{dy}{dx} = \frac{1}{\sec y \cdot \tan y} = \frac{1}{x\sqrt{x^2-1}}$ for all $x \in (1, \infty)$, and $\frac{dy}{dx} = \frac{1}{\sec y \cdot \tan y} = -\frac{1}{x\sqrt{x^2-1}}$ for all $x \in (-\infty, -1)$. That is,

$$\frac{d}{dx}g^{-1}(x) = \begin{cases} \frac{1}{x\sqrt{x^2-1}} & \text{for all } x \in (1, \infty) \\ -\frac{1}{x\sqrt{x^2-1}} & \text{for all } x \in (-\infty, -1) \end{cases}.$$

(**Note**: Depending on textbooks, $\sec^{-1} x$ is defined by either part (1) definition or part (2) definition. Please be careful. I **prefer** definition part (1) of $\sec^{-1} x$ for the simplicity of its derivative. Some authors like part (2) definition of $\sec^{-1} x$ since its domain is similar to the domain of $\cos^{-1} x$.)

Problem 5.1.3

(1) Express $\sec^{-1}(-3)$ under the definition of the inverse secant function in Example 5.1.3(1), in terms of $\cos^{-1}$.
(2) Express $\sec^{-1}(-3)$ under the definition of the inverse secant function in Example 5.1.3(2), in terms of $\cos^{-1}$.

(Answers: (1) $\sec^{-1}(-3) = 2\pi - \cos^{-1}\left(-\frac{1}{3}\right)$. (2) $\sec^{-1}(-3) = \cos^{-1}\left(-\frac{1}{3}\right)$. If $x > 0$, then $\sec^{-1}(x) = \cos^{-1}\left(\frac{1}{x}\right)$ by either definition of inverse secant function.)

Problem 5.1.4 Let $f : \left[\frac{\pi}{2}, \frac{3\pi}{2}\right] \to [-1, 1]$ be a function defined by $f(x) = \sin x$ for every $x \in \left[\frac{\pi}{2}, \frac{3\pi}{2}\right]$.
Show that $\frac{d}{dx}f^{-1}(x) = -\frac{1}{\sqrt{1-x^2}}$ for every $x \in (-1, 1)$.

Example 5.1.4

(1) Let $a > 1$. Let $f : \mathbb{R} \to (0, \infty)$ defined by $f(x) = a^x$ for every $x \in \mathbb{R}$. Then $f'(x) = (a^x)' = \left(e^{x \cdot \ln a}\right)' = e^{x \cdot \ln a} \cdot \ln a = a^x \cdot \ln a$. (See Theorem 4.3.11.)

(2) Let $g : (0, \infty) \to \mathbb{R}$ defined by $g(x) = \log_{10} x$ for every $x \in (0, \infty)$. Then

$$g'(x) = \left(\frac{\ln x}{\ln 10}\right)' = \frac{1}{x \cdot \ln 10}. \text{ (See Definition 4.3.2.)}$$

Remark 5.1.6 As an alternate to Example 5.1.4(1), we can let $y = a^x$. Hence, $\ln y = x \cdot \ln a$. By the chain rule, or by implicit differentiation, we have $\frac{1}{y} \cdot \frac{dy}{dx} = \ln a$. Hence, $\frac{dy}{dx} = y \cdot \ln a = a^x \cdot \ln a$. This method is called the logarithmic differentiation. In many elementary calculus textbooks, this logarithmic differentiation is used to show that $(a^x)' = a^x \cdot \ln a$. But this is slow and difficult to remember. I recommend the method explained in Example 5.1.4.

As a rule of thumb, whenever you see an exponential function a^x, think of it **always** as

$$a^x = e^{x \ln a}$$

as we explained in Theorem 4.3.11. This will also be very helpful when we want to integrate or to find limits related to exponential functions.

Example 5.1.5 For your information, let $f : (0, \infty) \times \mathbb{R} \to (\mathbb{R} - \{0\})$ be a function of two variables defined by $f(x, y) = x^y$ for every $(x, y) \in (0, \infty) \times \mathbb{R}$. Then the derivative of f with respect to x by treating y as a constant is called the partial derivative of f with respect to x, and it is denoted by $\frac{\partial}{\partial x} f(x, y)$. And it is given by $\frac{\partial}{\partial x} f(x, y) = \frac{\partial}{\partial x}(x^y) = x^{y-1}$ by Theorem 5.1.7.

Alternately, $\frac{\partial}{\partial x} f(x, y) = \frac{\partial}{\partial x}(x^y) = \frac{\partial}{\partial x}\left(e^{y \cdot \ln x}\right) = e^{y \cdot \ln x} \cdot \frac{1}{x} = x^y \cdot \frac{1}{x} = x^{y-1}$.
by Theorem 4.3.11. Also, we have $\frac{\partial}{\partial y} f(x, y) = \frac{\partial}{\partial y}(x^y) = \frac{\partial}{\partial y}\left(e^{y \cdot \ln x}\right) = e^{y \cdot \ln x} \cdot \ln x = x^y \cdot \ln x$.

I do not want to use logarithmic differentiation for finding these partial derivatives because logarithmic differentiation is too cumbersome and not practical.

5.2 The Mean Value Theorem

The Rolle's theorem is probably the most primitive form of the mean value theorem. So we'll start there. It is a powerful theorem as you will see. It is not only a lemma to prove the mean value theorem, but it is also used to prove Cauchy's mean value theorem. In turn, Cauchy's mean value theorem is used to prove L'Hospital's rule. And we will also use Rolle's theorem to prove Taylor's theorem in Theorem 7.5.1.

Theorem 5.2.1 (*Rolle's theorem*) Let $a < b$. Suppose $f : [a, b] \to \mathbb{R}$ is continuous and $f|_{(a,b)}$ is a differentiable function such that $f(a) = f(b)$. Then there exists a number $c \in (a, b)$ such that $f'(c) = 0$.

(Note that "$f|_{(a,b)}$ is differentiable" means that "f is a differentiable function on the interval (a, b)". This hypothesis is motivated by the type of function in Example 5.1.2 whose derivatives may not exists at a or b.)

Proof: Case 1: Assume that $f(x) > f(b)$ for some $x \in [a, b]$. By the extreme value theorem, there exists a number $c \in (a, b)$ such that $f(c) = \max f([a, b])$.
Then $f(c + h) - f(c) \le 0$ for all h such that $c + h \in [a, b]$. Hence, $f'(c) = \lim\limits_{h \to 0, h > 0} \frac{f(c+h)-f(c)}{h} \le 0$ and $f'(c) = \lim\limits_{h \to 0, h < 0} \frac{f(c+h)-f(c)}{h} \ge 0$.
Therefore, $f'(c) = 0$.

Case 2: Assume that $f(x) < f(b)$ for some $x \in [a, b]$. By the extreme value theorem, there exists a number $c \in (a, b)$ such that $f(c) = \min f([a, b])$.
Then $f(c + h) - f(c) \ge 0$ for all h such that $c + h \in [a, b]$. Hence,

$$f'(c) = \lim_{h \to 0, h > 0} \frac{f(c+h) - f(c)}{h} \geq 0 \text{ and } f'(c) = \lim_{h \to 0, h < 0} \frac{f(c+h) - f(c)}{h} \leq 0.$$

Therefore, $f'(c) = 0$.

Case 3: If $f(x) = f(b)$ for all $x \in [a, b]$, then f is a constant function and $f'(x) = 0$ for all $x \in [a, b]$.

These three cases prove this theorem.

The following mean value theorem is a generalization of Rolle's theorem. It differs from Rolle's theorem because it does not include the phrase "such that $f(a) = f(b)$".

Theorem 5.2.2 (*Mean Value Theorem*) Let $a < b$. Suppose $f : [a, b] \to \mathbb{R}$ is continuous and $f|_{(a,b)}$ is a differentiable function. Then there exists a number $c \in (a, b)$ such that $f'(c) = \frac{f(b) - f(a)}{b-a}$.

Proof The equation of the line through the points $(a, f(a))$ and $(b, f(b))$ is given by $y = \frac{f(b) - f(a)}{b-a}(x - a) + f(a)$. So let $F : [a, b] \to \mathbb{R}$ be a function defined by $F(x) = f(x) - \left[\frac{f(b) - f(a)}{b-a}(x - a) + f(a)\right]$ for every $x \in [a, b]$. This is to make $F(a) = F(b)$. To see this, note that we have

$$F(a) = f(a) - \left[\frac{f(b) - f(a)}{b-a}(a - a) + f(a)\right] = 0 \text{ and}$$

$$F(b) = f(b) - \left[\frac{f(b) - f(a)}{b - a}(b - a) + f(a)\right]$$
$$= f(b) - \left[f(b) - f(a) + f(a)\right] = 0.$$

Hence, $F(a) = F(b)$.

The function $F|_{(a,b)}$ is differentiable since $F'(x) = f'(x) - \frac{f(b) - f(a)}{b-a}$.

By Rolle's theorem, there exists a number $c \in (a, b)$ such that $F'(c) = 0$. But since $F'(x) = f'(x) - \frac{f(b) - f(a)}{b-a}$, we have $F'(c) = f'(c) - \frac{f(b) - f(a)}{b-a}$. Hence, $f'(c) - \frac{f(b) - f(a)}{b-a} = 0$, or we have $f'(c) = \frac{f(b) - f(a)}{b-a}$.

The next theorem is a generalization of the mean value theorem. It is derived from the proof of the mean value theorem using Rolle's theorem. The proof of Theorem 5.2.3 is a slight modification of the proof of Theorem 5.2.2.

Theorem 5.2.3 (*Cauchy's Mean Value Theorem*) Let $a \neq b$. Let $f, g : [a, b] \to \mathbb{R}$ be continuous functions such that $g(a) \neq g(b)$. Suppose $f|_{(a,b)}$ and $g|_{(a,b)}$ are differentiable. Then there exists $c \in (a, b)$ such that

$$\frac{f'(c)}{g'(c)} = \frac{f(b) - f(a)}{g(b) - g(a)}.$$

Proof Notice that if $g(x) = x$, we have the mean value theorem. So the idea of the proof of this theorem is to replace some of a by $g(a)$, b by $g(b)$, and x by $g(x)$ in the above proof of the mean value theorem.

Let $F : [a, b] \to \mathbb{R}$ be a function defined by

$$F(x) = f(x) - \left[\frac{f(b) - f(a)}{g(b) - g(a)} (g(x) - g(a)) + f(a) \right] \text{ for every } x \in [a, b].$$

Then, we have

$$F(a) = f(a) - \left[\frac{f(b) - f(a)}{g(b) - g(a)} (g(a) - g(a)) + f(a) \right] = 0 \text{ and}$$

$$F(b) = f(b) - \left[\frac{f(b) - f(a)}{g(b) - g(a)} (g(b) - g(a)) + f(a) \right]$$

$$= f(b) - \left[f(b) - f(a) + f(a) \right] = 0.$$

This shows that $F(a) = F(b)$.

The function $F|_{(a,b)}$ is differentiable since $F'(x) = f'(x) - \frac{f(b) - f(a)}{g(b) - g(a)} g'(x)$.

By Rolle's theorem, there exists a number. $c \in (a, b)$ such that $F'(c) = 0$. But then $F'(c) = f'(c) - \frac{f(b) - f(a)}{g(b) - g(a)} g'(c) = 0$. Hence, $f'(c) = \frac{f(b) - f(a)}{g(b) - g(a)} g'(c)$ or $\frac{f'(c)}{g'(c)} = \frac{f(b) - f(a)}{g(b) - g(a)}$.

Cauchy's mean value theorem is a lemma to prove L'Hospital's rule.

Theorem 5.2.4 (*L'Hospital's Rule*) Let $a < b$, and $f, g : (a, b) \to \mathbb{R}$ be differentiable functions such that $g(x) \neq 0 \neq f(x)$ for all $x \in (a, b)$.

(1) Suppose $\lim\limits_{x \to a} f(x) = 0 = \lim\limits_{x \to a} g(x)$. Then $\lim\limits_{x \to a} \frac{f(x)}{g(x)} = \lim\limits_{x \to a} \frac{f'(x)}{g'(x)}$.

   (Since $f, g : (a, b) \to \mathbb{R}$, all the limits are one-sided limits here. For example, $\lim\limits_{x \to a} f(x) = 0$ means that $\lim\limits_{x \to a, x > a} f(x) = 0$.)

(1') Suppose $\lim\limits_{x \to a} f(x) = \infty = \lim\limits_{x \to a} g(x)$. Then $\lim\limits_{x \to a} \frac{f(x)}{g(x)} = \lim\limits_{x \to a} \frac{f'(x)}{g'(x)}$.

(2) Suppose $\lim\limits_{x \to b} f(x) = 0 = \lim\limits_{x \to b} g(x)$. Then $\lim\limits_{x \to b} \frac{f(x)}{g(x)} = \lim\limits_{x \to b} \frac{f'(x)}{g'(x)}$.

(2') Suppose $\lim\limits_{x \to b} f(x) = \infty = \lim\limits_{x \to b} g(x)$. Then $\lim\limits_{x \to b} \frac{f(x)}{g(x)} = \lim\limits_{x \to b} \frac{f'(x)}{g'(x)}$.

Proof of (1): Since $\lim\limits_{x \to a} f(x) = 0 = \lim\limits_{x \to a} g(x)$, we can extend the functions f and g continuously to $\hat{f}, \hat{g} : [a, b] \to \mathbb{R}$ such that

$$\hat{f}(x) = \begin{cases} f(x) & \text{if } x \in (a, b) \\ 0 & \text{if } x = a \end{cases} \text{ and } \hat{g}(x) = \begin{cases} g(x) & \text{if } g \in (a, b) \\ 0 & \text{if } x = a \end{cases}.$$

Let $x \in (a, b)$. Then there exists c between a and x such that $\frac{f(x)}{g(x)} = \frac{\hat{f}(x)}{\hat{g}(x)} = \frac{\hat{f}(x) - \hat{f}(a)}{\hat{g}(x) - \hat{g}(a)} = \frac{\hat{f}'(c)}{\hat{g}'(c)} = \frac{f'(c)}{g'(c)}$ by Cauchy's mean value theorem.

This shows that $\lim\limits_{x \to a} \frac{f(x)}{g(x)} = \lim\limits_{c \to a} \frac{f'(c)}{g'(c)} = \lim\limits_{x \to a} \frac{f'(x)}{g'(x)}$ since $c \to a$ as $x \to a$.

Proof of (1'): Let $F, G : (a, b) \to \mathbb{R}$ be defined by $F(x) = \frac{1}{f(x)}$ and $G(x) = \frac{1}{g(x)}$ for every $x \in (a, b)$. Since $f(x) \neq 0 \neq g(x)$ for all $x \in (a, b)$, functions F and G are defined, and they are differentiable. Moreover, $\lim_{x \to a} f(x) = \infty = \lim_{x \to a} g(x)$ implies that $\lim_{x \to a} F(x) = 0 = \lim_{x \to a} G(x)$. So Part (1) applies to $\lim_{x \to a} \frac{F(x)}{G(x)}$, and we have

$$\lim_{x \to a} \frac{g(x)}{f(x)} = \lim_{x \to a} \frac{\frac{1}{f(x)}}{\frac{1}{g(x)}} = \lim_{x \to a} \frac{F(x)}{G(x)} = \lim_{x \to a} \frac{F'(x)}{G'(x)} \text{ by part (1)}$$

$$= \lim_{x \to a} \frac{-\frac{1}{f^2(x)} \cdot f'(x)}{-\frac{1}{g^2(x)} \cdot g'(x)} = \lim_{x \to a} \frac{f'(x)}{g'(x)} \cdot \left(\frac{g(x)}{f(x)} \right)^2$$

by the quotient rule, Theorem 5.1.4(5). Hence, we have

$$\lim_{x \to a} \frac{f(x)}{g(x)} = \lim_{x \to a} \frac{g(x)}{f(x)} \cdot \left(\frac{f(x)}{g(x)} \right)^2$$

$$= \lim_{x \to a} \left[\left\{ \frac{f'(x)}{g'(x)} \cdot \left(\frac{g(x)}{f(x)} \right)^2 \right\} \cdot \left(\frac{f(x)}{g(x)} \right)^2 \right] = \lim_{x \to a} \frac{f'(x)}{g'(x)}.$$

Proof of (2) and (2') are similar to (1) and (1'), respectively.

Corollary 5.2.5 (L'Hospital's Rule) Let $a > 0$. Let $f, g : (a, \infty) \to \mathbb{R}$ be differentiable functions such that

$$\lim_{x \to \infty} f(x) = 0 = \lim_{x \to \infty} g(x) \text{ or } \lim_{x \to \infty} f(x) = \infty = \lim_{x \to \infty} g(x).$$

Suppose $f(x) \neq 0 \neq g(x)$ for all $x \in (a, \infty)$. Then $\lim_{x \to \infty} \frac{f(x)}{g(x)} = \lim_{x \to \infty} \frac{f'(x)}{g'(x)}$.

Proof: Let $F, G : \left(0, \frac{1}{a}\right) \to \mathbb{R}$ be defined by $F(x) = \frac{1}{f(x)}$ and $G(x) = \frac{1}{g(x)}$ for every $x \in \left(0, \frac{1}{a}\right)$. Since $f(x) \neq 0 \neq g(x)$ for all $x \in (a, \infty)$, functions F and G are defined, and they are differentiable. And $G(x) \neq 0 \neq F(x)$ for all $x \in \left(0, \frac{1}{a}\right)$. If $\lim_{x \to a} f(x) = \infty = \lim_{x \to a} g(x)$, then $\lim_{x \to a} F(x) = 0 = \lim_{x \to a} G(x)$. If $\lim_{x \to a} f(x) = 0 = \lim_{x \to a} g(x)$. then $\lim_{x \to a} F(x) = \infty = \lim_{x \to a} G(x)$. In either case, by Theorem 5.2.4, we have

$$\lim_{x \to a} \frac{g(x)}{f(x)} = \lim_{x \to a} \frac{\frac{1}{f(x)}}{\frac{1}{g(x)}} = \lim_{x \to a} \frac{F(x)}{G(x)} = \lim_{x \to a} \frac{F'(x)}{G'(x)}$$

$$= \lim_{x \to a} \frac{-\frac{1}{f^2(x)} \cdot f'(x)}{-\frac{1}{g^2(x)} \cdot g'(x)} = \lim_{x \to a} \frac{f'(x)}{g'(x)} \cdot \left(\frac{g(x)}{f(x)} \right)^2.$$

Hence,

$$\lim_{x\to\infty} \frac{f(x)}{g(x)} = \lim_{x\to a} \frac{g(x)}{f(x)} \cdot \left(\frac{f(x)}{g(x)}\right)^2$$

$$= \lim_{x\to a} \left\{ \frac{f'(x)}{g'(x)} \cdot \left(\frac{g(x)}{f(x)}\right)^2 \right\} \cdot \left(\frac{f(x)}{g(x)}\right)^2$$

$$= \lim_{x\to a} \frac{f'(x)}{g'(x)}.$$

Remark 5.2.1 In order to better understand infinite series, I generally recommend using the comparison test to test the convergence of positive term series rather than the ratio test. Similarly, I recommend initially finding limits of functions without L'Hospital's rule. It will help you understand limits better.

Example 5.2.1

(1) We consider $\lim\limits_{x\to-\infty} \left(\frac{x}{1+x}\right)^{\pi x}$.

The part $x \to -\infty$ is troublesome. So let $x = -t$, or $t = -x$. Then $t \to \infty$ as $x \to -\infty$.

Hence, $\lim\limits_{x\to-\infty} \left(\frac{x}{1+x}\right)^{\pi x} = \lim\limits_{t\to\infty} \left(\frac{-t}{1-t}\right)^{-\pi t} = \lim\limits_{t\to\infty} \left(\frac{t}{t-1}\right)^{-\pi t} = \left\{ \lim\limits_{t\to\infty} \left(\frac{t-1}{t}\right)^t \right\}^\pi$

by Theorem 4.1.1 or by Theorem 4.3.12.

Thus, $\lim\limits_{t\to\infty} \left(\frac{t-1}{t}\right)^t = \lim\limits_{t\to\infty} e^{t\cdot\ln\left(\frac{t-1}{t}\right)}$ by Theorem 4.3.11.

$$\lim_{t\to\infty} t \cdot \ln\left(\frac{t-1}{t}\right) = \lim_{t\to\infty} \frac{\ln\left(1 - \frac{1}{t}\right)}{\frac{1}{t}} \quad \text{(apply L' Hospital's rule)}$$

$$= \lim_{t\to\infty} \frac{\left(\frac{1}{1-\frac{1}{t}}\right) \cdot \frac{1}{t^2}}{-\frac{1}{t^2}} = \lim_{t\to\infty} -\frac{1}{1-\frac{1}{t}} = -1.$$

Hence, $\lim\limits_{t\to\infty} \left(\frac{t-1}{t}\right)^t = \lim\limits_{t\to\infty} e^{t\cdot\ln\left(\frac{t-1}{t}\right)} = e^{-1}$ by Theorem 4.2.7.

Therefore, $\lim\limits_{x\to-\infty} \left(\frac{x}{1+x}\right)^{\pi x} = \left\{ \lim\limits_{t\to\infty} \left(\frac{t-1}{t}\right)^t \right\}^\pi = e^{-\pi}$.

(2) Next, we consider $\lim\limits_{x\to\infty} \frac{(3x-2)^3}{(5-7x^2)(5x+8)}$. Even though this problem can be solved using L'Hospital's rule three times, it is simpler not to use L'Hospital's rule as follows:

$$\lim_{x\to\infty} \frac{(3x-2)^3}{(5-7x^2)(5x+8)} = \lim_{x\to\infty} \frac{(3x-2)^3 \cdot \frac{1}{x^3}}{(5-7x^2) \cdot \frac{1}{x^2} \cdot (5x+8) \cdot \frac{1}{x}}$$

$$= \lim_{x \to \infty} \frac{\left(3 - \frac{2}{x}\right)^3}{\left(\frac{5}{x^2} - 7\right) \cdot \left(5 + \frac{8}{x}\right)} = -\frac{27}{35}.$$

(3) The limit $\lim_{x \to 0} \frac{\sin x}{x} = 1$ is a theorem we came across in Chap. 4. Using L'Hospital's rule to conclude $\lim_{x \to 0} \frac{\sin x}{x} = \lim_{x \to 0} \frac{\cos x}{1} = 1$ is not a proof of $\lim_{x \to 0} \frac{\sin x}{x} = 1$ since we used this limit to find the derivative of $\sin x$. So for a problem like $\lim_{x \to 0} \frac{\sin 2x}{\tan 3x}$, I prefer a solution without L'Hospital's rule such as the following:

$$\lim_{x \to 0} \frac{\sin 2x}{\tan 3x} = \lim_{x \to 0} \frac{\sin 2x}{\sin 3x} \cdot \cos 3x = \lim_{x \to 0} \frac{\sin 2x / 2x}{\sin 3x / 3x} \cdot \frac{2}{3} \cdot \cos 3x = 1 \cdot \frac{2}{3} \cdot 1 = \frac{2}{3}.$$

Problem 5.2.1 L'Hospital's rule is very powerful. But find the limits of the following functions without L'Hospital's rule if you can.

(1) $\lim_{x \to -3} \frac{(x-3)^2}{x^3+27}$.

(2) $\lim_{x \to -3} \frac{(x+3)^3}{(x^2-9)(3-x)}$.

(3) $\lim_{x \to 0} \frac{\sin 3x}{\tan 5x}$.

(4) $\lim_{x \to 0-} \frac{e^{3x}}{e^{2x}}$.

(5) $\lim_{x \to 0+} \frac{\ln x}{x}$.

(6) $\lim_{x \to \infty} \frac{\ln x}{x}$.

(7) $\lim_{x \to 0} \frac{3^x}{2^x}$.

(8) $\lim_{x \to 0+} \frac{3^x - 2^x}{x}$.

(9) $\lim_{x \to -\infty} \frac{3^x}{2^x}$.

(10) $\lim_{x \to -\infty} (3^x - 2^x)$.

(11) $\lim_{x \to \infty} \frac{3^x - 2^x}{x}$.

(12) $\lim_{x \to -\infty} \frac{3^x - 2^x}{x}$.

(13) $\lim_{x \to 0+} x^{3x}$.

(14) $\lim_{x \to 0+} \left(\frac{1}{x} - \frac{1}{\tan x}\right)$.

(15) $\lim_{x \to \frac{\pi}{2}-} \left(\frac{1}{x} - \frac{1}{\tan x}\right)$.

(16) $\lim_{x \to 1} \frac{\sin 3x}{\sin 5x}$.

Problem 5.2.2 Use mathematical induction to prove: $\lim\limits_{x \to 0+} x(\ln x)^n = 0$ for all integers $n \geq 1$.

Problem 5.2.3 Use mathematical induction to prove: $\lim\limits_{x \to \infty} \frac{(\ln x)^p}{x} = 0$ for all integers $p \geq 1$.

Example 5.2.2 This is an application of Problem 5.2.3.

For any positive integer p, $\sum\limits_{n=2}^{\infty} \frac{1}{(\ln n)^p}$ diverges.

Proof Suppose p is a positive integer. We know that $\lim\limits_{x \to \infty} \frac{(\ln x)^p}{x} = 0$. So there exists an integer N such that $\frac{(\ln x)^p}{x} = \left| \frac{(\ln x)^p}{x} - 0 \right| < 1$ for all $x \geq N$. Hence, $(\ln x)^p < x$ or $\frac{1}{(\ln x)^p} > \frac{1}{x}$ for all $x \geq N$. Therefore, $\sum\limits_{n=N}^{\infty} \frac{1}{(\ln n)^p} > \sum\limits_{n=N}^{\infty} \frac{1}{n}$. Since $\sum\limits_{n=N}^{\infty} \frac{1}{n}$ diverges, $\sum\limits_{n=2}^{\infty} \frac{1}{(\ln n)^p}$ diverges.

See Definition 4.3.1 for the definition of increasing/decreasing.

Theorem 5.2.6 (*Increasing/Decreasing Test*) Let $a < b$, and let $f : [a, b] \to \mathbb{R}$ be a continuous function such that $f|_{(a,b)}$ is differentiable.

(1) If, for every $a < x < b$, we have $f'(x) > 0$, then f is strictly increasing on $[a, b]$.
(2) If, for every $a < x < b$, we have $f'(x) < 0$, then f is strictly decreasing on $[a, b]$.

Proof of (1): Suppose $f'(x) > 0$ for every $a < x < b$. Let $a \leq \alpha < \beta \leq b$. Then there exists $\alpha < c < \beta$ such that $f(\beta) - f(\alpha) = f'(c) \cdot (\beta - \alpha)$ by the mean value theorem. But $f'(c) > 0$. So we have $f(\beta) - f(\alpha) = f'(c) \cdot (\beta - \alpha) > 0$, or $f(\alpha) < f(\beta)$. Therefore, f is strictly increasing on $[a, b]$.

Proof of (2) is left to the readers.

Remark 5.2.2 Even though most calculus textbooks include the first derivative test, I think the increasing/decreasing test is more important than the first derivative test. Since the first derivative test is a weaker statement than the increasing/decreasing test, I omit the first derivative test in this book.

Example 5.2.3 We prove that $\sin x < x$ for every $x \in (0, \infty)$.

Proof Let $f : \mathbb{R} \to \mathbb{R}$ defined by $f(x) = x - \sin x$ for every $x \in \mathbb{R}$. Then $f'(x) = 1 - \cos x > 0$ for every $x \in \left(0, \frac{\pi}{2}\right)$. By Theorem 5.2.6, f is increasing. So $f(x) =$

$x - \sin x > f(0) = 0$, or we have $\sin x < x$ for every $x \in \left(0, \frac{\pi}{2}\right)$. If $x \geq \frac{\pi}{2}$, then $x \geq \frac{\pi}{2} > 1 \geq \sin x$. Therefore, $\sin x < x$ for every $x \in (0, \infty)$.

Problem 5.2.4

(1) Prove that $\cos x > 1 - \frac{x^2}{2!}$ for all $x \in \mathbb{R}$.

(2) Prove that $\sin x \geq x - \frac{x^3}{6}$ for every $x \in \left[0, \frac{\pi}{2}\right]$.

(3) Prove that $\sin x \leq x - \frac{x^3}{6} + \frac{x^5}{120}$ for every $x \in \left[0, \frac{\pi}{2}\right]$.

Problem 5.2.5 Use Theorem 5.2.6 and appropriate limits to graph the following equations.

(1) $y = \frac{x}{1-x^2}$.

(2) $y = \frac{1}{(1-x)(1+x)}$. Note that $x \in (-\infty, -1) \cup (-1, 1) \cup (1, \infty)$.

(3) $y = \frac{1}{\sqrt{(1-x)(1+x)}}$.

(4) $y = \frac{1}{1+e^{-x}}$.

(5) $y = 1 + xe^{-x}$.

Solution to (1): First, note that $x \in (-\infty, -1) \cup (-1, 1) \cup (1, \infty)$. $\frac{dy}{dx} = \frac{1+x^2}{(1-x^2)^2}$. Hence, the increasing/decreasing test gives us the following table:

$$
\begin{array}{lllll}
x & : \ldots & -1 & \ldots & 1 & \ldots \\
\frac{dy}{dx} & : + & \times & + & \times & + \, . \\
y & : \nearrow & \times & \nearrow & \times & \nearrow
\end{array}
$$

The first row indicates the x-axis. The second row indicates the sign of $\frac{dy}{dx}$ on the intervals $(-\infty, -1), (-1, 1), (1, \infty)$. The sign "$\times$" under -1, for example, indicates that $\frac{dy}{dx}$ **does not exist** at -1. The arrows on the third row indicates the graph of y is increasing on the intervals $(-\infty, -1), (-1, 1), (1, \infty)$ after applying the increasing/decreasing test. The sign "$\times$" under -1 on this row indicates that y value does not exist at -1.

Now, let us consider the limits.

$$\lim_{x \to -\infty} y = \lim_{x \to -\infty} \frac{x}{1 - x^2} = \lim_{x \to -\infty} \frac{1}{-2x} = 0,$$

$$\lim_{x \to -1-} y = \lim_{x \to -1-} \frac{x}{1-x^2} = \lim_{x \to -1-} \frac{x}{(1-x)(1+x)} = \infty \text{ and}$$

$$\lim_{x \to -1+} y = \lim_{x \to -1+} \frac{x}{1 - x^2} = \lim_{x \to -1+} \frac{x}{(1 - x)(1 + x)} = -\infty.$$

$$\lim_{x \to 1-} y = \lim_{x \to 1-} \frac{x}{1-x^2} = \lim_{x \to 1-} \frac{x}{(1-x)(1+x)} = \infty, \text{ and}$$

Fig. 5.1 The graph of
$y = \frac{x}{1-x^2}$

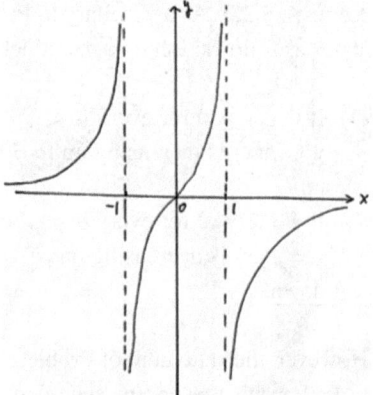

$$\lim_{x \to 1+} y = \lim_{x \to 1+} \frac{x}{1 - x^2} = \lim_{x \to 1+} \frac{x}{(1 - x)(1 + x)} = -\infty.$$

And, finally, we have $\lim\limits_{x \to \infty} y = \lim\limits_{x \to \infty} \frac{x}{1-x^2} = \lim\limits_{x \to \infty} \frac{1}{-2x} = 0$. I used "$\lim\limits_{x \to} y$" because y is the subject. Concavity of the graph helps little to improve the graph, so I ignored it. Summarizing all this information gives us the graph Fig. 5.1.

Note: Let $f : (-1, 1) \to \mathbb{R}$ be a function defined by $f(x) = \frac{x}{1-x^2}$ for all $x \in (-1, 1)$. We can use Solution to (1) as an alternate solution to Example 1.3.6(7) with the help of Theorem 4.3.6. Note that the $f^{-1} : \mathbb{R} \to (-1, 1)$ is given explicitly in Example 4.3.1as

$$f^{-1}(y) = \begin{cases} \frac{-1+\sqrt{1+4y^2}}{2y} & \text{if } y \neq 0 \\ 0 & \text{if } y = 0 \end{cases} \quad \text{for every } y \in \mathbb{R}.$$

Problem 5.2.6

(1) Let $f : (0, \infty) \to \mathbb{R}$ be a function defined by $f(x) = \left(1 + \frac{1}{x}\right)^x$ for every $x \in (0, \infty)$. Prove that f is increasing.
(2) Find $\lim\limits_{x \to 0, x > 0} \left(1 + \frac{1}{x}\right)^x$, $\lim\limits_{x \to 0, x < 0} \left(1 + \frac{1}{x}\right)^x$, and $\lim\limits_{x \to -\infty} \left(1 + \frac{1}{x}\right)^x$.

We know that $\lim\limits_{x \to \infty} \left(1 + \frac{1}{x}\right)^x = e$ from Lemma 4.4.2.

Problem 5.2.7 Let $a < b$, and let $f : [a, b] \to \mathbb{R}$ be a twice differentiable function.

(1) If we have $f'(a) \geq 0$ and $f''(x) \geq 0$ for every $a < x < b$, then, prove that f' is increasing on $[a, b]$.
(2) If we have $f'(a) \leq 0$ and $f''(x) \leq 0$ for every $a < x < b$, then, prove that f' is decreasing on $[a, b]$.

Remark 5.2.3 As a corollary to Problem 5.2.7, we have **the second derivative test**, which states as follows: Let $a < b$, and let $f : [a, b] \to \mathbb{R}$ be a twice differentiable function.

(1) If $f''(x) \geq 0$ for every $a < x < b$, then the slope $f'(x)$ of the tangent line to the curve $y = f(x)$ is increasing on $[a, b]$. In this case, we say that the graph of $f(x)$ is <u>concave up</u>.
(2) If $f''(x) \leq 0$ for every $a < x < b$, then the slope $f'(x)$ of the tangent line to the curve $y = f(x)$ is decreasing on $[a, b]$. In this case, we say that the graph of $f(x)$ is <u>concave down</u>.

However, the statement of Problem 5.2.7 is an application of the increasing/decreasing test to f'. For this reason, the statement of Problem 5.2.7 seems more important to me than the second derivative test.

Definition 5.2.1 Suppose $f : \mathbb{R} \to \mathbb{R}$ is a twice differentiable function. Suppose $f''(p) = 0$ and the sign of $f''(x)$ changes from (positive to negative) or (negative to positive) at $x = p$ as x increases, then the point p is said to be an <u>inflection</u> point of f, a point at which the concavity of f changes from concave down to concave up, or vice versa.

Problem 5.2.8 Let $\sigma > 0$ and μ be constants. Let $f : \mathbb{R} \to (0, \infty)$ be a function defined by $f(x) = \frac{1}{\sigma\sqrt{2\pi}} e^{-\frac{1}{2\sigma^2}(x-\mu)^2}$ for all $x \in \mathbb{R}$. This function f is called a <u>normal distribution function</u>, and it plays an important role in probability and statistics.

(1) Prove that $f'(x) = 0$ only when $x = \mu$. (μ is the <u>mean</u> of the distributive function f.)
(2) The function f is increasing on $(-\infty, \mu)$, and decreasing on (μ, ∞).
(3) $\mu + \sigma$ and $\mu - \sigma$ are inflection points of f. (σ is the <u>standard deviation</u> of the distributive function f.)
(4) Prove that the graph of $y = f(x)$ is symmetric with respect to the vertical line $x = \mu$.

(Note: The area bounded by the graph of $y = f(x)$ and the x-axis is 1.)

Next, we introduce antiderivatives of functions and then in the next chapter, we will discuss integrations of functions of one variable.

Definition 5.2.2 Let X be a subset of $\mathbb{R}$, and let $f : X \to \mathbb{R}$ be a function. If $F : X \to \mathbb{R}$ is a differentiable function such that $F'(x) = f(x)$ for every $x \in X$, we say that F is an <u>anti-derivative</u> of f, and we write $F(x) = \int f(x)dx$. (We read $\int f(x)dx$ as the anti-derivative of f. Reading as "the integral of f" is misleading since we have not introduced integrations. Integrations will be discussed in the next chapter.)

Example 5.2.4

(1) $\int 0 dx = C$ for any constant C, so the anti-derivative of a function is not unique.
(2) $\int e^x dx = e^x + C$,
(3) $\int x^r dx = \frac{1}{r+1} x^{r+1} + C$ if $r \neq -1$ (See Theorem 5.1.7 and Remark 5.1.5.)

In order to see this, $\frac{d}{dx} x^{r+1} = (r+1)x^r$ by Theorem 5.1.6(2). Hence, $\frac{d}{dx} \frac{1}{r+1} x^{r+1} = x^r$. Therefore, $\int x^r dx = \frac{1}{r+1} x^{r+1} + C$.
(4) $\int x^{-1} dx = \int \frac{1}{x} dx = \ln|x| + C$ by Theorem 5.1.6. This is the case when $r = -1$ in the left side of part (3).
(5) $\int \sin x dx = -\cos x + C$, $\int \cos x dx = \sin x + C$.
(6) Let $t = \sin x$. Then $\frac{dt}{dx} = \cos x$ so that $dt = \cos x dx$, and $\int \sin x \cos x dx = \int t dt = \frac{1}{2}t^2 + C = \frac{1}{2}\sin^2 x + C$.

This method is called integration by substitution, and this method works because of the chain rule. Once you do the substitution $t = \sin x$, often **keeping** the variable t in the answer may be better.

Alternately, since $2\sin x \cos x = \sin 2x$, we have

$$\int \sin x \cos x dx = \int \frac{1}{2} \sin 2x dx = -\frac{1}{4} \cos 2x + C.$$

Does $\frac{1}{2}\sin^2 x = -\frac{1}{4}\cos 2x$?

Problem 5.2.9 Find the following anti-derivatives.

(1) $\int 3x\cos(x^2)dx$.
(2) $\int \sin x \cos^2 x dx$.
(3) $\int \cos^2 x dx$.
(4) $\int \cos^3 x dx$.
(5) $\int \sec x dx$.
(6) $\int \sec^2 x dx$.
(7) $\int \sec^3 x dx$.
(8) $\int 2^x dx$.
(9) $\int \ln x dx$.
(10) $\int \log_{10} x dx$

For hints and solutions, look at calculus textbooks, and review anti-derivatives.

Theorem 5.2.7 Let $f : \mathbb{R} \to \mathbb{R}$ be a function. Suppose $F, G : \mathbb{R} \to \mathbb{R}$ are anti-derivatives of f. Then there is a constant C such that $F(x) - G(x) = C$ for every $x \in \mathbb{R}$.

Proof Let $g : \mathbb{R} \to \mathbb{R}$ be a function defined by $g(x) = F(x) - G(x)$ for every $x \in \mathbb{R}$. Let a be a number, and let $x \in \mathbb{R}$. By the mean value theorem, there exists a number c between a and x such that $g(a) - g(x) = g'(c) \cdot (a - x)$. But $g'(t) = F'(t) - G'(t) = f'(t) - f'(t) = 0$ for every $t \in \mathbb{R}$. So $g(a) - g(x) = g'(c) \cdot (a - x) = 0$ or $g(x) = g(a)$ for every $x \in \mathbb{R}$. Let $C = g(a)$. Hence, $F(x) - G(x) = g(a) = C$ for every $x \in \mathbb{R}$. This proves this theorem.

Corollary 5.2.8 If $F, G : \mathbb{R} \to \mathbb{R}$ are differentiable functions such that $F'(x) = G'(x)$ for every $x \in \mathbb{R}$. Then there is a constant C such that $F(x) = G(x) + C$ for every $x \in \mathbb{R}$.

Example 5.2.5 (This is an alternate approach to Example 1.4.4.)

(1) Suppose you deposit \$$P$/year "continuously[1]" into an account that earns the annual interest rate of $100 \cdot i\%$/year continuously. Let \$$A(t)$ be the amount of money you have in the account t years after the first deposit. Then the differential equation for $A(t)$ is given by $\frac{dA}{dt} = i \cdot A + P$ with the initial condition $A(0) = 0$. Solve this differential equation. This is similar to the <u>future value</u> calculation in Example 1.4.4.

(2) In part (1), if your goal is to have \$1 million for your retirement after 40 years at the age 65, and if the bank gives you 6%/year interest rate, what is your annual deposit should be?

(3) Suppose you borrowed \$$F$ from a bank that charges you the annual interest rate of $100 \cdot i\%$/year continuously. You are to amortize this loan by making the annual deposits of equal amount (mortgage payments) of \$$P$/year over 30 years. If the amount of money you owe to the bank is \$$A(t)$ after t years after you borrowed the money, then the differential equation for $A(t)$ is given by $\frac{dA}{dt} = i \cdot A - P$ with the initial condition $A(0) = F$ and the terminal condition $A(30) = 0$. Find the yearly payment \$$P$. (This is called a 30-year mortgage. This is similar to the <u>present value</u> calculation in Example 1.4.4.)

(4) Suppose you borrowed \$1,000,000 for a purchase of a house, and want to pay it back in 30 years with an equal amount of payment each month at an interest rate of 6%/year compounded continuously. You are to amortize this loan with an equal yearly payment. What is your yearly payment?

(5) After 20 years, you want to pay back the loan in a lump sum in order to buy a new house. How much money do you have to pay to the bank?

Solution to (1): The differential equation
$$\frac{dA}{dt} = i \cdot A + P,$$

---

[1] It is impossible to deposit money continuously into an account. But by thinking this way as a model, one can obtain approximate solutions to Problem 1.4.5. We made Problem 1.4.5 into a differential equation problem.

where i and P are constants, can be thought of as either a separable differential equation or a linear differential equation. Hence, there are at least two elementary methods to solve this problem. Here, we solve this differential equation as separable. By re-writing it as $\frac{1}{i \cdot A + P} \cdot dA = dt$, it becomes

(a) $\int \frac{1}{i \cdot A + P} \cdot dA = \int dt$.

But $\int \frac{1}{i \cdot A + P} \cdot dA = \frac{1}{i} \ln|i \cdot A + P| + C$. So from (a), we have
$\frac{1}{i} \ln|i \cdot A + P| = t + C$, or $\ln|i \cdot A + P| = it + C$. (Here, we replaced iC by C since iC is still a constant.) Hence, we have $iA + P = \pm e^{it + C} = \pm e^C e^{it} = K e^{it}$, where $K = \pm e^C$. So,

(b) $A = \frac{1}{i}\left(K e^{it} - P\right)$.

The initial condition $A(0) = 0$ gives us $K = P$. Hence,

$$A = \frac{P}{i}\left(e^{it} - 1\right).$$

(2) Solving $A = \frac{P}{i}\left(e^{it} - 1\right)$ for P, we have $P = \frac{Ai}{e^{it} - 1}$. Since we want $A(40) = 1,000,000$ when $i = 0.06$, we have

$$P = \frac{1,000,000 \cdot 0.06}{e^{0.06 \cdot 40} - 1} \approx 5987.$$

So if you deposit $\frac{\$5987}{12} \approx \$500$ per month, most likely you will have \$1,000,000 at the time of your retirement. It is "likely" because the interest rate fluctuates.

(3) From (b), by changing P to $-P$, we have $A = \frac{1}{i}\left(K e^{it} + P\right)$.

The initial condition $A(0) = F$, and the terminal condition $A(30) = 0$ gives us $Fi = K + P$ and $0 = K e^{30i} + P$. Hence, $P = -K e^{30} = -(Fi - P)e^{30i}$.
   Hence,

(c) $P = \frac{Fi \cdot e^{30i}}{e^{30i} - 1}$, and
(d) $A = \frac{1}{i}\left((Fi - P)e^{it} + P\right)$.
(4) Use the formulas in (3) for this calculation. From the equation (c), we have

$$P = \frac{1,000,000 \cdot 0.06 \cdot e^{30 \cdot 0.06}}{e^{30 \cdot 0.06} - 1} \approx 71,882.$$

Hence, your annual payment is approximately \$71,882. So it is about $\frac{\$71882}{12} \approx \$5990$ per month.

(5) We use the equation (d) for this. After 20 years of payment, the amount you have is

$$A(20) = \frac{1}{0.06}\big((1000000 \cdot 0.06 - 71882)e^{20 \cdot 0.06} + 71882\big) \approx \$540,540.$$

So you have to pay about \$540,540 to the bank.

Note: An alternate way to look at problems (4) and (5) is the following: Suppose when you are 65 years old, you have accumulated \$1 million in a bank for your retirement, and suppose you are expecting to live for another 30 years. If the bank continues to pay 6% of interest rate compounded continuously, then you will receive approximately \$71,882 per year. If you die at the age 85, then you leave an inheritance of \$540,540 for your loved ones.

5.3   Continuous but Nowhere-Differentiable Functions

Weierstrass constructed the first continuous function that is not differentiable at any points on the domain (0, 1). Since then, many such examples were discovered. I also constructed one in my paper Continuous Nowhere-Differentiable Functions—an Application of Contraction Mappings, *The American Mathematical Monthly*, Vol. 98, no. 5, 411–416, 1991 [2]. Let me explain this next.

Example 5.3.1 The continuous but nowhere-differentiable function $f : [0, 1] \rightarrow [0, 1]$ will be obtained as the limit of the following functions $f_i : [0, 1] \rightarrow [0, 1], i = 1, 2, 3, \dots$ constructed as follow: First, divide the unit square $[0, 1] \times [0, 1] = [0, 1]^2$ by four lines $x = \frac{1}{3}, x = \frac{2}{3}, y = \frac{1}{3}$, and $y = \frac{2}{3}$. And we let $R_1 = \left[0, \frac{1}{3}\right] \times \left[0, \frac{2}{3}\right], R_2 = \left[\frac{1}{3}, \frac{2}{3}\right] \times \left[\frac{1}{3}, \frac{2}{3}\right]$, and $R_3 = \left[\frac{2}{3}, 1\right] \times \left[\frac{2}{3}, 1\right]$ as in Fig. 5.2.

Fig. 5.2 Pictures of $[0,1]^2$ and $R_1 \cup R_2 \cup R_3$

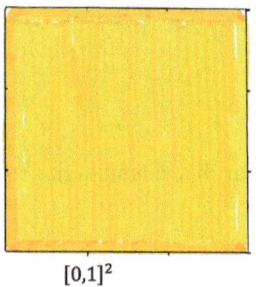

$[0,1]^2$ $R_1 \cup R_2 \cup R_3$

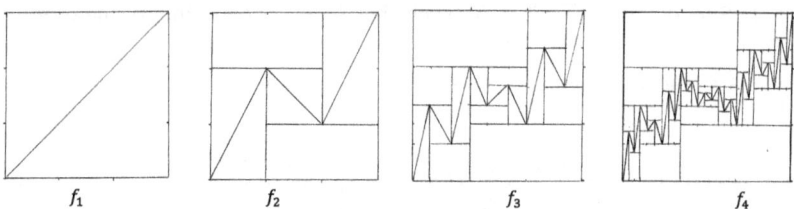

Fig. 5.3 Graphs of f_1, f_2, f_3 and f_4

Let $f_1 : [0, 1] \to [0, 1]$ be the identity function defined by $f_1(x) = x$ for all $x \in [0, 1]$.

The graph of $f_2 : [0, 1] \to [0, 1]$ is obtained by filling the regions R_1 and R_3 by the shrunk graph of $y = f_1(x)$, and the region R_2 being filled by the flipped and shrunk graph of f_1 about $x = \frac{1}{2}$.

The graph of $f_3 : [0, 1] \to [0, 1]$ by filling the regions R_1 and R_3 are filled by the shrunk graph of $y = f_2(x)$, and the region R_2 being filled by the flipped and shrunk graph of $y = f_2(x)$ about $x = \frac{1}{2}$. See Fig. 5.3.

Inductively, suppose n is an integer such that $f_n : [0, 1] \to [0, 1]$ is defined. We place the shrunk graph of $y = f_n(x)$ into the regions R_1 and R_3, and then fill the region R_2 with the shrunk graph of $y = f_n(x)$ flipped about $x = \frac{1}{2}$. Then the result is the graph of a continuous function $f_{n+1} : [0, 1] \to [0, 1]$.

By continuing this process indefinitely, we obtain a sequence of continuous functions $\langle f_n \rangle_{n=1}^{\infty}$. I think you can see that the sequence $\langle f_n \rangle_{n=1}^{\infty}$ converges plausibly to a continuous function, say $f : [0, 1] \to [0, 1]$. We will prove this sequence $\langle f_n \rangle_{n=1}^{\infty}$ is "uniformly Cauchy" in Example 7.2.6, and finally we will prove that $\langle f_n \rangle_{n=1}^{\infty}$ converges to a **continuous** function f in Example 7.3.1. This provides a nice example of uniform convergence.

Proof that this function $f : [0, 1] \to [0, 1]$ is nowhere-differentiable is rather complicated, so I will leave it to the article of mine mentioned above in the *American Mathematical Monthly*. However, I hope you can see that it is plausible that f is nowhere differentiable as it is approximated by f_n.

Since the Cantor set C has no isolated point, our definition of differentiation Definition 5.1.1 works. So let us prove the following theorem.

Theorem 5.3.1 The Cantor function $f : C \to [0, 1]$ is continuous but nowhere-differentiable. (See the figure in Theorem 3.6.1 for the graph of f.)

Proof By Theorem 4.1.2, we know that the Cantor function is continuous. We prove that f is not differentiable at $s = \sum\limits_{n=1}^{\infty} \frac{s_n}{3^n} \in C$.

Claim 1: Suppose $s = \sum_{n=1}^{\infty} \frac{s_n}{3^n} \in C$ such that $s_n = 2$ for infinitely many $n \in \mathbb{N}$. Let $t_n = \sum_{k=1}^{n} \frac{s_k}{3^k} \in C$ for every integer $n \geq 1$. Then $t_n \to s$. Then

$$f'(s) = \lim_{n \to \infty} \frac{f(s) - f(t_n)}{s - t_n} = \lim_{n \to \infty} \frac{\sum_{k=1}^{\infty} \frac{s_k}{2^{k+1}} - \sum_{k=1}^{n} \frac{s_k}{2^{k+1}}}{\sum_{k=1}^{\infty} \frac{s_k}{3^{k+1}} - \sum_{k=1}^{n} \frac{s_k}{3^{k+1}}}$$

$$= \lim_{n \to \infty} \frac{\sum_{k=n+1}^{\infty} \frac{s_k}{2^{k+1}}}{\sum_{k=n+1}^{\infty} \frac{s_k}{3^{k+1}}}$$

For each $n \geq 1$, let M_n be the smallest integer such that $M_n \geq n$, and $s_{M_n} = 2$. Then

$$f'(s) = \lim_{n \to \infty} \frac{\sum_{k=n+1}^{\infty} \frac{s_k}{2^{k+1}}}{\sum_{k=n+1}^{\infty} \frac{s_k}{3^{k+1}}} > \lim_{n \to \infty} \frac{\frac{s_{M_n}}{2^{M_n+2}}}{\sum_{k=M_n+1}^{\infty} \frac{2}{3^{k+1}}}$$

$$= \lim_{n \to \infty} \frac{s_{M_n}}{2^{M_n+2}} \cdot \frac{1}{\frac{2}{3^{M_n+2}} \sum_{k=0}^{\infty} \frac{1}{3^k}}$$

$$= \lim_{n \to \infty} \frac{2}{2^{M_n+2}} \cdot \frac{3^{M_n+2}}{2} \cdot \left(1 - \frac{1}{3}\right)$$

$$= \lim_{n \to \infty} \left(\frac{3}{2}\right)^{M_n+2} \cdot \frac{2}{3} = \lim_{n \to \infty} \left(\frac{3}{2}\right)^{M_n+1} = \infty$$

since $M_n \to \infty$ as $n \to \infty$.

Hence, this shows that f is not differentiable at s.

Claim 2: Let $s = \sum_{n=1}^{\infty} \frac{s_n}{3^n} \in C$ such that $s_n = 2$ for only finitely many integers $n \geq 1$. Then there is the smallest integer $N \geq 0$ such that $s_k = 0$ for all integers $k \geq N + 1$. Then $s = \sum_{n=1}^{N} \frac{s_n}{3^n}$. Since $s < 1$, there is an integer $M > N$ such that $s \leq 1 - \frac{1}{3^M}$. Then $s + \frac{1}{3^M} = s + \sum_{k=M+1}^{\infty} \frac{2}{3^k} \in C$. Let $t_n = s + \sum_{k=M+n}^{\infty} \frac{2}{3^k} \in C$ for every integer $n \geq 1$. Then $t_n \to s$. Then

$$f'(s) = \lim_{n \to \infty} \frac{f(t_n) - f(s)}{t_n - s}$$

$$= \lim_{n \to \infty} \frac{\left\{\sum_{k=1}^{N} \frac{s_k}{2^{k+1}} + \sum_{k=M+n}^{\infty} \frac{2}{2^{k+1}}\right\} - \sum_{k=1}^{N} \frac{s_k}{2^{k+1}}}{\left\{s + \sum_{k=M+n}^{\infty} \frac{2}{3^{k+1}}\right\} - s}$$

$$= \lim_{n \to \infty} \frac{\sum_{k=M+n}^{\infty} \frac{2}{2^{k+1}}}{\sum_{k=M+n}^{\infty} \frac{2}{3^{k+1}}}$$

$$= \lim_{n \to \infty} \frac{\frac{2}{2^{k+1}} \cdot \frac{1}{\left(1-\frac{1}{2}\right)}}{\frac{2}{3^{k+1}} \cdot \frac{1}{\left(1-\frac{1}{3}\right)}} = \frac{4}{3} \cdot \lim_{n \to \infty} \left(\frac{3}{2}\right)^n = \infty.$$

Hence, this shows that f is not differentiable at s.

From Claims 1 and 2, f is shown to be nowhere-differentiable.

References

1. Euclid; *The Elements*, at about 295 B.C.
2. H. Katsuura, Continuous nowhere-differentiable functions—an application of contraction mappings. Am. Math. Mon. **98**(5), 411–416 (1991)

Integration

6.1   Introduction

The origin of integration is very old. By piling up rectangular figures as in Fig. 6.1, we can obtain a pyramid like figure. This may be the original idea of finding volumes of solids by the slicing method. In fact, the pyramid in Saqqara, Egypt, resembles Fig. 6.1. Looking at a closeup picture of the pyramids in Giza, we can see that they were built with rectangular stones by making layers of rectangular bases like the ones in Fig. 6.1. This shows the idea of triple integrals since we can approximate the volume of these pyramids by adding the volumes of the rectangular stones. Hence, one might say that this is the origin of integral calculus. These pyramids were built almost 5000 years ago. However, the usual credit for the concept of integration goes to either Eudoxus or Democritus, perhaps both, who invented the idea called the **method of exhaustion** at about 350 B.C. in Greece. This method is typically applied to find the area of a two-dimensional figure. The most famous application of the method of exhaustion is its application to proving that the area of a circle of radius $r > 0$ is πr^2. Archimedes' argument to find the area of a circle did not use $\lim_{x \to 0} \frac{\sin x}{x} = 1$, but his argument was similar to the proof of Theorem 4.2.9. Archimedes[1] (287–212 B.C.) is also famous for discovering the volume formula $\frac{4\pi r^3}{3}$ and the surface area formula $4\pi r^2$ for a sphere of radius $r > 0$. The method of exhaustion is considered to be the start of integral calculus.

We introduce the definition of Darboux integration, which is equivalent to Riemann integration. The main focus of this chapter is to prove that a continuous function on a closed interval is integrable in Theorem 6.2.2 and the fundamental theorem of calculus.

---

[1] Archimedes discovered this formula for the volume of a sphere, but it was not by the method of exhaustion. It was done by the application of a lever. For an interesting read, see *Mathematical Methods in Science* by George Polya, The Mathematical Association of America, 1977, 70–75 [1].

© The Author(s), under exclusive license to Springer Nature Switzerland AG 2025   173
H. Katsuura, *Introduction to Analysis*, Synthesis Lectures on Mathematics & Statistics,
https://doi.org/10.1007/978-3-031-67954-4_6

Fig. 6.1 This object was built and photographed by the author. This resembles to the pyramid in Saqqara, built in the 27th century BC. It is older than the pyramids in Giza. The pyramids in Giza have thinner layers of rectangular slices compared to the one in Saqqara.

Theorem 6.2.2 is rather difficult to prove because it requires Theorem 6.2.1 which says that a continuous function on a finite closed interval is uniformly continuous. We discuss uniform continuity further in Sect. 6.3.

It seems that ancient Greek mathematics originated with Thales of Miletus[2], who was followed by Pythagoras, Plato, and Aristotle. Greek mathematics culminated in the establishment of the Library of Alexandria by Ptolemy I. Its construction was completed in about 290 B.C. by Ptolemy II. However, its function as a library and as a university started even before its completion. The first director of the library was Euclid, and he wrote *The Elements* [2] in about 295 B.C. commissioned by Ptolemy II. Archimedes came to Alexandria in about 250 B.C. when Eratosthenes was the head librarian there. Eratosthenes[3] was the first to calculate the radius of the earth. Apollonius of Perga investigated and named the three types of the conic sections by hyperbolas, parabolas, and ellipses in about 200 B.C. (Conics: Books I–IV, Green Lion Press, 2013 [3]). The Ptolemy dynasty of Egypt was ended by the invasion of Romans shortly after 1 A.D. in the Julian Calendar, which is about 40 B.C. Cleopatra was the Pharaoh at that time. However, though the Ptolemy dynasty ended, the Library of Alexandria continued to flourish. Heron of Alexandria discovered the famous Heron's formula[4] for the area of a triangle in about 60 A.D. Another famous mathematician/astronomer there was Claudius Ptolemy of Alexandria (no relation to the Ptolemy dynasty) who wrote the Almagest on planetary motion, which contains

---

[2] Theorem of Thales: If AB is the diameter of a circle, and if C is a point on the circle different from A and B, then the angle $\angle ACB$ is always a right angle.

[3] Incidentally, at the time of Eratosthenes, people thought the earth was spherical, and he calculated the earth radius to be about eight times the distance between Alexandria and Syene (current name is Aswan), Egypt. Looking at a map, the distance between Alexandrea and Syene is about $800km$. So the radius of the earth is $6400km$ according to Eratosthenes. The modern measurement of the earth radius is $6378.1km$. See page 12 of [1].

[4] Heron's Formula: Let a, b, c be the edge lengths of a triangle. Let $s = \frac{1}{2}(a + b + c)$. Then the area of the triangle is given by $\sqrt{s(s-a)(s-b)(s-c)}$.

his famous Ptolemy's theorem[5] on a cyclic quadrilateral. The last official librarian was Theon, 335–405 AD. But unofficially, his daughter, Hypatia, probably the first female scientist to be remembered in history, continued to lead the library until she was assassinated by a Christian mob in about 415 AD. The Library of Alexandria was said to have been destroyed at that time. But as I said before, Greek mathematicians were also reaching the limit of what they could do without zero and without algebra.

6.2   Integration

There are different approaches to integration, starting with the method of exhaustion. I was first introduced to Riemann's definition of an integration as a student, and learned to use Leibniz's notations $\int f(x)dx$ and $\int_a^b f(x)dx$. Computationally, Riemann's definition works well, but it is rather difficult to use to prove theorems. In contrast, it seems that Darboux gave an equivalent definition of the integral to that found in Riemann's integration and Darboux's definition is more suited for proving theorems. So, we will use the Darboux definition of integration. We will prove that Riemann's and Darboux definitions of the integral are equivalent for a continuous function. Throughout this textbook, when we say **integration**, we mean Darboux integration throughout this textbook.

What is curious about learning integrations in my experience is that none of my teachers told me about how Newton and Leibniz defined integrations. So, I consulted an English translation of NEWTON'S PRINCIPIA; MOTTE's TRANSLATION REVISED, University of California Press, Berkeley, Los Angeles, London, 1934 [4]. (Originally, Newton wrote it in Latin.) In it, there are a few figures that describe integrations between pages 29 and 31. It is difficult to decipher what he meant, but it seems closer to Darboux's definition than to Riemann's from his figures.

Let us begin our introduction to integrations.

Definition   6.2.1 A   partition   of   an   interval $[a, b]$ is   a   finite   set $P =$ $\{a = a_0 < a_1 < a_2 < \cdots < a_n = b\}$ for some positive integer n. Let $Q =$ $\{a = b_0 < b_1 < b_2 < \cdots < b_m = b\}$, for some integer $m \geq n$, be another partition of $[a, b]$. Then Q is said to be a refinement of P if $Q \supset P$.

Definition 6.2.2 Let $P = \{a = a_0 < a_1 < a_2 < \cdots < a_m = b\}$ be a partition of $[a, b]$ for some positive integer m. Let $\|P\| = \max\{(a_i - a_{i-1}) : i = 1, 2, 3, \ldots, m\}$. Then $\|P\|$ is said to be the norm of the partition P.

If $\langle P_n \rangle_{n=1}^{\infty}$ is a sequence of partitions of the interval $[a, b]$ such that $\lim_{n \to \infty} \|P_n\| = 0$, then $\langle P_n \rangle_{n=1}^{\infty}$ is said to be a partition sequence of $[a, b]$.

---

[5] Ptolemy's Theorem: If a quadrilateral $ABCD$ is inscribed in a circle, then $|AB||CD| + |AD||BC| = |AC||BD|$.

If $\langle P_n \rangle_{n=1}^{\infty}$ is a partition sequence of the interval $[a, b]$ such that $P_{n+1} \supset P_n$ for every positive integer n, then $\langle P_n \rangle_{n=1}^{\infty}$ is said to be a <u>refinement-partition sequence</u>, or simply an <u>R-partition sequence</u> of $[a, b]$.

Example 6.2.1

(1) Let $n > 0$ be an integer. Let $P_n = \{0\} \cup \left\{ \frac{1}{k} \right\}_{k=1}^{n}$. Then P_n is a partition of the interval $[0, 1]$, and P_{n+1} is a refinement of P_n for each integer $n \geq 1$. However, $\|P_n\| = \frac{1}{2}$ if $n \geq 2$ so that $\lim_{n \to \infty} \|P_n\| \neq 0$. Thus, $\langle P_n \rangle_{n=1}^{\infty}$ is **not** a partition sequence of $[0, 1]$.

(2) Let n be a positive integer. Let $P_n = \left\{ \frac{k}{n} \right\}_{k=0}^{n}$. Then $\langle P_n \rangle_{n=1}^{\infty}$ is a partition sequence of $[0,1]$ since $\lim_{n \to \infty} \|P_n\| = \lim_{n \to \infty} \frac{1}{n} = 0$. But $\langle P_n \rangle_{n=1}^{\infty}$ is **not** an R-partition sequence since P_{n+1} is not a refinement of P_n for some n. For example, $P_4 \not\subset P_5$.

However, $P_4 \subset P_8 \subset P_{16} \subset P_{32}$ gives the idea for the next example in (3).

(3) Let n be a positive integer. Let $P_n = \left\{ \frac{k}{2^n} \right\}_{k=0}^{2^n}$. Then P_n is a partition of the interval $[0,1]$, and P_{n+1} is a refinement of P_n for each n. Moreover, $\lim_{n \to \infty} \|P_n\| = \lim_{n \to \infty} \frac{1}{2^n} = 0$. Thus, the sequence $\langle P_n \rangle_{n=1}^{\infty}$ is an R-partition sequence of $[0, 1]$.

Problem 6.2.1

(a) Find a partition-sequence of $[-2, 3]$ that is not an R-partition sequence.
(b) Find an R-partition sequence of $[-2, 3]$

Definition 6.2.3 Let $a < b$, and let $f : [a, b] \to \mathbb{R}$ be a bounded function. Let $P = \{a = a_0 < a_1 < a_2 < \cdots < a_k = b\}$ be a partition of $[a, b]$ for some positive integer k. Let
$m_i = \inf\{f(x) : x \in [a_{i-1}, a_i]\} = \inf f([a_{i-1}, a_i])$ and
$\quad M_i = \sup\{f(x) : x \in [a_{i-1}, a_i]\} = \sup f([a_{i-1}, a_i])$ for each $i = 1, 2, \ldots, k$.
Let $m(f, P) = \sum_{i=1}^{k} m_i \cdot (a_i - a_{i-1})$,
and $M(f, P) = \sum_{i=1}^{k} M_i \cdot (a_i - a_{i-1})$.
(See $m(f, P)$ for Fig. 6.2, $M(f, P)$ for Fig. 6.3, where $P = \{a, a_1, a_2, a_3, a_4, a_5 = b\}$)

Integration: If, for any R-partition sequence $\langle P_n \rangle_{n=1}^{\infty}$ of the interval $[a, b]$, $\lim_{n \to \infty} m(f, P_n)$ and $\lim_{n \to \infty} M(f, P_n)$ exist, and if $\lim_{n \to \infty} m(f, P_n) = \lim_{n \to \infty} M(f, P_n)$, then we say that f is <u>integrable</u> on $[a, b]$, and we denote this limit by $\int_a^b f(x)dx$ and $\int_a^b f(x)dx$ is said to be the <u>integral</u> of f on $[a, b]$. Hence, if we have $\lim_{n \to \infty} m(f, P_n) = \int_a^b f(x)dx = \lim_{n \to \infty} M(f, P_n)$, then the value $\int_a^b f(x)dx$ is said to be the <u>integration</u> of f from a to b.

Fig. 6.2 The sum of the yellow region is $m(f, P)$.

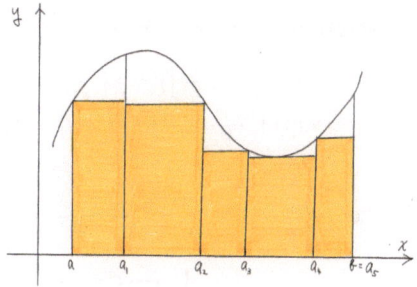

Fig. 6.3 The sum of the yellow region is $M(f, P)$.

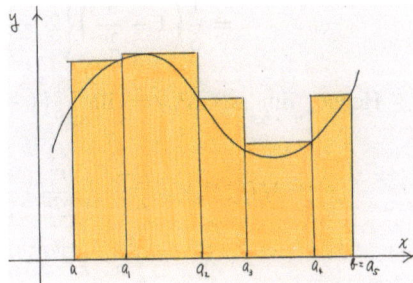

Definition 6.2.4 Riemann Integration: Let $a < b$, and let $f : [a, b] \to \mathbb{R}$ be a bounded function. Let $\langle P_n \rangle_{n=1}^{\infty}$ be a partition-sequence of the interval $[a, b]$. Let $P_n = \{a = a_{n,0} < a_{n,1} < a_{n,2} < \cdots < a_{n,k_n} = b\}$ for some integer $k_n > 0$ for each integer $n > 0$, and let $c_{n,i}$ be an arbitrary element of $[a_{n,i-1}, a_{n,i}]$ for each $i = 1, 2, 3, \ldots, k_n$ for each n.

If $\lim\limits_{n \to \infty} \sum_{i=1}^{k_n} f(c_{n,i}) \cdot (a_{n,i} - a_{n,i-1})$ exists, then f is said to be <u>Riemann integrable</u> on $[a, b]$, and we denote this limit by $\mathcal{R}_a^b f(x) dx$, i.e.,

$$\mathcal{R}_a^b f(x) dx = \lim_{n \to \infty} \sum_{i=1}^{k_n} f(c_{n,i}) \cdot (a_{n,i} - a_{n,i-1}).$$

And $\mathcal{R}_a^b f(x) dx$ is said to be the <u>Riemann integral</u> of f on $[a, b]$ or the <u>Riemann integration</u> of f from a to b.

Remark 6.2.1 Our definition of the integral and Riemann's definition of integral are equivalent. However, we will show that $\int_a^b f(x) dx = \mathcal{R}_a^b f(x) dx$ only when f is continuous in Theorem 6.2.3 since we are interested mostly in continuous functions and since continuity makes this proof easier.

Problem 6.2.2 Let $f : [0, 1] \to \mathbb{R}$ be a function defined by $f(x) = x^2$ for all $x \in [0, 1]$. Let n be a positive integer. Let P_n be the partition of $[0,1]$ given in Example 6.2.1(3). Evaluate $m(f, P_n)$, $M(f, P_n)$ and $\int_0^1 x^2 dx$.

Solution: Let n be a positive integer. Let $P_n = \left\{\frac{k}{2^n}\right\}_{k=0}^{2^n}$. Then

$$m_i = \inf f\left(\left[\frac{i-1}{2^n}, \frac{i}{2^n}\right]\right) = f\left(\frac{i-1}{2^n}\right) = \left(\frac{i-1}{2^n}\right)^2 = \frac{(i-1)^2}{2^{2n}}, \text{ and}$$

$$M_i = \sup f\left(\left[\frac{i-1}{2^n}, \frac{i}{2^n}\right]\right) = f\left(\frac{i}{2^n}\right) = \left(\frac{i}{2^n}\right)^2 = \frac{i^2}{2^{2n}} \text{ for each } i = 1, 2, \ldots, 2^n.$$

Hence, $m(f, P_n) = \sum_{i=1}^{2^n} \frac{(i-1)^2}{2^{2n}} \cdot \frac{1}{2^n} = \frac{1}{2^{3n}} \sum_{i=1}^{2^n} (i-1)^2$

$$= \frac{1}{2^{3n}} \sum_{i=1}^{2^n-1} i^2 = \frac{1}{2^{3n}} \cdot \frac{1}{6}(2^n-1)2^n\{2(2^n-1)+1\} \text{ by Problem 1.4.1(1),}$$

$$= \frac{1}{6}\left(1 - \frac{1}{2^n}\right)\left\{2 - \frac{1}{2^n}\right\}.$$

Hence, $\lim_{n\to\infty} m(f, P_n) = \lim_{n\to\infty} \frac{1}{6}\left(1 - \frac{1}{2^n}\right)\left\{2 - \frac{1}{2^n}\right\} = \frac{2}{6} = \frac{1}{3}$.

$$M(f, P_n) = \sum_{i=1}^{2^n} \frac{i^2}{2^{2n}} \cdot \frac{1}{2^n} = \frac{1}{2^{3n}} \sum_{i=1}^{2^n} i^2$$

$$= \frac{1}{2^{3n}} \cdot \frac{1}{6} 2^n(2^n+1)\{2 \cdot 2^n + 1\} \text{ by Problem 1.4.1(1),}$$

$$= \frac{1}{6}\left(1 + \frac{1}{2^n}\right)\left\{2 + \frac{1}{2^n}\right\}.$$

Hence, $\lim_{n\to\infty} M(f, P_n) = \lim_{n\to\infty} \frac{1}{6}\left(1 + \frac{1}{2^n}\right)\left\{2 + \frac{1}{2^n}\right\} = \frac{2}{6} = \frac{1}{3}$.

Since $\lim_{n\to\infty} m(f, P_n) = \frac{1}{3} = \lim_{n\to\infty} M(f, P_n)$, we have $\int_0^1 x^2 dx = \frac{1}{3}$.

The next Lemma is an application of Problem 2.5.6 from Chap. 2.

Lemma 6.2.1 Let $a < b$ and let $f : [a, b] \to \mathbb{R}$ be a bounded function. Let $P = \{a = a_0 < a_1 = b\}$ and $Q = \{a = a_0 < b_1 < a_1 = b\}$ be partitions of $[a, b]$. (Note that Q is a refinement of P.) Then $m(f, P) \leq m(f, Q) \leq M(f, Q) \leq M(f, P)$.

Proof We will show $m(f, P) \leq m(f, Q)$.

$$m(f, P) = \inf f([a_0, a_1]) \cdot (a_1 - a_0)$$

$$= \inf f([a_0, a_1]) \cdot (b_1 - a_0) + \inf f([a_0, a_1]) \cdot (a_1 - b_1)$$

$$\leq \inf f([a_0, b_1]) \cdot (b_1 - a_0) + \inf f([b_1, a_1]) \cdot (a_1 - b_1) \text{ by Problem 2.5.6,}$$

$$= m(f, Q).$$

This proves that $m(f, P) \leq m(f, Q)$.

Next, $m(f, Q) = \inf f([a_0, b_1]) \cdot (b_1 - a_0) + \inf f([b_1, a_1]) \cdot (a_1 - b_1)$

$$\leq \sup f([a_0, b_1]) \cdot (b_1 - a_0) + \sup f([b_1, a_1]) \cdot (a_1 - b_1)$$
$$= M(f, Q)$$

Next, we will show that $M(f, Q) \leq M(f, P)$.

$$M(f, Q) = \sup f([a_0, b_1]) \cdot (b_1 - a_0) + \sup f([b_1, a_1]) \cdot (a_1 - b_1)$$
$$\leq \sup f([a_0, a_1]) \cdot (b_1 - a_0) + \sup f([a_0, a_1]) \cdot (a_1 - b_1) \text{ by Problem 2.5.6}$$
$$= \sup f([a_0, a_1]) \cdot (a_1 - a_0)$$
$$= M(f, P).$$

This proves this lemma.

The next lemma is a generalization of Lemma 6.2.2.

Lemma 6.2.3 Let $a < b$, and $k > n > 0$ are integers. Let $f : [a, b] \to \mathbb{R}$ be a bounded function. Let $P = \{a = a_0 < a_1 < a_2 < \cdots < a_n = b\}$ and $Q = \{a = a_0 < b_1 < b_2 < \cdots < b_k = b\}$ be partitions of $[a, b]$ such that $P \subset Q$. Then

$$m(f, P) \leq m(f, Q) \leq M(f, Q) \leq M(f, P).$$

Proof is omitted.

Lemma 6.2.4 Let $a < b$ and let $f : [a, b] \to \mathbb{R}$ be a bounded function. Then, for any R-partition sequence $\langle P_n \rangle_{n=1}^{\infty}$ of the interval $[a, b]$, $\lim_{n \to \infty} m(f, P_n)$ and $\lim_{l \to \infty} M(f, P_l)$ exist. Moreover, $\lim_{n \to \infty} m(f, P_n) \leq \lim_{l \to \infty} M(f, P_l)$

Proof We have $m(f, P_n) \leq M(f, P_l)$ for any positive integers n and l by Lemma 6.2.3. So the sequence $\langle m(f, P_n) \rangle_{n=1}^{\infty}$ is increasing and bounded above by $M(f, P_l)$ for any positive integer l, $\lim_{n \to \infty} m(f, P_n)$ exists by Theorem 3.2.1, and we have $\lim_{n \to \infty} m(f, P_n) \leq M(f, P_l)$ for any positive integer l. Hence, the sequence $\langle M(f, P_l) \rangle_{l=1}^{\infty}$ is decreasing by Lemma 6.2.3, and it is bounded below by $\lim_{n \to \infty} m(f, P_n)$. Therefore, $\lim_{l \to \infty} M(f, P_l)$ exists by Theorem 3.2.1 and $\lim_{n \to \infty} m(f, P_n) \leq \lim_{l \to \infty} M(f, P_l)$.

Problem 6.2.3

(1) Prove that $\int_a^b 0 dx = 0$.
(2) Prove that $\int_a^b x dx = \frac{1}{2}(b - a)$.

The key to prove the integrability of a continuous function on a closed interval is to realize that a continuous function defined on a finite closed interval is uniformly continuous, which is a new concept defined below, and to apply Lemma 6.2.4.

Definition 6.2.5 Let X be an interval. A function $f : X \to \mathbb{R}$ is <u>uniformly continuous</u> if for every $\varepsilon > 0$, there exist a $\delta > 0$ such that
   $$\sup f\left([p, q]\right) - \inf f\left([p, q]\right) < \varepsilon \text{ for every } p < q \in X \text{ such that } q - p < \delta.$$
   (Since X is an interval of real numbers, the interval $[p, q]$ in this definition is a subset of X.)

Remark 6.2.2 Let X be an interval. Suppose a function $f : X \to \mathbb{R}$ is uniformly continuous. We will prove that f is continuous in the next section.

Theorem 6.2.1 Let $a < b$. Suppose $f : [a, b] \to \mathbb{R}$ is a continuous function. Then f is uniformly continuous. That is, a continuous function defined on a **finite closed interval is uniformly continuous.**

Proof Suppose, on the contrary, f is not uniformly continuous. Then there exists an $\varepsilon > 0$ such that for every $\delta > 0$ such that there exists an interval $[p, q] \subset [a, b]$ such that $q - p < \delta$ and

$$\sup f\left([p, q]\right) - \inf f\left([p, q]\right) \geq \varepsilon.$$

Hence, for every integer $n > 0$, there exists an interval $[p_n, q_n] \subset [a, b]$ such that $q_n - p_n < \frac{1}{n}$ and $\sup f\left([p_n, q_n]\right) - \inf f\left([p_n, q_n]\right) \geq \varepsilon$.
   By the extreme value theorem, there are $s_n, t_n \in [p_n, q_n]$ such that
   $f(s_n) = \max f\left([p_n, q_n]\right) = \sup f\left([p_n, q_n]\right)$ and
   $f(t_n) = \min f\left([p_n, q_n]\right) = \inf f\left([p_n, q_n]\right)$ for each n.
   Now, we have sequences $\langle s_n \rangle$ and $\langle t_n \rangle$ in the closed interval $[a, b]$. By the Bolzano-Weierstrass theorem, $\langle s_n \rangle$ has a convergent subsequence $\langle s_{n_k} \rangle$. Again, by the same theorem, $\langle t_{n_k} \rangle$ has a convergent subsequence $\langle t_{n_{k_i}} \rangle$.
   (Note that $\langle t_{n_k} \rangle$ is a subsequence of $\langle t_n \rangle$ but we do not know its convergence. So we are taking a convergent subsequence $\langle t_{n_{k_i}} \rangle$ of $\langle t_{n_k} \rangle$.)
   Since $\langle s_{n_k} \rangle$ is convergent, its subsequence $\langle s_{n_{k_i}} \rangle$ is also convergent. Let $\lim\limits_{i \to \infty} s_{n_{k_i}} = s$ and $\lim\limits_{i \to \infty} t_{n_{k_i}} = t$. Since $q_{n_{k_i}} - p_{n_{k_i}} < \frac{1}{n_{k_i}} \leq \frac{1}{i}$ and $s_{n_{k_i}}, t_{n_{k_i}} \in \left[p_{n_{k_i}}, q_{n_{k_i}}\right]$ for every $i > 0$, we must have $\lim\limits_{i \to \infty} \left| s_{n_{k_i}} - t_{n_{k_i}} \right| \leq \lim\limits_{i \to \infty} \left| q_{n_{k_i}} - p_{n_{k_i}} \right| \leq \lim\limits_{i \to \infty} \frac{1}{i} = 0$. Hence, $s = t$. Since the function f is continuous, we have

$$\lim_{i \to \infty} f\left(s_{n_{k_i}}\right) = f(s) = f(t) = \lim_{i \to \infty} f\left(t_{n_{k_i}}\right).$$

On the other hand, we have
$$f\left(s_{n_{k_i}}\right) = \sup f\left(\left[p_{n_{k_i}}, q_{n_{k_i}}\right]\right) \text{ and } f\left(t_{n_{k_i}}\right) = \inf f\left(\left[p_{n_{k_i}}, q_{n_{k_i}}\right]\right) \text{ so that}$$
$f\left(s_{n_{k_i}}\right) - f\left(t_{n_{k_i}}\right) \geq \varepsilon$ for each i. This implies that

$$\lim_{i \to \infty} f\left(s_{n_{k_i}}\right) = f(s) \neq f(t) = \lim_{i \to \infty} f\left(t_{n_{k_i}}\right).$$

This is a contradiction to $f(s) = f(t)$. Therefore, f is uniformly continuous.

Now, we are ready to prove the integrability of a continuous function on a finite closed interval.

Theorem 6.2.2 Let $a < b$. If $f : [a, b] \to \mathbb{R}$ is a continuous function, then $\int_a^b f(x)dx$ exists. In other words, a continuous function on a finite closed interval is integrable.

Proof Let $\langle P_n \rangle_{n=1}^{\infty}$ be a R-partition sequence of the interval $[a, b]$. By Lemma 6.2.4, since a continuous function on a closed interval is bounded by the extreme value theorem, $\lim_{n \to \infty} m(f, P_n)$ and $\lim_{n \to \infty} M(f, P_n)$ exist. So, we have to show that $\lim_{n \to \infty} m(f, P_n) = \lim_{n \to \infty} M(f, P_n)$.

Let $\varepsilon > 0$. By Theorem 6.2.1, there exists a $\delta > 0$ such that such that if $[p, q] \subset [a, b]$ and $q - p < \delta$, then

$$\sup f\left([p, q]\right) - \inf f\left([p, q]\right) < \frac{\varepsilon}{b - a}.$$

Since $\lim_{n \to \infty} \|P_n\| = 0$, there exists an integer N such that $\|P_n\| < \delta$ for every integer $n \geq N$. Let $n \geq N$, and let $P_n = \{a = a_0 < a_1 < a_2 < \cdots < a_k = b\}$ for some k. Let $m_i = \inf f\left([a_{i-1}, a_i]\right)$ and $M_i = \sup f\left([a_{i-1}, a_i]\right)$ for each $i = 1, 2, \ldots, k$. Then

$$M_i - m_i = \sup f\left([a_{i-1}, a_i]\right) - \inf f\left([a_{i-1}, a_i]\right) < \frac{\varepsilon}{b - a}.$$

Since $\|P_n\| < \delta$ implies that $a_i - a_{i-1} < \delta$ for each $i = 1, 2, \ldots, k$.
Hence, we have

$$M(f, P_n) - m(f, P_n) = \sum_{i=1}^{n} M_i \cdot (a_i - a_{i-1}) - \sum_{i=1}^{n} m_i \cdot (a_i - a_{i-1})$$
$$= \sum_{i=1}^{n} (M_i - m_i) \cdot (a_i - a_{i-1})$$
$$< \sum_{i=1}^{n} \frac{\varepsilon}{b - a} \cdot (a_i - a_{i-1}) = \frac{\varepsilon}{b - a} \cdot \sum_{i=1}^{n} (a_i - a_{i-1}) = \frac{\varepsilon}{b - a} \cdot (b - a) = \varepsilon.$$

So $0 \leq M(f, P_n) - m(f, P_n) < \varepsilon$. This proves that $\lim_{n \to \infty} [M(f, P_n) - m(f, P_n)] = 0$. Since f is a bounded function, $\lim_{n \to \infty} m(f, P_n)$ and $\lim_{n \to \infty} M(f, P_n)$ exists by Lemma

6.2.4. Hence, $\lim_{n\to\infty} M(f, P_n) - \lim_{n\to\infty} m(f, P_n) = \lim_{n\to\infty} [M(f, P_n) - m(f, P_n)] = 0$ by Theorem 3.1.4 so that $\lim_{n\to\infty} m(f, P_n) = \lim_{n\to\infty} M(f, P_n)$.

Therefore, this proves Theorem 6.2.2.

Theorem 6.2.3 Let $a < b$. Let $f : [a, b] \to \mathbb{R}$ be a continuous function. Then our definition of integral and Riemann integral of f on $[a, b]$ are equivalent. That is, $\{$if $\int_a^b f(x)dx$ exists, then $\int_a^b f(x)dx = \mathcal{R}_a^b f(x)dx\}$, and $\{$if $\mathcal{R}_a^b f(x)dx$ exists, then $\mathcal{R}_a^b f(x)dx = \int_a^b f(x)dx\}$.

Proof Suppose $\int_a^b f(x)dx$ exists. We will show that $\int_a^b f(x)dx = \mathcal{R}_a^b f(x)dx$. Let $\langle P_n \rangle_{n=1}^{\infty}$ be a partition sequence of $[a, b]$. Let $Q_n = \bigcup_{i=1}^n P_i$ for each integer $n > 0$. Then $\langle Q_n \rangle_{n=1}^{\infty}$ is an R-partition sequence. Hence, $\lim_{n\to\infty} m(f, Q_n)$ and $\lim_{n\to\infty} M(f, Q_n)$ exist, and $\lim_{n\to\infty} m(f, Q_n) = \int_a^b f(x)dx = \lim_{n\to\infty} M(f, Q_n)$ by Theorem 6.2.2.

Let $n > 0$ be an integer. By Lemma 6.2.3, we have

$$m(f, Q_n) \leq m(f, P_n) \leq M(f, P_n) \leq M(f, Q_n).$$

Let $P_n = \{a = a_{n,0} < a_{n,1} < a_{n,2} < \cdots < a_{n,k_n} = b\}$ for some integer $k_n > 0$, and we let $c_{n,i}$ be an arbitrary element of $[a_{n,i-1}, a_{n,i}]$ for each $i = 1, 2, 3, \ldots, k_n$.

Then $m(f, Q_n) \leq m(f, P_n) \leq \sum_{i=1}^{k_n} f(c_{n,i}) \cdot (a_{n,i} - a_{n,i-1}) \leq M(f, P_n) \leq M(f, Q_n)$.

Since $\lim_{n\to\infty} m(f, Q_n) = \int_a^b f(x)dx = \lim_{n\to\infty} M(f, Q_n)$, we have $\int_a^b f(x)dx = \lim_{n\to\infty} \sum_{i=1}^n f(c_i) \cdot (a_i - a_{i-1}) = \mathcal{R}_a^b f(x)dx$ by the sandwich theorem.

Conversely, suppose $\mathcal{R}_a^b f(x)dx$ exists. Let $\langle P_n \rangle_{n=1}^{\infty}$ is an R-partition sequence of $[a, b]$. Then $\langle P_n \rangle_{n=1}^{\infty}$ is a partition sequence of $[a, b]$. Let $n > 0$ be an integer. Let $P_n = \{a = a_0 < a_1 < a_2 < \cdots < a_k = b\}$ for some k.

Since f is continuous, there are $c_i \in [a_{i-1}, a_i]$ and $\widehat{c}_i \in [a_{i-1}, a_i]$ for each $i = 1, 2, 3, \ldots, k$ such that

$$f(c_i) = \min\{f(x) : x \in [a_{i-1}, a_i]\} = \inf\{f(x) : x \in [a_{i-1}, a_i]\}, \text{ and}$$

$$f(\widehat{c}_i) = \max\{f(x) : x \in [a_{i-1}, a_i]\} = \sup\{f(x) : x \in [a_{i-1}, a_i]\}.$$

Since f is Riemann integrable, we have

$$\lim_{n\to\infty} m(f, P_n) = \lim_{n\to\infty} \sum_{i=1}^n f(c_i) \cdot (a_i - a_{i-1}) = \mathcal{R}_a^b f(x)dx$$

$$= \lim_{n\to\infty} \sum_{i=1}^n f(\widehat{c}_i) \cdot (a_i - a_{i-1}) = \lim_{n\to\infty} M(f, P_n).$$

Therefore, $\mathcal{R}_a^b f(x)dx = \int_a^b f(x)dx$.

This proves this theorem.

Problem 6.2.4

(1) Let $f : [0, 2] \to \mathbb{R}$ be a function defined by $f(x) = 1$ if $x = 1$ and $f(x) = 2$ for every $x \in [0, 2]$ such that $x \neq 1$. First, show that $\lim_{x \to 1} f(x)$ does not exist. (Note that $\lim_{x \to 1} f(x) \neq 2$.) Hence, f is not continuous. Second, show that it is integrable by showing $\int_0^2 f(x)dx = 4$.

This problem shows that the converse of Theorem 6.2.2 is not true.

(2) Let $g : [0, 2] \to \mathbb{R}$ be a function defined by $g(x) = 1$ for every rational number $x \in [0, 2]$ and $g(x) = 2$ for every irrational number $x \in [0, 2]$. Prove that g is not continuous and not integrable.

Remark 6.2.3

(1) Theorem 6.2.3 can be improved to $\int_a^b f(x)dx = \mathcal{R}_a^b f(x)dx$ for any integrable function f on $[a, b]$.
(2) As you can see from Problem 6.2.4, **some discontinuous functions are integrable**, while some discontinuous functions are not integrable.

Question: Is there a nice characterization of an integrable function?

An answer to this question is difficult. One answer is in the study of Lebesgue measure and Lebesgue integrations that some of you may study in the future.

Definition 6.2.6 Let X be a subset of $\mathbb{R}$. Let $f : X \to \mathbb{R}$ be a function. Let $a < b$. If $[a, b] \subset X$, and $f|_{[a,b]}$ is integrable on $[a, b]$, we simply say that f is <u>integrable</u> on $[a, b]$, and we write $\int_a^b f(x)dx$ in place of $\int_a^b f|_{[a,b]}(x)dx$. We define $\int_b^a f(x)dx$ to be $-\int_a^b f(x)dx$, i.e., $\int_b^a f(x)dx = -\int_a^b f(x)dx$.

Proposition 6.2.1 Suppose $a < b < c < d$. Let $f : [a, d] \to \mathbb{R}$ be an integrable function. Then f is integrable on $[b, c]$.

Proof of this proposition is omitted.

Proposition 6.2.2 Let $a < c$. Let $f, g : [a, c] \to \mathbb{R}$ be integrable functions. Let $b, p, q \in [a, c]$ and k be any constant. Then the following are true.

(1) f is integrable on $[a, b]$ and on $[b, c]$, and $\int_a^c f(x)dx = \int_a^b f(x)dx + \int_b^c f(x)dx$.
(2) The function f is integrable on $[p, q]$ if $p < q$.
(3) $\int_p^q f(x)dx = \int_p^b f(x)dx + \int_b^q f(x)dx$. (Here, we are not assuming $p < q$.)
(4) $\int_p^q (f + g)(x)dx = \int_p^q f(x)dx + \int_p^q g(x)dx$.

(5) $\int_p^q k \cdot f(x)dx = k \cdot \int_p^b f(x)dx$.

Proof of this proposition is omitted.

Theorem 6.2.4 Let $a < c$. Let $f : [a, c] \to \mathbb{R}$ be integrable function such that $f(x) \geq 0$ for every $x \in [a, c]$. Then $\int_a^c f(x)dx \geq 0$.

Proof Let P be a partition of the interval $[a, c]$. Since f is integrable, we have that $\int_a^c f(x)dx \geq m(f, P) \geq m(0, P) = 0$.

Theorem 6.2.5 Let $a < c$. Let $f, g : [a, c] \to \mathbb{R}$ be integrable functions such that $f(x) \leq g(x)$ for every $x \in [a, c]$. Then $\int_a^c f(x)dx \leq \int_a^c g(x)dx$.

Proof This is a corollary of Theorem 6.2.4 and Proposition 6.2.2(4). The detail is left as an exercise.

Theorem 6.2.6 Let $a < c$. Let $f : [a, c] \to \mathbb{R}$ be a continuous function. Then $g : [a, c] \to \mathbb{R}$ defined by $g(x) = |f(x)|$ for every $x \in [a, c]$ is also integrable, and $\left| \int_a^c f(x)dx \right| \leq \int_a^c g(x)dx$. We usually write this as $\left| \int_a^c f(x)dx \right| \leq \int_a^c |f(x)|dx$.

Proof Let $F : \mathbb{R} \to \mathbb{R}$ be a function defined by $F(x) = |x|$ for every $x \in \mathbb{R}$. Then the function g is a composition of two continuous function f and F, i.e., $g = F \circ f$. Hence, g is continuous. This proves that g is integrable. Since $g(x) \geq f(x)$ for every $x \in [a, c]$, the rest follows from Theorem 6.2.6.

Theorem 6.2.7 (The **Fundamental Theorem of Calculus** by Newton and Leibniz) Let $a < c$. Let $f : [a, c] \to \mathbb{R}$ be a continuous function on $[a, c]$.

(1) Let $p \in [a, c]$. Let $F : [a, c] \to \mathbb{R}$ be a function defined by $F(t) = \int_p^t f(x)dx$ for every $t \in [a, c]$. Then $\frac{d}{dt}F(t) = f(t)$ for every $t \in [a, c]$, i.e., F is an anti-derivative of f.
(2) The function F is continuous.
(3) Let $G : [a, c] \to \mathbb{R}$ be an anti-derivative of f, i.e., $\frac{d}{dt}G(t) = f(t)$ for every $t \in [a, c]$. Then $\int_a^c f(x)dx = G(c) - G(a)$. We write $G(c) - G(a)$ by $G(t)|_a^c$.

Proof of (1): Let $t \in [a, c]$ and let $h \neq 0$ such that $t + h \in [a, c]$. Then

$$F(t+h) - F(t) = \int_p^{t+h} f(x)dx - \int_p^t f(x)dx = \int_t^p f(x)dx + \int_p^{t+h} f(x)dx$$

$$= \int_t^{t+h} f(x)dx.$$

Let $U(t, h) = \sup\{f(x) : x \text{ is between } t \text{ and } t + h\}$ and

$$L(t, h) = \inf\{f(x) : x \text{ is between } t \text{ and } t + h\}.$$

These numbers $U(t, h)$ and $L(t, h)$ exists by the Extreme Value theorem.
Since $L(t, h) \le U(t, h)$, we have
$L(t, h) \cdot h \le \int_t^{t+h} f(x)dx \le U(t, h) \cdot h$ if $h > 0$, and
$U(t, h) \cdot h \le \int_t^{t+h} f(x)dx \le L(t, h) \cdot h$ if $h < 0$.
Regardless of whether $h > 0$ or $h < 0$, we have

$$L(t, h) \le \frac{F(t + h) - F(t)}{h} \le U(t, h).$$

But $L(t, h) \le f(t) \le U(t, h)$ so that $\lim_{h \to 0} L(t, h) = f(t) = \lim_{h \to 0} U(t, h)$. This proves that
$\lim_{h \to 0} \frac{F(t+h) - F(t)}{h} = f(t)$. Therefore, we have shown that $\frac{d}{dt} F(t) = f(t)$.

Proof of (2): Since F is differentiable, it is continuous by Theorem 5.1.1.

Proof of (3): Let $p = a$ in (1). Then $F(t) = \int_a^t f(x)dx$ and $\frac{d}{dt} F(t) = f(t)$ for every $t \in [a, c]$.
In particular, we have $\int_a^c f(x)dx = F(c)$. By Theorem 5.2.7, $F(t) - G(t) = C$ for some
constant C for every $t \in [a, c]$. So $0 = F(a) = G(a) + C$ or $C = -G(a)$. And $F(c) - G(c) = C = -G(a)$ or $F(c) = G(c) - G(a)$. Therefore, $\int_a^c f(x)dx = F(c) = G(c) - G(a)$.

6.3   Uniform Continuity

For us, the only purpose of introducing uniform continuity is to prove Theorem 6.2.2. However, because uniform continuity of a function plays an important role in other branches of mathematics, we will give some additional discussion on uniform continuity in this section.

We used Definition 6.2.5 for uniform continuity because it was easier to apply to the proof of Theorem 6.2.2. However, following is an equivalent form of uniform continuity.

Theorem 6.3.1 Let X be an interval. A function $f : X \to \mathbb{R}$ is uniformly continuous if and only if for every $\varepsilon > 0$, there exists a $\delta > 0$ such that $|f(p) - f(q)| < \varepsilon$ **for every** $p, q \in X$ and $|p - q| < \delta$.

Proof Suppose function $f : X \to \mathbb{R}$ is uniformly continuous. Let $\varepsilon > 0$. Then there exist a $\delta > 0$ such that if $p < q \in X$ and $q - p < \delta$, then $\sup f([p, q]) - \inf f([p, q]) < \varepsilon$. But $|f(p) - f(q)| \le \sup f([p, q]) - \inf f([p, q])$. Therefore, for every $\varepsilon > 0$, there exists a $\delta > 0$ such that $|f(p) - f(q)| < \varepsilon$ for every $p, q \in X$ and $|p - q| < \delta$.

Conversely, suppose, for every $\varepsilon > 0$, there exists a $\delta > 0$ such that $|f(p) - f(q)| < \varepsilon$ for every $p, q \in X$ and $|p - q| < \delta$. Suppose f is not uniformly continuous. Then there is an $\varepsilon > 0$ such that, for every $\delta > 0$, there are $p < q \in X$ such that $q - p < \delta$ and

$\sup f ([p, q]) - \inf f ([p, q]) \geq 3\varepsilon$. By the definition of sup and inf, there are $s, t \in [p, q]$ such that $f(s) > \sup f ([p, q]) - \varepsilon$ and $f(t) < \inf f ([p, q]) + \varepsilon$.

Hence, $-f(t) > -\inf f ([p, q]) - \varepsilon$ so that

$$f(s) - f(t) > \sup f ([p.q]) - \inf f ([p, q]) - 2\varepsilon.$$

But $\sup f ([p, q]) - \inf f ([p, q]) \geq 3\varepsilon$.

Thus, $f(s) - f(t) > \sup f ([p.q]) - \inf f ([p, q]) - 2\varepsilon \geq 3\varepsilon - 2\varepsilon = \varepsilon$

Since $s, t \in [p, q]$, we have $|s - t| \leq q - p < \delta$. But we have $f(s) - f(t) > \varepsilon$. This is a contradiction to our assumption. Therefore, f is uniformly continuous.

Remark 6.3.1 Some authors use Theorem 6.3.1 as the definition of uniform continuity.

Theorem 6.3.2 Let X be an interval. Suppose a function $f : X \rightarrow \mathbb{R}$ is uniformly continuous. Then the function f is continuous.

Proof Let $\varepsilon > 0$. Since f is uniformly continuous, there is a $\delta > 0$ such that if $p, q \in X$ and $|p - q| < \delta$, then $|f(p) - f(q)| < \varepsilon$ by Theorem 6.3.1.

Let $p \in X$. Let $x \in X$ such that $|p - x| < \delta$. then $|f(p) - f(x)| < \varepsilon$. Therefore, the function f is continuous at p. That is, f is continuous everywhere in X.

Uniform continuity is important mainly because of Theorem 6.2.1, which says that a continuous function defined on a **finite closed interval** is uniformly continuous. So this gives us the following **question**: Is there a continuous function defined on a finite interval that is not uniformly continuous? In order to answer this question, Theorem 6.3.1 is useful as in the next examples.

Example 6.3.2 Let X be an interval. From Theorem 6.3.1, we have the following:

A function $f : X \rightarrow \mathbb{R}$ is <u>not uniformly continuous</u> if there exists an $\varepsilon > 0$ such that, for every $\delta > 0$, there are $p, q \in X$ such that $|p - q| < \delta$ and $|f(p) - f(q)| \geq \varepsilon$.

Example 6.3.2 Let $f : (0, 2) \rightarrow \mathbb{R}$ be a function defined by $f(x) = \frac{1}{x}$ for every $x \in (0, 2)$. We know that f is continuous. However, f is not uniformly continuous.

In order to see this, let $\delta > 0$. Let n be a positive integer such that $\frac{1}{n} < \delta$. Then $\frac{1}{n+1}, \frac{1}{n} \in (0, 1)$ and $\left| \frac{1}{n+1} - \frac{1}{n} \right| < \frac{1}{n} < \delta$ yet $f \left(\frac{1}{n+1} \right) - f \left(\frac{1}{n} \right) = (n+1) - n = 1$. Therefore, the function f is not uniformly continuous.

Problem 6.3.1 Let $f : (1, \infty) \rightarrow \mathbb{R}$ be a function defined by $f(x) = \frac{1}{x}$ for every $x \in (1, \infty)$. Prove that f is uniformly continuous.

Problem 6.3.2 Some functions can be uniformly continuous even though their domain is not finite closed interval. Let $f : (1, \infty) \to \mathbb{R}$ be a function defined by $f(x) = 2x$ for every $x \in (1, \infty)$. Prove that f is uniformly continuous.

Example 6.3.3 Some bounded functions can be non-uniformly continuous. Let $f : (0, 1) \to \mathbb{R}$ is a function defined by $f(x) = \cos \frac{1}{x}$ for every $x \in (0, 1)$. Then f is not uniformly continuous. Let $\delta > 0$. Then there is an integer $n > 0$ such that

$0 < \frac{1}{(2n)\pi} - \frac{1}{(2n+1)\pi} < \delta$. Then $0 < \frac{1}{(2n)\pi}, \frac{1}{(2n+1)\pi} < 1$ and

$\cos \frac{1}{\frac{1}{(2n)\pi}} - \cos \frac{1}{\frac{1}{(2n+1)\pi}} = \cos(2n)\pi - \cos(2n+1)\pi = 1 - (-1) = 2 > 0$. Therefore, f is not uniformly continuous.

6.4 A Brief Review of the Techniques of Integrations

Example 6.4.1

(1) $\int_{-2}^{-1} \frac{1}{x} dx = [\ln|x|]_{-2}^{-1} = -\ln 2.$
(2) $\int_{1}^{\pi} \sin x dx = [-\cos x]_{1}^{\pi} = 1 + \cos 1.$
(3) $\int_{1}^{\pi} \cos x dx = [\sin x]_{1}^{\pi} = -\sin 1.$
(4) $\int_{1}^{2} \sin x \cos x dx.$

Let $t = \sin x$. Then $\frac{dt}{dx} = \cos x$ so that $dt = \cos x dx$. When $x = 1$, we have $t = \sin 1$. When $x = 2$, we have $t = \sin 2$. Hence,

$$\int_{1}^{2} \sin x \cos x dx = \int_{\sin 1}^{\sin 2} t dt = \left[\frac{1}{2}t^2\right]_{\sin 1}^{\sin 2} = \frac{1}{2}[\sin^2 2 - \sin^2 1].$$

When the substitution method is used to integrate, **please do not forget to change the limits of integrations**.

Alternate solution.

$$\int_{1}^{2} \sin x \cos x dx = \int_{1}^{2} \frac{1}{2} \sin 2x dx = -\frac{1}{4}[\cos 2x]_{1}^{2} = -\frac{1}{4}(\cos 4 - \cos 2).$$

Definition 6.4.1 Let $a < b$, and let $f : (a, b] \to \mathbb{R}$ and $g : [a, b) \to \mathbb{R}$ be functions. Then we define $\int_{a}^{b} f(x)dx = \lim_{t \to a+} \int_{t}^{b} f(x)dx$ and $\int_{a}^{b} g(x)dx = \lim_{t \to b-} \int_{a}^{t} g(x)dx.$

Let $F : (-\infty, b] \to \mathbb{R}$ and $G : [a, \infty) \to \mathbb{R}$ be functions. Then we define $\int_{-\infty}^{b} f(x)dx =$

$\lim_{t \to -\infty} \int_{t}^{b} F(x)dx$ and $\int_{a}^{\infty} G(x)dx = \lim_{t \to \infty} \int_{a}^{t} G(x)dx.$

Problem 6.4.1 Evaluate the following integrals.

(1) $\int_{-1}^{2} 3x\sin(x^2)dx$ by substitution.

(2) $\int_{-\frac{\pi}{3}}^{\frac{\pi}{3}} \cos^2 x dx$.

(3) $\int_{-\frac{\pi}{6}}^{\frac{\pi}{3}} \sin^2 x dx$.

(4) $\int_{-1}^{2} e^{2x}dx$.

(5) $\int_{-1}^{2} 2^{-2x}dx$

(6) $\int_{-10}^{0} \frac{1}{2-3x}dx$.

(7) $\int_{0}^{1} \frac{1}{\sqrt{1-x^2}}dx$ by letting $x = \sin\theta$.

(8) $\int_{0}^{1} \frac{1}{\sqrt{1+x^2}}dx$ by letting $x = \tan\theta$.

(9) $\int_{2}^{6} \frac{1}{\sqrt{x^2-1}}dx$ by letting $x = \sec\theta$.

(Notice the trigonometric substitutions in (7)–(9) are based on the relations

$$1 - x^2 \Leftrightarrow 1 - \sin^2\theta, \qquad 1 + x^2 \Leftrightarrow 1 + \tan^2\theta, \qquad \text{and } x^2 - 1 \Leftrightarrow \sec^2\theta - 1.)$$

(10) $\int_{0}^{2} \frac{1}{\sqrt{4-x^2}}dx$.

(11) $\int_{0}^{2} \frac{1}{\sqrt{4+x^2}}dx$.

(12) $\int_{3}^{5} \frac{1}{\sqrt{x^2-4}}dx$.

(13) $\int_{1}^{2} \ln x dx$.

(14) $\int_{0}^{1} \sec x dx$.

(15) $\int_{0}^{1} \sec^2 x dx$.

(16) $\int_{0}^{1} \sec^3 x dx$.

Look at any calculus book, and review integrations.

Example 6.4.2 We will prove that the area $\mathcal{A}$ of a circle of radius $r > 0$ is πr^2. (Compare this to Theorem 4.2.9.)

The equation $x^2 + y^2 = r^2$ is the equation of the circle centered at $(0, 0)$ with radius $r > 0$. Then $\mathcal{A} = 4\int_{0}^{r} \sqrt{r^2 - x^2}dx$. Let $x = r\sin\theta$. The $dx = r\cos\theta d\theta$. When $x = 0$, $r\sin\theta = 0$ implies that $\theta = 0$. When $x = r$, $r\sin\theta = r$ implies that $\theta = \frac{\pi}{2}$. Hence, we have $\mathcal{A} = 4\int_{0}^{r} \sqrt{r^2 - x^2}dx = 4\int_{0}^{\frac{\pi}{2}} \sqrt{r^2 - r^2\sin^2\theta} \cdot r \cdot \cos\theta d\theta = 4r^2 \int_{0}^{\frac{\pi}{2}} \cos^2\theta d\theta$

$$= 4r^2 \int_{0}^{\frac{\pi}{2}} \frac{1}{2}(1 + \cos 2\theta)d\theta = 2r^2\left[\theta + \frac{1}{2}\sin 2\theta\right]_{0}^{\frac{\pi}{2}} = \pi r^2.$$

(Note that the antiderivative of $\cos 2\theta$ being $\frac{1}{2}\sin 2\theta$ is obtained from $\frac{d}{dx}\sin x = \cos x$. As we mentioned in Remark 4.2.6, since used the limit $\lim_{x\to 0} \frac{\sin x}{x} = 1$ to prove $\frac{d}{dx}\sin x = \cos x$, and therefore, we used $\int \cos x \, dx = \sin x + C$ to derive the area of a circle, using area of a circle to derive $\lim_{x \to 0} \frac{\sin x}{x} = 1$ is not a derivation at all.)

Problem 6.4.2 Prove that the volume of a sphere of radius $r > 0$ is $\frac{4\pi r^3}{3}$ by using the technique of the solid of revolution, or by double integrals, or by triple integrals.

Problem 6.4.3 Prove that the surface area of a sphere of radius $r > 0$ is $4\pi r^2$ by technique using the solid of revolution, or by double integrals.

Problem 6.4.4 Use mathematical induction to prove: $\int_0^1 (\ln x)^n dx = (-1)^n \cdot n!$ for all integers $n \geq 1$. Here, the meaning of $\int_0^1 (\ln x)^n dx$ is $\int_0^1 (\ln x)^n dx = \lim_{a \to 0, a > 0} \int_a^1 (\ln x)^n dx$. (Use Problem 5.2.3).

References

1. George Polya; *Mathematical Methods in Science*, The Mathematical Association of America (1977)
2. Euclid, *The Elements*, at about 295 B.C.
3. Apollonius of Perga; *Conics: Books I–IV*, translated and reproduced by Green Lion Press (2013)
4. Newton; *Principia*, Motte's Translation Revised, University of California Press, Berkeley, Los Angeles, London (1934)

Sequences and Series of Functions

7

7.1 Introduction of Sequences of Functions

In Chap. 3, we talked about sequences of numbers. In this chapter, we are to study sequences of functions. Let X be a nonempty set of real numbers. Instead of having one function, suppose we have a collection of functions $f_n : X \to \mathbb{R}$ for all positive integers n. Let $x \in X$ be fixed. Then since $\langle f_n(x) \rangle_{n=1}^{\infty}$ is a sequence of numbers, we can determine the convergence of this sequences of numbers. If it does not converge, we are not interested. So, we will be dealing here with convergent sequences $\langle f_n(x) \rangle_{n=1}^{\infty}$ for each $x \in X$. This is the motivation of the following definition:

Definition 7.1.1 Let X be a nonempty subset of $\mathbb{R}$ **throughout this chapter**. Let $f_n : X \to \mathbb{R}$ be a function for each positive integer n. If the sequence $\langle f_n(x) \rangle_{n=1}^{\infty}$ converges for each $x \in X$, then the sequence $\langle f_n \rangle_{n=1}^{\infty}$ is said to <u>converge pointwise</u>. Suppose that the sequence $\langle f_n \rangle_{n=1}^{\infty}$ converges pointwise. Then a function $f : X \to \mathbb{R}$ can be defined by $f(x) = \lim_{n \to \infty} f_n(x)$ for each $x \in X$. We say that the function f is the <u>pointwise limit</u> of the sequence $\langle f_n \rangle_{n=1}^{\infty}$.

Suppose $\langle f_n \rangle_{n=1}^{\infty}$ converges pointwise to a function f. Our main purposes in this chapter are related to the following questions:

Question 1: If each function f_n is continuous, when is f continuous?

Question 2: If each function f_n is differentiable, when is f differentiable?

Question 3: If each function f_n is integrable, when is f integrable?

© The Author(s), under exclusive license to Springer Nature Switzerland AG 2025
H. Katsuura, *Introduction to Analysis*, Synthesis Lectures on Mathematics & Statistics,
https://doi.org/10.1007/978-3-031-67954-4_7

Example 7.1.1 For each positive integer n, let $f_n : [0, 1] \to \mathbb{R}$ be a function defined by $f_n(x) = \frac{x}{n}$ for every $x \in [0, 1]$. Let $f : [0, 1] \to \mathbb{R}$ be defined by $f(x) = 0$ for every $x \in [0, 1]$. Then f is the pointwise limit of $\langle f_n \rangle_{n=1}^{\infty}$ since $\lim_{n \to \infty} \frac{x}{n} = 0$ for every $x \in [0, 1]$. Note that f_n and f are continuous.

Example 7.1.2 For each positive integer n, let $f_n : [0, 1] \to \mathbb{R}$ be a function defined by $f_n(x) = x^n$ for every $x \in [0, 1]$. Let $f : [0, 1] \to \mathbb{R}$ be defined by $f(x) = 0$ for every $x \in [0, 1)$ and $f(1) = 1$. Then f is the pointwise limit of $\langle f_n \rangle_{n=1}^{\infty}$ (see Theorem 3.2.3). Note that f_n is continuous for each n, while f is not continuous at 1.

Example 7.1.3 For each positive integer n, let $f_n : \mathbb{R} \to \mathbb{R}$ be a function defined by $f_n(x) = \sum_{k=0}^{n} \frac{x^k}{k!}$ for every $x \in \mathbb{R}$. Let $f : \mathbb{R} \to \mathbb{R}$ be defined by $f(x) = e^x$ for every $x \in \mathbb{R}$. Then f is the pointwise limit of $\langle f_n \rangle_{n=1}^{\infty}$ by Theorem 4.4.1. The function f_n is continuous for each n since it is a polynomial, and f is continuous by Theorem 4.2.7.

Problem 7.1.1 For each positive integer n, let $f_n : \mathbb{R} \to \mathbb{R}$ be a function defined by $f_n(x) = \frac{x}{n}$ for every $x \in \mathbb{R}$. Find the pointwise limit of $\langle f_n \rangle_{n=1}^{\infty}$ if it exists.

Example 7.1.4 Let $\mathbb{Q} \cap [0, 2]$ be the set of all rational number in $[0, 2]$. Then $\mathbb{Q} \cap [0, 2]$ is a countable set, which means that there is an **onto** function $a : \mathbb{N} \to \mathbb{Q} \cap [0, 2]$, i.e., $\mathbb{Q} \cap [0, 2] = \{a(n) : n = 1, 2, 3, \ldots\}$. Without loss of generality, let $a(1) = 1$. Let $f_1 : [0, 2] \to \mathbb{R}$ be a function defined by $f_1(x) = 1$ if $x = a(1)$ and $f(x) = 2$ for every $x \in [1, 2]$ such that $x \neq a(1)$. In Problem 6.2.4, you showed that f_1 is integrable by showing $\int_0^2 f_1(x) dx = 4$. For each integer $n \geq 1$, let $f_n : [0, 2] \to \mathbb{R}$ be a function defined by $f_n(x) = 1$ if $x \in \{a(1), a(2), \ldots, a(n)\}$ and $f_n(x) = 2$ for every $x \in [1, 2]$ such that $x \notin \{a(1), a(2), \ldots, a(n)\}$. Then it can be shown that $\int_0^2 f_n(x) dx = 4$ for all integers $n \geq 1$. Hence, $\lim_{n \to \infty} \left(\int_0^2 f_n(x) dx \right) = 4$.

Let $g : [0, 2] \to \mathbb{R}$ be a function defined by $g(x) = 1$ for every rational number $x \in [0, 2]$ and $g(x) = 2$ for every irrational number $x \in [0, 2]$. In Problem 6.2.4, you showed that g is not integrable. However, $\lim_{n \to \infty} f_n(x) = g(x)$ for every $x \in [0, 2]$. Hence, $\int_0^2 \left(\lim_{n \to \infty} f_n(x) \right) dx$ does not exist.

Therefore, $\int_0^2 \left(\lim_{n \to \infty} f_n(x) \right) dx \neq \lim_{n \to \infty} \left(\int_0^2 f_n(x) dx \right)$.

Example 7.1.5

(1) For each positive integer n, let $f_n : (-1, 1) \to \mathbb{R}$ be a function defined by $f_n(x) = \sum_{k=0}^{n} x^k$ for every $x \in (-1, 1)$. Then the function $f : (-1, 1) \to \mathbb{R}$ defined by $f(x) = \frac{1}{1-x}$ for every $x \in (-1, 1)$ is the limit function of $\langle f_n \rangle_{n=1}^{\infty}$ by Theorem 3.3.2 on the geometric series.

(2) For each positive integer n, let $F_n : (-1, 1) \to \mathbb{R}$ be a function defined by $F_n(x) = \sum_{k=0}^{n} \frac{x^{k+1}}{k+1}$ for every $x \in (-1, 1)$. Then $\frac{d}{dx} F_n(x) = f_n(x)$. So let $F : (-1, 1) \to \mathbb{R}$ be a function defined by $F(x) = \int_0^x \frac{1}{1-x} dx = -\ln(1-x)$ for every $x \in (-1, 1)$.

Question: Is F the pointwise limit of $\langle F_n \rangle_{n=1}^{\infty}$?
   We will answer this question in Example 7.4.3.

7.2   Uniform Convergence

As you saw in Example 7.1.2, even though f_n is continuous for each positive integer n, the pointwise limit of $\langle f_n \rangle_{n=1}^{\infty}$ may not be continuous. The purpose of this section and the next section is to answer Question 1 in the previous section. One answer is the uniform convergence in Definition 7.2.2. We need some preliminaries.

Definition 7.2.1 If $f : X \to \mathbb{R}$, let $\|f\| = \sup\{|f(x)| : x \in X\}$, and it is called the <u>sup norm</u> of the function f, or simply the <u>norm</u> of f. The norm $\|f\|$ can be zero, a positive number, or ∞.

   (The words "uniform" and "norm" are used very frequently in many branches of mathematics. Sometimes they can be very confusing. Please be careful.)

   The value $\|f\|$ depends on the domain of the function f as you can see in the next two examples.

Example 7.2.1 Let $f : [0, 1) \to \mathbb{R}$ be defined by $f(x) = x$ for every $x \in [0, 1)$. Then $\|f\| = 1$.

Example 7.2.2 Let $f : (-3, 1) \to \mathbb{R}$ be defined by $f(x) = x$ for every $x \in (-3, 1)$. Then $\|f\| = 3$.

   The norm of functions **defined on the same domain** satisfies the triangle inequality, and the norm behaves like the absolute value of numbers.

Theorem 7.2.1 (Triangle Inequality) Let $f, g : X \to \mathbb{R}$ be bounded functions. Then $\|f + g\| \leq \|f\| + \|g\|$.

Proof $\|f + g\| = \sup\{|f(x) + g(x)| : x \in X\} \leq \sup\{|f(x)| + |g(x)| : x \in X\}$ by the triangle inequality Theorem 1.1.1.
   Let $x \in X$. Then

$$|f(x)| + |g(x)| \leq \sup\{|f(x)| : x \in X\} + \sup\{|g(x)| : x \in X\} = \|f\| + \|g\|.$$

Hence, $\sup\{|f(x)| + |g(x)| : x \in X\} \leq \|f\| + \|g\|$.

Therefore, $\|f + g\| \leq \|f\| + \|g\|$.

Problem 7.2.1 Let $f, g, h : X \to \mathbb{R}$ be bounded functions. Prove that $\|f - g\| \leq \|f - h\| + \|h - g\|$. (This is also called the triangle inequality.)

Remark 7.2.1 Let $B(X)$ be the set of all **bounded** functions from X into $\mathbb{R}$. For every $f, g \in B(X)$, we define the distance between f and g, denoted by $d(f, g)$, defined by $d(f, g) = \|f - g\| = \sup\{|f(x) - g(x)| : x \in X\}$. Then d is a metric of the metric space $B(X)$. (See Sect. 3.9 for the definition of a metric space.) This metric d is sometimes called the sup-metric on $B(X)$.

Definition 7.2.2 Different textbooks give different definitions of uniform convergence of a sequence of functions and they all look different. So let us list equivalent definitions of uniform convergence.

For each positive integer n, let $f_n : X \to \mathbb{R}$ be a function. Suppose $f : X \to \mathbb{R}$ is a function.

(1) A sequence of functions $\langle f_n \rangle_{n=1}^{\infty}$ converges uniformly to a function f if for every $\varepsilon > 0$, there exists a positive integer M such that $\|f_n - f\| < \varepsilon$ for each integer $n \geq M$. (This resembles to Definition 3.1.2 of the convergence of a sequence. Hence, **we primarily use this as our definition of the uniform convergence**.)

(2) The sequence of functions $\langle f_n \rangle_{n=1}^{\infty}$ converges uniformly to a function $f : X \to \mathbb{R}$ if, for every $\varepsilon > 0$, there is an integer M such that $|f_n(x) - f(x)| < \varepsilon$ for each integer $n \geq M$ and **for every** $x \in X$.

(3) Let $B(X)$ be the set of all **bounded** functions from X into $\mathbb{R}$. For each $\varepsilon > 0$, let $N_\varepsilon(f) = \{g \in B(X) : \|f - g\| < \varepsilon\}$.

Then, a sequence of functions $\langle f_n \rangle_{n=1}^{\infty}$ converges uniformly to a function f if, for every $\varepsilon > 0$, there exists a positive integer M such that $f_n \in N_\varepsilon(f)$ for each integer $n \geq M$. (This is related to the metric spaces in Sect. 3.11 with the sup-metric in Remark 7.2.1.)

(4) For each $\varepsilon > 0$, let $G_\varepsilon(f) = \{(x, y) : f(x) - \varepsilon < y < f(x) + \varepsilon, x \in X\}$.

When X is an interval $[a, b]$, then the yellow trimmed region in Fig. 7.1 is $G_\varepsilon(f)$. Let us denote the graph of f_n by $\mathbf{f}_n$. That is,

$$\mathbf{f}_n = \{(x, y) : y = f_n(x), x \in X\}.$$

Fig. 7.1 $G_\varepsilon(f)$: Note that ε width is measured vertically from the graph of $y = f(x)$. So the place where the inclination of the graph of $y = f(x)$ is steep, the width of $G_\varepsilon(f)$ looks narrow

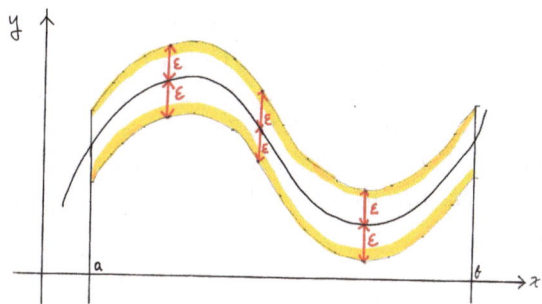

Then, a sequence of functions $\langle f_n\rangle_{n=1}^{\infty}$ <u>converges uniformly</u> to a function f if, for every $\varepsilon > 0$, there exists a positive integer M such that $f_n \subset G_\varepsilon(f)$ for each integer $n \geq M$. (This graph may help you visualize this concept.)

Remark 7.2.2 For each positive integer n, let $f_n : X \to \mathbb{R}$ be a function.

The sequence of functions $\langle f_n\rangle_{n=1}^{\infty}$ <u>converges pointwise</u> to a function $f : X \to \mathbb{R}$ if, **for every** $x \in X$ and for every $\varepsilon > 0$, there is an integer $M(x)$, specific to the choice of x, such that $|f_n(x) - f(x)| < \varepsilon$ for each integer $n \geq M(x)$.

Compare this definition of pointwise convergence to Definition 7.2.2(2) of uniform convergence. The only difference between these two is the position of "for every $x \in X$". Since many textbooks use Definition 7.2.2(2) as the definition of uniform convergence, let us prove the next proposition.

Proposition 7.2.1 In Definition 7.2.2, statements (1) and (2) are equivalent.

Proof Suppose the sequence of functions $\langle f_n\rangle_{n=1}^{\infty}$ converges uniformly as in (1) to a function $f : X \to \mathbb{R}$. Let $\varepsilon > 0$. Then there exists a positive integer M such that $\|f_n - f\| < \varepsilon$ for each integer $n \geq M$. Let $x \in X$. Then $|f_n(x) - f(x)| \leq \|f_n - f\|$. Therefore, $|f_n(x) - f(x)| < \varepsilon$ for each integer $n \geq M$ and for every $x \in X$. Hence, $\langle f_n\rangle_{n=1}^{\infty}$ converges uniformly as in (2) to a function f.

Conversely, suppose $\langle f_n\rangle_{n=1}^{\infty}$ converges uniformly as in (2) to a function f. Let $\varepsilon > 0$. Then there is an integer M such that $|f_n(x) - f(x)| < \frac{\varepsilon}{2}$ for each integer $n \geq M$ and for every $x \in X$. Hence, if $n \geq M$, we have

$$\|f_n - f\| = \sup\{|f_n(x) - f(x)| : x \in X\} \leq \frac{\varepsilon}{2} < \varepsilon.$$

Therefore, $\langle F_n\rangle_{n=1}^{\infty}$ converges uniformly as in (1) to a function $f : X \to \mathbb{R}$.

Remark 7.2.3 We **primarily** use Definition 7.2.2(1) for uniform convergence since it is very similar to Definition 3.1.2 of the convergence of a sequence.

Example 7.2.3 For each positive integer n, let $f_n : [0, 1] \to \mathbb{R}$ be a function defined by $f_n(x) = \frac{x}{n}$ for every $x \in [0, 1]$. Let $f : [0, 1] \to \mathbb{R}$ be defined by $f(x) = 0$ for every $x \in [0, 1]$. We prove that $\langle f_n \rangle_{n=1}^{\infty}$ converges uniformly to f.

We saw in Example 7.1.1 that f is the pointwise limit of $\langle f_n \rangle_{n=1}^{\infty}$. We show that f is the uniform limit of $\langle f_n \rangle_{n=1}^{\infty}$. Let $\varepsilon > 0$. There exists a positive integer M such that $\frac{1}{M} < \varepsilon$. So if $n \geq M$, then $\|f_n - f\| = \|f_n\| = \frac{1}{n} \leq \frac{1}{M} < \varepsilon$. Therefore, $\langle f_n \rangle_{n=1}^{\infty}$ converges uniformly to f.

Problem 7.2.2 For each positive integer n, let $f_n : X \to \mathbb{R}$ be a function. Let $f : X \to \mathbb{R}$. If f Is a uniform limit of $\langle f_n \rangle_{n=1}^{\infty}$, then show that f is necessarily the pointwise limit of $\langle f_n \rangle_{n=1}^{\infty}$.

Example 7.2.4 Let X be a set of real numbers. For each positive integer n, let $f_n : X \to \mathbb{R}$. Suppose $\langle f_n \rangle_{n=1}^{\infty}$ converges pointwise to $f : X \to \mathbb{R}$.

Question: What do we mean by f is not the uniform limit of $\langle f_n \rangle_{n=1}^{\infty}$?

Let us negate Definition 7.2.2 **(1)**. The sequence of functions $\langle f_n \rangle_{n=1}^{\infty}$ does not converge uniformly to f if there exists an $\varepsilon > 0$ such that for any positive integer M, there exists an integer $n \geq M$ such that $\|f_n - f\| \geq \varepsilon$. Compare this to Remark 3.1.5.

Example 7.2.5 For each positive integer n, let $f_n : \mathbb{R} \to \mathbb{R}$ be a function defined by $f_n(x) = \frac{x}{n}$ for every $x \in \mathbb{R}$. Let $f : \mathbb{R} \to \mathbb{R}$ be defined by $f(x) = 0$ for every $x \in \mathbb{R}$. Then f is a pointwise limit of $\langle f_n \rangle_{n=1}^{\infty}$ (see Problem 7.1.1). We will show that $\langle f_n \rangle_{n=1}^{\infty}$ does not converge uniformly to f.

Let M be a positive integer. Then $M \in \mathbb{R}$. And we have $|f_M(M) - f(M)| = \left|\frac{M}{M} - 0\right| = 1$. This shows that $\|f_M - f\| \geq 1$. Therefore, $\langle f_n \rangle_{n=1}^{\infty}$ does not converge uniformly to f.

It looks too short to be a proof of the statement. But it is. Please compare this example to Example 7.2.3. The only difference between these two examples is the domain of these functions.

Problem 7.2.3 This is related to Problem 7.1.2. For each positive integer n, let $f_n : [0, 1] \to \mathbb{R}$ be a function defined by $f_n(x) = x^n$ for every $x \in [0, 1]$. Let $f : [0, 1] \to \mathbb{R}$ be defined by $f(x) = 0$ for every $x \in [0, 1)$ and $f(1) = 1$. Prove that $\langle f_n \rangle_{n=1}^{\infty}$ does not converge uniformly to f.

Problem 7.2.4 For each positive integer n, let $f_n : [1, \infty) \to \mathbb{R}$ be a function defined by $f_n(x) = x^{\frac{1}{n}}$ for every $x \in [1, \infty)$. Let $f : [1, \infty) \to \mathbb{R}$ be defined by $f(x) = 1$ for every $x \in [1, \infty)$.

(1) Prove that f is the pointwise limit of $\langle f_n \rangle_{n=1}^{\infty}$. (An answer is in Lemma 3.8.1.)
(2) Prove that $\langle f_n \rangle_{n=1}^{\infty}$ does not converge uniformly to f.

(Hint: Perhaps using $(2^n)^{\frac{1}{n}} = 2$ may be useful.)

Problem 7.2.5 Let $R > 1$. For each positive integer n, let $f_n : [-R, R] \to \mathbb{R}$ be a function defined by $f_n(x) = \sum_{k=0}^{n} \frac{x^k}{k!}$ for every $x \in [-R, R]$. Let $f : [-R, R] \to \mathbb{R}$ be a function defined by $f(x) = e^x$ for every $x \in [-R, R]$. Prove that $\langle f_n \rangle_{n=1}^{\infty}$ converge uniformly to f. Note that $f(x) = \sum_{k=0}^{\infty} \frac{x^k}{k!}$ by Theorem 4.4.1.

Definition 7.2.3 If a sequence of functions $\langle f_n \rangle_{n=1}^{\infty}$ <u>converges uniformly</u> to a function f : $X \to \mathbb{R}$, then the function f is said to be the <u>uniform limit</u> of $\langle f_n \rangle_{n=1}^{\infty}$. If $\langle f_n \rangle_{n=1}^{\infty}$ converges uniformly to **some** function, we say that $\langle f_n \rangle_{n=1}^{\infty}$ <u>converges uniformly</u>.

Problem 7.2.6 Let $0 < R < 1$. For each positive integer n, let $f_n : [-R, R] \to \mathbb{R}$ be a function defined by $f_n(x) = \sum_{k=0}^{n} x^k$ for every $x \in [-R, R]$. Find the limit function $f : [-R, R] \to \mathbb{R}$, and prove that $\langle f_n \rangle_{n=1}^{\infty}$ converge uniformly to f.

Sometimes, the uniform limit of $\langle f_n \rangle_{n=1}^{\infty}$ is not known and it can be illusive. Recall how we defined the number like $2^{\sqrt{2}}$ in Chap. 3. When $\langle r_n \rangle_{n=1}^{\infty}$ is a sequence of rational numbers that converges to an irrational number r, we proved the convergence of $\langle a^{r_n} \rangle_{n=1}^{\infty}$ in Theorem 3.9.1(1) so that we could define the number a^r to be the limit of $\langle a^{r_n} \rangle_{n=1}^{\infty}$. In that proof, we could not use that $\langle a^{r_n} \rangle_{n=1}^{\infty}$ converges to a^r since a^r was not defined yet. Similar situations exist when we deal with a sequence of functions. Thus, we need the idea of a Cauchy sequence for a sequence of functions in terms of uniform convergence.

Definition 7.2.4 For each positive integer n, let $f_n : X \to \mathbb{R}$ be a function. A sequence of functions $\langle f_n \rangle_{n=1}^{\infty}$ is said to be <u>uniformly Cauchy</u> or a <u>uniform Cauchy sequence</u> if for every $\varepsilon > 0$, there exists a positive integer M such that $\|f_n - f_m\| < \varepsilon$ for any integers $n, m \geq M$. (Definition 7.2.4 resembles to Definition 3.7.2 because of the use of the norm $\|\|$.)

Problem 7.2.7 For each positive integer n, let $f_n : \mathbb{R} \to \mathbb{R}$ be a function defined by $f_n(x) = \frac{x}{n}$ for every $x \in \mathbb{R}$. Prove that $\langle f_n \rangle_{n=1}^{\infty}$ is not uniformly Cauchy.

(This problem is related to Problem 7.1.1 and Example 7.2.5.)

In Chap. 3, Problem 3.7.3(1) asks to prove that a convergent sequence is Cauchy. Here, we prove in the next lemma that a uniformly convergent sequence of functions is uniformly Cauchy.

Lemma 7.2.1 For each positive integer n, let $f_n : X \to \mathbb{R}$ be a function. Let $f : X \to \mathbb{R}$ be also a function. Suppose the sequence of functions $\langle f_n \rangle_{n=1}^{\infty}$ converges uniformly to a function f. Then the sequence $\langle f_n \rangle_{n=1}^{\infty}$ is uniformly Cauchy.

Proof Let $\varepsilon > 0$. Since $\langle f_n \rangle_{n=1}^{\infty}$ converges uniformly to a function f, there exists a positive integer M such that $\|f_n - f\| < \frac{\varepsilon}{2}$ for every integer $n \geq M$. Let $m, n \geq M$ be integers. Then

$\|f_n - f_m\| \le \|f_n - f\| + \|f - f_m\|$ by the triangle inequality Theorem 7.2.1

$$< \frac{\varepsilon}{2} + \frac{\varepsilon}{2} = \varepsilon.$$

Therefore, $\langle f_n \rangle_{n=1}^{\infty}$ is a uniform Cauchy sequence.

Problem 7.2.8 Let $\langle f_n \rangle_{n=1}^{\infty}$ be a sequence of functions defined in Example 7.1.4. Prove that $\langle f_n \rangle_{n=1}^{\infty}$ is not uniformly Cauchy. (Therefore, $\langle f_n \rangle_{n=1}^{\infty}$ does not converge uniformly to g defined in Example 7.1.4.)

Remark 7.2.4 Just like all convergent sequences are Cauchy, uniformly convergent sequences of functions are uniformly Cauchy by Lemma 7.2.1. But is there an interesting uniformly Cauchy sequence with an unknown limit function? We will give such an example in Example 7.2.7. For this, we need Theorem 7.2.2. We need a short lemma to prove Theorem 7.2.2.

Lemma 7.2.2 Let $\langle a_n \rangle_{n=1}^{\infty}$ be a sequence of real numbers such that $\lim_{n \to \infty} a_n = a$ for some $a \in \mathbb{R}$. Suppose $a_n < \varepsilon$ for every $n > 0$. Then $a \le \varepsilon$.

Proof This is by the sandwich theorem, Theorem 3.2.2(1).

Theorem 7.2.2 Suppose $f_n : X \to \mathbb{R}$ is a function for each integer $n > 0$ such that the sequence $\langle f_n \rangle_{n=1}^{\infty}$ is uniformly Cauchy. Then we have the following:

(1) For each $t \in X$, the sequence of numbers $\langle f_n(t) \rangle_{n=1}^{\infty}$ is Cauchy (as in Definition 3.7.2).
(2) By Theorem 3.7.2, $\lim_{n \to \infty} f_n(t)$ exists for each $t \in X$. Let $f : X \to \mathbb{R}$ be a function defined by $f(t) = \lim_{n \to \infty} f_n(t)$ for every $t \in X$. Then the sequence $\langle f_n \rangle_{n=1}^{\infty}$ converges uniformly to f.

Proof of (1): Let $t \in X$ and $\varepsilon > 0$. Since $\langle f_n \rangle_{n=1}^{\infty}$ is uniformly Cauchy, there is a positive integer M such that $\|f_n - f_m\| < \varepsilon$ for each integer $n, m \ge M$. But by definition of the norm of a function, we have

$$|f_n(t) - f_m(t)| \le \sup\{|f_n(x) - f_m(x)| : x \in X\} = \|f_n - f_m\|.$$

Hence, $|f_n(t) - f_m(t)| < \varepsilon$ for each integer $n, m \ge M$. Therefore, $\langle f_n(t) \rangle_{n=1}^{\infty}$ is a Cauchy sequence.

Proof of (2): Let $\varepsilon > 0$. Since $\langle f_n \rangle_{n=1}^{\infty}$ is uniformly Cauchy, there is a positive integer M such that $\|f_n - f_m\| < \frac{\varepsilon}{2}$. for each integer $n, m \ge M$. Let $n, m \ge M$. Then

$$\|f_n - f_m\| < \frac{\varepsilon}{2}.$$

Hence $|f_n(t) - f_m(t)| \le \|f_n - f_m\| < \frac{\varepsilon}{2}$ for every $t \in X$.

By fixing n, and by letting $m \to \infty$, we have

$|f_n(t) - f(t)| = \lim_{m \to \infty} |f_n(t) - f_m(t)| \le \frac{\varepsilon}{2}$ for every $t \in X$ by Lemma 7.2.2.

Hence, for every integer $n \ge M$,

$$\|f_n - f\| = \sup\{|f_n(x) - f(x)| : x \in X\} \le \frac{\varepsilon}{2} < \varepsilon.$$

Therefore, the sequence $\langle f_n \rangle_{n=1}^{\infty}$ converges uniformly to f.

Example 7.2.6 Let $\langle f_n \rangle_{n=1}^{\infty}$ be the sequence of functions $f_n : [0, 1] \to [0, 1]$ for every integer $n > 0$ defined in Example 5.3.1. We will prove this sequence $\langle f_n \rangle_{n=1}^{\infty}$ is uniformly Cauchy, and therefore, we can conclude that $\langle f_n \rangle_{n=1}^{\infty}$ converges to a function $f : [0, 1] \to [0, 1]$ by Theorem 7.2.2. (We do not know this functon $f: [0,1] \to [0,1]$ yet. Hence, we cannot prove that $\langle f_n \rangle_{n=1}^{\infty}$ converges to f directly. So we use Theorem 7.2.2. In the next section, we will prove the continuity of f.) What do we know about $\|f_n - f_m\|$? We can calculate $\|f_1 - f_2\|$ to be $\frac{2}{3} - \frac{1}{3} = \frac{1}{3}$. But for larger integers n and m, calculating exact value of $\|f_n - f_m\|$ seems very difficult. It seems too difficult to do, and it is. So we need a plan B. The definition of uniform Cauchy does not ask for calculating $\|f_n - f_m\|$ exactly. It asks to show that $\|f_n - f_m\|$ gets closer to 0 as two integers m and n get larger. Let us look at the construction of f_n more carefully. For a simplicity, let $S_1 = [0, 1] \times [0, 1]$.

   (Step 1) Notice that the graph of f_1 is contained in the square S_1, and the graph of f_2 is contained in the square. For that matter, the graph of f_n for each $n \ge 1$ is contained in the square S_1. Since the height of S_1 is 1, we have $\|f_1 - f_m\| \le 1$ for all $m \ge 1$.

   (Step 2) Notice that the graph of f_2 is contained in the square $R_1 \cup R_2 \cup R_3$, and the graph of f_3 is contained in $R_1 \cup R_2 \cup R_3$. (We are using the notations In Example 5.3.1.) For that matter, the graph of f_n for each $n \ge 2$ is contained in $R_1 \cup R_2 \cup R_3$. Since the tallest rectangle among R_1, R_2, R_3 is R_1, the maximum vertical height of $R_1 \cup R_2 \cup R_3$ is $\frac{2}{3}$. Hence, we have $\|f_2 - f_m\| \le \frac{2}{3}$ for all $m \ge 2$.

   (Step 3) It is not easy to describe the region that contains f_3 as we did for Steps 1 and 2. See graph of f_3 in Fig. 7.3. This requires a little thinking. Let w_i; $i = 1, 2, 3$ be the operations that

   (i) w_1 shrinks the square S_1. and places its images perfectly onto R_1 so that $R_1 = w_1(S_1)$,
   (ii) w_2 flips the square S_1 about the vertical line $x = \frac{1}{2}$, shrinks it, and then places the image perfectly onto R_2 so that $w_2(S_1) = R_2$, and
   (iii) w_3 shrinks the square S_1 and places its image perfectly onto R_3 so that $R_3 = w_3(S_1)$.

Let $S_2 = w_1(S_1) \cup w_2(S_1) \cup w_3(S_1)$. So, $S_2 = R_1 \cup R_2 \cup R_3$. In general, if $n \ge 1$ is an integer such that S_n is defined, we let $S_{n+1} = w_1(S_n) \cup w_2(S_n) \cup w_3(S_n)$. This defines S_n for all integer $n \ge 1$ by mathematical induction. The pictures of S_1, S_2, S_3, and S_4 are drawn in Fig. 7.2.

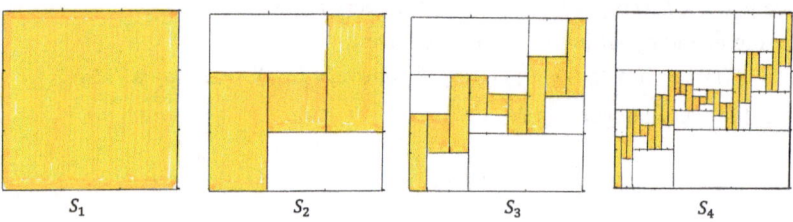

Fig. 7.2 Pictures of S_1, S_2, S_3 and S_4

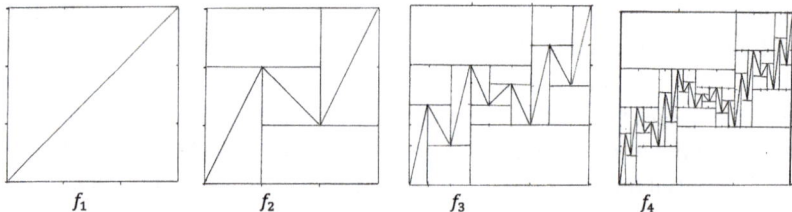

Fig. 7.3 Graphs of f_1, f_2, f_3 and f_4

Now, we can see that the graph of f_3 is contained in S_3. For that matter, the graph of f_m is contained in S_3 for all $m \geq 3$. Note that S_3 is a union of nine rectangles, and the tallest rectangles among these nine is $w_1(w_1(R_1)) = (w_1)^2(R_1)$. In the picture of S_3 in Fig. 7.2, $(w_1)^2(R_1)$ is the one at the left most corner rectangle with the height $\left(\frac{2}{3}\right)^2$. This shows that $\|f_3 - f_m\| \leq \left(\frac{2}{3}\right)^2$ for all $m \geq 3$.

(Step n) Let $n \geq 2$ be an integer. In a similar way, we can see that the graph of f_m is contained in S_n for all $m \geq n$. Since S_n is a union of 3^{n-1} many rectangles, and the tallest rectangle is $(w_1)^{n-1}(S_1)$ at the left most corner of S_n, which is $\left(\frac{2}{3}\right)^{n-1}$. This shows that $\|f_n - f_m\| \leq \left(\frac{2}{3}\right)^{n-1}$ for all $m \geq n$. Since $\lim_{n \to \infty} \left(\frac{2}{3}\right)^{n-1} = 0$, we can see that the sequence $\langle f_n \rangle_{n=1}^{\infty}$ is uniformly Cauchy.

7.3   Basic Theorems on Uniform Convergence

The next two theorems are some of the reasons why the uniform convergence is important.

Theorem 7.3.1 Let X be a nonempty subset of $\mathbb{R}$. For each integer $n > 0$, let $f_n : X \to \mathbb{R}$ be a **continuous** function. Suppose $\langle f_n \rangle_{n=1}^{\infty}$ converges uniformly to $f : X \to \mathbb{R}$. Then f is continuous. In short, the uniform limit of continuous functions is continuous.

Proof Let $p \in X$. We will show that f is continuous at p. Let $\varepsilon > 0$. Since $\langle f_n \rangle_{n=1}^{\infty}$ converge uniformly to f, there exists a positive integer M such that $\|f_n - f\| < \frac{\varepsilon}{3}$ for each integer $n \geq M$. Also, since $f_M : X \to \mathbb{R}$ is continuous at p, there exists a $\delta > 0$ such that if $x \in X$ and $|x - p| < \delta$, then $|f_M(x) - f_M(p)| < \frac{\varepsilon}{3}$. Let $x \in X$ such that $|x - p| < \delta$. Then we have

(1) $|f_M(x) - f_M(p)| < \frac{\varepsilon}{3}$ by the continuity of f_M.

By the uniform convergence, we have

(2) $|f_M(x) - f(x)| \leq \|f_M - f\| < \frac{\varepsilon}{3}$, and
(3) $|f_M(p) - f(p)| \leq \|f_M - f\| < \frac{\varepsilon}{3}$.

Therefore, by the triangle inequality, we have

$$
\begin{aligned}
|f(x) - f(p)| &= |(f(x) - f_M(x)) + (f_M(x) - f_M(p)) + (f_M(p) - f(p))| \\
&\leq |f(x) - f_M(x)| + |f_M(x) - f_M(p)| + |f_M(p) - f(p)| \\
&\leq \|f - f_M\| + |f_M(x) - f_M(p)| + \|f_M - f\| \\
&< \frac{\varepsilon}{3} + \frac{\varepsilon}{3} + \frac{\varepsilon}{3} = \varepsilon.
\end{aligned}
$$

This proves the continuity of f.

Remark 7.3.1 Suppose $\langle f_n \rangle_{n=1}^{\infty}$ and f are the ones in Problem 7.2.7. Prove that $\langle f_n \rangle_{n=1}^{\infty}$ is not uniformly Cauchy. This shows that even though $\langle f_n \rangle_{n=1}^{\infty}$ does not uniformly converge to f, f can be continuous.

Problem 7.3.1 Let $f : \mathbb{R} \to \mathbb{R}$ be a function defined by $f(x) = e^x$ for every $x \in \mathbb{R}$. By Theorem 4.2.7, we know that f is continuous. Alternately, use Problem 7.2.5 and the above theorem to prove that f is continuous.

Example 7.3.1 Let $\langle f_n \rangle_{n=1}^{\infty}$ be the sequence of functions defined in Example 5.3.1. By Example 7.2.6, it is uniformly Cauchy. By Theorem 7.2.2, the sequence $\langle f_n \rangle_{n=1}^{\infty}$ converges uniformly to a function $f : [0, 1] \to [0, 1]$. Hence, by Theorem 7.3.1, we can conclude that the function f is continuous. In order to see that f is **nowhere differentiable**, please see my paper; *Continuous Nowhere-Differentiable Functions—an Application of Contraction Mappings*, The American Mathematical Monthly, Vol. 98, no. 5, 411–416, 1991 [1].

Example 7.3.2 The need for uniform convergence in the next theorem can be shown in this example. For every integer $n \geq 2$, let $f_n : [0, 1] \to \mathbb{R}$ be a function defined by
$f_n(x) = n^2 x$ when $x \in \left[0, \frac{1}{n}\right], f_n(x) = -n^2 x + 2n$ when $x \in \left[\frac{1}{n}, \frac{2}{n}\right]$, and
$f_n(x) = 0$ when $x \in \left[\frac{2}{n}, 1\right]$.

Check that f_n is continuous for every integer $n \geq 2$.

Let $f : [0, 1] \to \mathbb{R}$ be a function defined by $f(x) = 0$ for every $x \in [0, 1]$. Then the sequence $\langle f_n \rangle$ converges pointwise to the **continuous** function f. Yet $\langle f_n \rangle$ does not converge uniformly to f (**prove this**). Moreover, $\lim_{n \to \infty} \int_0^1 f_n(x)dx = 1$ and $\int_0^1 f(x)dx = 0$. Hence,

$$\lim_{n \to \infty} \int_0^1 f_n(x)dx \neq \int_0^1 \lim_{n \to \infty} f_n(x)dx.$$

Theorem 7.3.2 Let $a < b$. Let $\langle f_n \rangle$ be a sequence of continuous functions on $[a, b]$ that converges uniformly to a function $f : [a, b] \to \mathbb{R}$. Let $F_n(t) = \int_a^t f_n(x)dx$ and $F(t) = \int_a^t f(x)dx$ for every $t \in [a, b]$ and for every positive integer n. Then the sequence of functions $\langle F_n \rangle_{n=1}^{\infty}$ converges to F uniformly on $[a, b]$. In particular, we have
$\int_a^t \lim_{n \to \infty} f_n(x)dx = \lim_{n \to \infty} \int_a^t f_n(x)dx$ for every $t \in [a, b]$.
(Functions F_n and F are continuous by Theorem 6.2.7.)

Proof Since both f and f_n are continuous on $[a, b]$, F_n and F are defined for every positive integer n by Theorems 6.3.1 and 7.3.1. Let $\varepsilon > 0$. Since $\langle f_n \rangle_{n=1}^{\infty}$ converges uniformly to f on $[a, b]$, there exists a positive integer M such that $\langle f - f_n \rangle < \frac{\varepsilon}{2(b-a)}$ for every integer $n \geq M$. (Here, $\|f - f_n\| = \sup\{|f(x) - f_n(x)| : x \in [a, b]\}$.) Let $n \geq M$ be an integer and $t \in [a, b]$. Then $|F(t) - F_n(t)| = \left| \int_a^t f_n(x)dx - \int_a^t f(x)dx \right| \leq \int_a^t |f_n(x) - f(x)|dx \leq \int_a^t \|f - f_n\|dx \leq \|f - f_n\|(b - a)$. This shows that $\|F - F_n\| = \sup\{|F(t) - F_n(t)| : t \in [a, b]\} \leq \|f - f_n\|(b - a) < \frac{\varepsilon}{2(b-a)}(b - a) = \varepsilon$.
This proves that $\langle F_n \rangle$ converges to F uniformly on $[a, b]$.

Example 7.3.2 Let $0 < R < 1$. For each positive integer n, let $f_n : [-R, R] \to \mathbb{R}$ be a function defined by $f_n(x) = \sum_{k=0}^n x^k$ for every $x \in [-R, R]$. From Problem 7.2.6, we know that $\langle f_n \rangle_{n=1}^{\infty}$ converge uniformly. Let $t \in [-R, R]$. Then

$$\lim_{n \to \infty} \int_0^t f_n(x)dx = \lim_{n \to \infty} \int_0^t \sum_{k=0}^n x^k dx = \lim_{n \to \infty} \sum_{k=0}^n \frac{t^{k+1}}{k+1} = \sum_{k=0}^{\infty} \frac{t^{k+1}}{k+1}.$$

On the other hand,

$$\int_0^t \lim_{n \to \infty} f_n(x)dx = \int_0^t \sum_{k=0}^{\infty} x^k dx = \int_0^t \frac{1}{1-x}dx = [-\ln|1 - x|]_0^t = -\ln(1 - t).$$

Therefore, $\sum_{k=0}^{\infty} \frac{t^{k+1}}{k+1} = -\ln(1 - t)$ for every $t \in [-R, R]$, $R < 1$, by Theorem 7.3.2.

Question: Can we prove that $\sum_{k=0}^{\infty} \frac{t^{k+1}}{k+1} = -\ln(1 - t)$ for every $t \in (-1, 1)$?
The answer is YES. This is what we are going to answer in the next section.

Problem 7.3.2 Let $0 < R < 1$. Let $f_n : [-R, R] \to \mathbb{R}$ be a function defined by $f_n(x) = \sum_{k=0}^{n} \frac{x^{k+1}}{k+1}$ for every $x \in [-R, R]$.

(1) Find the limit function of the sequence of functions $\langle f_n \rangle$.

(2) Let $x \in [-R, R]$. Find $\int_0^x \lim_{n \to \infty} f_n(x) dx$ and $\lim_{n \to \infty} \int_0^x f_n(x) dx$.

Example 7.3.3 Let $\langle f_n \rangle$ and f be the sequence of continuous functions and the continuous function in Example 7.2.6, respectively. Then $\langle f_n \rangle$ converges uniformly to f. Note that $\int_0^1 f_n(x) dx$ is the area under the curves $y = f_n(x)$. Since $f_n\left(\frac{1}{2}\right) = \frac{1}{2}$ and the graphs of $y = f_n(x)$ are symmetric with respect to the point $\left(\frac{1}{2}, \frac{1}{2}\right)$, we can see that $\int_0^1 f_n(x) dx = \frac{1}{2}$ for every integer $n > 0$. Hence, by Corollary 7.3.2, we have $\int_0^1 f(x) dx = \int_0^1 \lim_{n \to \infty} f_n(x) dx = \lim_{n \to \infty} \int_0^1 f_n(x) dx = \lim_{n \to \infty} \frac{1}{2} = \frac{1}{2}$.

Problem 7.3.3 Let $f_n : \mathbb{R} \to \mathbb{R}$ be a function defined by $f_n(x) = \frac{\sin(n^2 x)}{n^2}$ for every $x \in \mathbb{R}$ for every positive integer n. And let $f : \mathbb{R} \to \mathbb{R}$ be a function defined by $f(x) = 0$ for every $x \in \mathbb{R}$.

(1) Prove that f_n converges uniformly to f on $\mathbb{R}$.

(2) Verify $\int_0^t \lim_{n \to \infty} f_n(x) dx = \lim_{n \to \infty} \int_0^t f_n(x) dx$ for every $t \in \mathbb{R}$.

(3) Prove that $\langle f_n' \rangle$ does **not** converge uniformly to f' on $\mathbb{R}$. (This shows that even if $\langle f_n \rangle$ is a sequence of differentiable functions that converges uniformly to a differentiable function f, the sequence of the derivatives $\langle f_n' \rangle$ may not converge uniformly to the function f'.)

We conclude this section by quoting the next theorem for your information. See "*An old result on sequences and differentiability—a concise proof*" by D. P. Minassian, The American Mathematical Monthly, 109(2), 2002, page 172 [2].

Theorem 7.3.3 Let $a < b$. Let $\langle f_n \rangle$ be a sequence of differentiable functions on $[a, b]$. Let $\langle f_n' \rangle$ converge uniformly to some function $g : [a, b] \to \mathbb{R}$, and let $\langle f_n(c) \rangle$ converge for some $c \in [a, b]$. Then

(i) $\langle f_n \rangle$ converges uniformly to some function $f : [a, b] \to \mathbb{R}$, and

(ii) $f' = g$.

7.4   Power Series

Definition 7.4.1 Let $\langle a_n \rangle_{n=1}^{\infty}$ be a sequence of real numbers. Let X be a subset of $\mathbb{R}$. For each non-negative integer n, let $f_n : X \to \mathbb{R}$ be a function defined by $f_n(x) = \sum_{k=0}^{n} a_k x^k$ for every $x \in X$. Then we have a sequence of functions $\langle f_n \rangle_{n=1}^{\infty}$. We simply say that $\sum_{n=0}^{\infty} a_n x^n$ is a _power series_, treating x as a variable, and treating $f(x) = \sum_{n=0}^{\infty} a_n x^n$ to be the pointwise limit of $\langle f_n(x) \rangle_{n=1}^{\infty}$ for every $x \in X$. **Recall** Theorem 3.5.2: (The ratio test) If $a_n \neq 0$ for all integer n, and $\lim\limits_{n \to \infty} \frac{|a_{n+1}|}{|a_n|} < 1$, then $\sum_{n=1}^{\infty} a_n$ converges absolutely.

Theorem 7.4.1 Let $\sum_{n=0}^{\infty} a_n x^n$ be a power series such that $r = \lim\limits_{n \to \infty} \left| \frac{a_{n+1}}{a_n} \right|$ exists. If $r \neq 0$, then the power series $\sum_{n=0}^{\infty} a_n x^n$ converges absolutely for all $x \in \left(-\frac{1}{r}, \frac{1}{r} \right)$. If $r = 0$, then $\sum_{n=0}^{\infty} a_n x^n$ converges absolutely for every $x \in (-\infty, \infty)$.

Proof Let $b_n = a_n x^n$. Suppose $r = \lim\limits_{n \to \infty} \left| \frac{a_{n+1}}{a_n} \right|$ exists. Then $\lim\limits_{n \to \infty} \left| \frac{b_{n+1}}{b_n} \right| = \lim\limits_{n \to \infty} \left| \frac{a_{n+1} x^{n+1}}{a_n x^n} \right| = |x| \lim\limits_{n \to \infty} \left| \frac{a_{n+1}}{a_n} \right| = r|x|$. So if $r \neq 0$ and if $r|x| < 1$, or if $|x| < \frac{1}{r}$, then the series $\sum_{n=0}^{\infty} a_n x^n$ converges absolutely by the ratio test. That is, if $x \in \left(-\frac{1}{r}, \frac{1}{r} \right)$, the series $\sum_{n=0}^{\infty} a_n x^n$ converges absolutely. If $r = 0$, then $\sum_{n=0}^{\infty} a_n x^n$ converges absolutely for any $x \in (-\infty, \infty)$ by the ratio test.

Definition 7.4.2 Suppose $r = \lim\limits_{n \to \infty} \left| \frac{a_{n+1}}{a_n} \right|$ exists. Let $R = \frac{1}{r}$ if $r \neq 0$, and $R = \infty$ if $r = 0$. This number R is called the _radius of convergence_ of the _power series_ $\sum_{n=0}^{\infty} a_n x^n$. If we say that "$R$ is the radius of convergence of a power series", then R is a positive number or $R = \infty$. The interval $(-R, R)$ is called the _interval of absolute convergence_ for the power series $\sum_{k=0}^{\infty} a_k x^k$ by Theorem 7.4.1.

Note that $R = \lim\limits_{n \to \infty} \left| \frac{a_n}{a_{n+1}} \right|$ since $R = \frac{1}{r}$.

Note that if $b \in \mathbb{R}$, then the _interval of absolute convergence_ for the power series $\sum_{n=0}^{\infty} a_n (x - b)^n$ is $(b - R, b + R)$.

Remark 7.4.1 If $R = \lim\limits_{n \to \infty} \left| \frac{a_n}{a_{n+1}} \right| = 0$, then the power series $\sum_{n=0}^{\infty} a_n x^n$ converges only for $x = 0$ and diverges for all $x \neq 0$. If R is the radius of convergence, and if $x \in \mathbb{R} - [-R, R]$, then the power series $\sum_{n=0}^{\infty} a_n x^n$ does not converge, but the series may or may not converge at $x = \pm R$.

Example 7.4.1

(1) $\sum_{n=0}^{\infty} \frac{x^n}{n+1}$ is a power series. Let $a_n = \frac{1}{n+1}$. Then $R = \lim\limits_{n \to \infty} \left| \frac{a_n}{a_{n+1}} \right| = \lim\limits_{n \to \infty} \left| \frac{\frac{1}{n+1}}{\frac{1}{n+2}} \right| = \lim\limits_{n \to \infty} \frac{n+2}{n+1} = \lim\limits_{n \to \infty} \frac{1 + \frac{2}{n}}{1 + \frac{1}{n}} = 1$. Hence, $R = 1$ is the radius of convergence for this power series $\sum_{n=0}^{\infty} \frac{x^n}{n+1}$. The **interval of absolute convergence** for this power series is $(-1, 1)$. Note that $\sum_{n=0}^{\infty} \frac{1}{n+1}$ diverges while $\sum_{n=0}^{\infty} \frac{(-1)^n}{n+1}$ converges. So $[-1, 1)$ is the **interval of convergence** for the power series $\sum_{n=0}^{\infty} \frac{x^n}{n+1}$.

(2) $\sum_{n=0}^{\infty} \frac{x^n}{n!}$ is a power series. Let $a_n = \frac{1}{n!}$. Then $R = \lim_{n \to \infty} \left| \frac{a_n}{a_{n+1}} \right| = \lim_{n \to \infty} \left| \frac{\frac{1}{n!}}{\frac{1}{(n+1)!}} \right| =$

$\lim_{n \to \infty} (n+1) = \infty$ is the radius of convergence of the power series $\sum_{n=0}^{\infty} \frac{x^n}{n!}$. That is, $\sum_{n=0}^{\infty} \frac{x^n}{n!}$ converges for all $x \in \mathbb{R}$. Of course, we knew that $\sum_{n=0}^{\infty} \frac{x^n}{n!}$ for all $x \in \mathbb{R}$, from Sect. 4.4, but we refrained from using this fact here.

Let us re-state Definition 7.4.1.

Definition 7.4.3 Let $\langle a_n \rangle_{n=1}^{\infty}$ be a sequence of real numbers. Let $R > 0$ be the radius of convergence of the power series $\sum_{n=0}^{\infty} a_n x^n$. For each non-negative integer n, let $f_n :$ $(-R, R) \to \mathbb{R}$ be a function defined by $f_n(x) = \sum_{k=0}^{n} a_k x^k$ for every $x \in (-R, R)$. Then we have a sequence of polynomial functions $\langle f_n \rangle_{n=1}^{\infty}$. Let $f : (-R, R) \to \mathbb{R}$ be a function defined by $f(x) = \sum_{n=0}^{\infty} a_n x^n$ for every $x \in (-R, R)$. Then f is the pointwise limit of $\langle f_n \rangle_{n=1}^{\infty}$. We simply say that $\sum_{n=0}^{\infty} a_n x^n$ is a <u>power series</u> to indicate the sequence $\langle f_n \rangle_{n=1}^{\infty}$, and at the same time, by treating x as a variable, we treat $f(x) = \sum_{n=0}^{\infty} a_n x^n$ to be the pointwise limit of $\langle f_n \rangle_{n=1}^{\infty}$.

Example 7.4.2

(1) Let us consider the power series $\sum_{n=0}^{\infty} 2^{\frac{1}{n}} (2x)^n = \sum_{n=0}^{\infty} 2^{n+\frac{1}{n}} x^n$. Let $a_n = 2^{n+\frac{1}{n}}$.

Hence, $R = \lim_{n \to \infty} \left| \frac{a_n}{a_{n+1}} \right| = \lim_{n \to \infty} \left| \frac{2^{n+\frac{1}{n}}}{2^{n+1+\frac{1}{n+1}}} \right| = \lim_{n \to \infty} 2^{-1+\frac{1}{n}-\frac{1}{n+1}} = \lim_{n \to \infty} \frac{2^{\frac{1}{n(n+1)}}}{2} = \frac{1}{2}$.

Therefore, $\sum_{n=0}^{\infty} 2^{\frac{1}{n}} \left(\frac{x+2}{2} \right)^n$ converges absolutely for all $x \in \left(-\frac{1}{2}, \frac{1}{2} \right)$. The interval of convergence is also $\left(-\frac{1}{2}, \frac{1}{2} \right)$.

(2) Let us consider the power series $\sum_{n=0}^{\infty} 2^{\frac{1}{n}} \left(\frac{x+2}{2} \right)^n = \sum_{n=0}^{\infty} 2^{\frac{1}{n}-n} (x+2)^n$. Let $a_n = 2^{\frac{1}{n}-n}$. Hence, $R = \lim_{n \to \infty} \left| \frac{a_n}{a_{n+1}} \right| = \lim_{n \to \infty} \left| \frac{2^{\frac{1}{n}-n}}{2^{\frac{1}{n+1}-n-1}} \right| = \lim_{n \to \infty} 2^{1+\frac{1}{n}-\frac{1}{n+1}} =$

$\lim_{n \to \infty} 2 \cdot 2^{\frac{1}{n(n+1)}} = 2$. Therefore, $\sum_{n=0}^{\infty} 2^{\frac{1}{n}} \left(\frac{x+2}{2} \right)^n$ converges absolutely for all $x \in (-2-2, -2+2) = (-4, 0)$. The interval of absolute convergence is $(-4, 0)$. The interval of convergence is also $(-4, 0)$.

Problem 7.4.1 Show that each of the following power series below has the radius of convergence 1.

(1) $\sum_{n=0}^{\infty} \frac{x^n}{n+1}$.
(2) $\sum_{n=0}^{\infty} x^n$.
(3) $\sum_{n=0}^{\infty} (n+1) x^n$

By generalizing Theorem 7.4.1, we have the next theorem.

Theorem 7.4.2 Suppose $R > 0$ is the radius of convergence for the power series $\sum_{n=0}^{\infty} a_n x^n$. For each non-negative integer n, let $f_n : (-R, R) \to \mathbb{R}$ be a function defined by $f_n(x) = \sum_{k=0}^{n} a_k x^k$ for every $x \in (-R, R)$. Let $f : (-R, R) \to \mathbb{R}$ be a function defined by $f(x) = \sum_{n=0}^{\infty} a_n x^n$ for every $x \in (-R, R)$.

(1) Let $0 < S < R$. Then $\langle f_n|_{[-S,S]} \rangle_{n=1}^{\infty}$ converges uniformly to $f|_{[-S,S]}$.

(We usually say that "$\langle f_n \rangle_{n=1}^{\infty}$ converges uniformly to f on $[-S, S]$" for this.)

(2) The function $f : (-R, R) \to \mathbb{R}$ is continuous.

Proof

(1) Let $\varepsilon > 0$. We know that $\sum_{n=0}^{\infty} a_n x^n$ converges absolutely for each $x \in [-S, S]$. In particular, we have $\sum_{n=0}^{\infty} |a_n| S^n$ converges. Thus, there exists a positive integer M such that

$$\left| \sum_{k=0}^{\infty} |a_k| S^k - \sum_{k=0}^{M} |a_k| S^k \right| = \sum_{k=M+1}^{\infty} |a_k| S^k < \varepsilon.$$

Let $n \geq M$ be an integer. Then, by restricting the domain of $f - f_n$ to the interval $[-S, S]$, we have

$$\|f - f_n\| = \sup_{x \in [-S,S]} |f(x) - f_n(x)| = \sup_{x \in [-S,S]} \left| \sum_{k=0}^{\infty} a_k x^k - \sum_{k=0}^{n} a_k x^k \right|$$

$$= \sup_{x \in [-S,S]} \left| \sum_{k=n+1}^{\infty} a_k x^k \right| \leq \sup_{x \in [-S,S]} \sum_{k=n+1}^{\infty} |a_k| |x|^k$$

$$\leq \sum_{k=n+1}^{\infty} |a_k| S^k \leq \sum_{k=M+1}^{\infty} |a_k| S^k < \varepsilon.$$

This proves that $\langle f_n \rangle_{n=1}^{\infty}$ converges uniformly to $f|_{[-S,S]}$.

(2) Let $p \in (-R, R)$. Then $p \in [-S, S]$ for some $0 < S < R$. By Theorem 7.3.1, the function $f|_{[-S,S]}$ is continuous. Since $f|_{[-S,S]}$ is continuous at p, f is continuous at p. Therefore, $f : (-R, R) \to \mathbb{R}$ is continuous.

Problem 7.4.2 In order to clarify the proof of part (2) in the above theorem, let me ask the following: Suppose $f : (-R, R) \to \mathbb{R}$ is a function for some $R > 0$. Let $0 < S < R$. Let $-S < p < S$. Suppose the function $f|_{[-S,S]}$ is continuous at p. Prove that f is continuous at p.

By realizing that the radii of convergence of the power series $\sum_{n=0}^{\infty} a_n x^n$, $\sum_{n=0}^{\infty} \frac{a_n}{n+1} x^{n+1}$, and $\sum_{n=1}^{\infty} n a_n x^{n-1}$ are the same, we have the following Theorem. This next theorem is the reason why we are interested in uniform convergence and power series.

Theorem 7.4.3 Suppose $R > 0$ is the radius of convergence for the power series $\sum_{n=0}^{\infty} a_n x^n$. For each non-negative integer n, let $f_n : (-R, R) \to \mathbb{R}$ be a function defined by $f_n(x) = \sum_{k=0}^{n} a_k x^k$ for every $x \in (-R, R)$. Let $f : (-R, R) \to \mathbb{R}$ be a function defined by $f(x) = \sum_{n=0}^{\infty} a_n x^n$ for every $x \in (-R, R)$. Then we have the following. Let $x \in (-R, R)$.

(1) $\int_0^x f(x) dx = \lim_{n \to \infty} \int_0^x f_n(x) dx = \sum_{n=0}^{\infty} \frac{a_n}{n+1} x^{n+1}$.

(2) $\frac{d}{dx} f(x) = \lim_{n \to \infty} \frac{d}{dx} f_n(x) = \sum_{n=1}^{\infty} n a_n x^{n-1}$.

Proof of (1): Let $x \in (-R, R)$. Then there is a number $S > 0$ such that $x \in [-S, S]$. Hence, by Theorem 7.3.2, we have $\int_0^x f(x) dx = \int_0^x \lim_{n \to \infty} f_n(x) dx = \lim_{n \to \infty} \int_0^x f_n(x) dx$.

Proof of (2): The radius of convergence for $\sum_{k=1}^{\infty} k a_k x^{k-1}$ is also R. Let $g(x) = \sum_{k=1}^{\infty} k a_k x^{k-1}$ for every $x \in (-R, R)$. Then we have

$$f(x) = \sum_{k=1}^{\infty} a_k x^k = \lim_{n \to \infty} \sum_{k=1}^{n} \frac{k a_k}{k} x^k = \lim_{n \to \infty} \sum_{k=1}^{n} k a_k \int_0^x x^{k-1} dx$$

$$= \lim_{n \to \infty} \int_0^x \sum_{k=1}^{n} k a_k x^{k-1} dx = \int_0^x \left\{ \lim_{n \to \infty} \sum_{k=1}^{n} k a_k x^{k-1} \right\} dx \quad \text{by (1)}$$

$$= \int_0^x g(x) dx.$$

Hence, the function $f(x)$ is an antiderivative of $g(x)$. Therefore, we have

$$\frac{d}{dx} f(x) = \frac{d}{dx} \sum_{n=0}^{\infty} a_n x^n = \frac{d}{dx} \sum_{n=1}^{\infty} a_n x^n = g(x) = \lim_{n \to \infty} \frac{d}{dx} f_n(x).$$

Remark 7.4.2 As we saw in Example 7.3.4, the derivatives of a uniformly convergent sequence of differentiable functions may not behave as nicely as other derivatives. However, Theorem 7.4.3 shows that a power series behaves nicely on the interval of absolute convergence. In Remark 5.1.4, we concluded plausibly that $\frac{d}{dx} e^x = e^x$, but now we can **finally** prove this in Theorem 7.4.4!

Theorem 7.4.4 $\frac{d}{dx} e^x = e^x$ for every $x \in \mathbb{R}$.

Proof By Theorem 7.4.3(2),

$$\frac{d}{dx} e^x = \frac{d}{dx} \sum_{k=0}^{\infty} \frac{x^k}{k!} = \sum_{k=0}^{\infty} \frac{k x^{k-1}}{k!} = \sum_{k=1}^{\infty} \frac{x^{k-1}}{(k-1)!} = \sum_{k=0}^{\infty} \frac{x^k}{k!} = e^x.$$

Example 7.4.3 From Problem 7.4.1, we know that the radius of convergence of $\sum_{k=0}^{\infty} x^k$ is 1. But by Theorem 3.3.1, we know that $\sum_{k=0}^{\infty} r^k = \frac{1}{1-r}$ for all $|r| < 1$. Hence, we have

(1) $\sum_{k=0}^{\infty} x^k = \frac{1}{1-x}$ for all $|x| < 1$.

(2) By integrating $\int_0^t \frac{1}{1-x} dx = [-\ln(1-x)]_0^t = -\ln(1-t) = \ln \frac{1}{1-t}$. So

$$\sum_{k=0}^{\infty} \frac{x^{k+1}}{k+1} = \ln \frac{1}{1-x} = -\ln(1-x) \text{ for all } |x| < 1.$$ See Theorem 7.3.2 and Example 7.3.2. This is re-stated by Theorem 7.4.3(1). This answers the question in Example 7.1.5.

(3) By Theorem 7.4.3, we can differentiate both sides of (1) to obtain

$$\sum_{k=1}^{\infty} k x^{k-1} = \frac{1}{(1-x)^2} \text{ for all for all } |x| < 1.$$

(4) By replacing x by $-x$ in (2). We have $\ln(x + 1) = x - \frac{1}{2}x^2 + \cdots + (-1)^{n+1}\frac{1}{n}x^n + \cdots$ for all $|x| < 1$.

Remark 7.4.3 If we let $x = 1$ in Example 7.4.3(4), the lefthand side becomes $\ln 2$, while the right-hand side becomes $1 - \frac{1}{2} + \cdots + (-1)^{n+1}\frac{1}{n} + \cdots$. Therefore, Leibniz thought $1 - \frac{1}{2} + \cdots + (-1)^{n+1}\frac{1}{n} + \cdots = \ln 2$. was "obvious", and this expansion is known as Leibniz's identity.

   However, note that the continuity of $\sum_{k=0}^{\infty} \frac{x^{k+1}}{k+1}$ is only guaranteed when $x \in (-1, 1)$ by Theorem 7.4.2. Hence, this leads us to the following question:

   Question: Is it **really** true that $1 - \frac{1}{2} + \cdots + (-1)^{n+1}\frac{1}{n} + \cdots = \ln 2$?

   We answered YES in Example 3.4.1. But Example 3.4.1 was given in late 20th century. Many people after Leibniz thought it was "obviously" yes. But it was **left unproven**. Abel finally **questioned** this type of equalities derived from power series. As a result, he was able to establish Theorem 7.4.5, and it became a published result. This is instructive to me and, perhaps, it is also instructive to the readers on how to conduct mathematical research.

   After graduation when I had to stand on my own as a mathematical researcher without my thesis advisor, I was worried and concerned. To me, the most difficult part of mathematical research is to find research problems of my own that are new, interesting, and also ones I can solve. There are many known unsolved problems in the mathematical literature and many have tried to solve them. As you might guess, they are usually very difficult to solve. It is good to tackle these unsolved problems, and if you solve them, you get a moment of glory. But I do not recommend that anyone try to solve **only** these published unsolved problems. I have my own unsolved math problems, and keep them mostly secret until I solve them. I think many mathematicians are secretive about their problems until they either solve them or give them up! What Abel did was very instructive as a way to find **your** unsolved problem. (And proving Theorem 7.4.5 is not easy!) Yes, I often find mathematics problems for my research in similar ways from reading textbooks and math papers inquisitively, and I always look out for hidden questions that can be derived from them. Often, the problems I found and solved were thought "obvious" by some authors.

Theorem 7.4.5 (*Abel*) Let $0 < R < \infty$ be the radius of convergence of the power series $\sum_{n=0}^{\infty} a_n x^n$. For each non-negative integer n, let $f_n : (-R, R) \to \mathbb{R}$ be a function defined by $f_n(x) = \sum_{k=0}^{n} a_k x^k$ for every $x \in (-R, R)$.

(1) Suppose $\sum_{n=0}^{\infty} a_n R^n$ converges. Then $\lim_{x \to R, x < R} \sum_{n=0}^{\infty} a_n x^n = \sum_{n=0}^{\infty} a_n R^n$.

(In other words, if we let $F : (-R, R] \to \mathbb{R}$ be a function defined by $F(x) = \sum_{n=0}^{\infty} a_n x^n$ for every $x \in (-R, R]$, then F is continuous at R.)

(2) Suppose $\sum_{n=0}^{\infty} a_n (-R)^n$ converges. Then $\lim_{x \to -R, x > R} \sum_{n=0}^{\infty} a_n x^n = \sum_{n=0}^{\infty} a_n (-R)^n$.

(Again, if we let $\hat{F} : [-R, R) \to \mathbb{R}$ be a function defined by $\hat{F}(x) = \sum_{n=0}^{\infty} a_n x^n$ for every $x \in [-R, R)$, then $\hat{F}$ is continuous $-R$.)

We omit the proof. If you are interested, please see page 174 of Principle of Mathematical Analysis by W. Rudin, McGraw-Hill, Inc., 1976 [3] for a proof.

Example 7.4.4 Because of Abel's theorem, Example 7.4.2(4) can be improved to
$$\ln(x + 1) = x - \tfrac{1}{2}x^2 + \cdots + (-1)^{n+1}\tfrac{1}{n}x^n + \cdots \text{ for all } x \in (-1, 1].$$
In particular, when $x = 1$, it says that $\sum_{k=0}^{\infty} \frac{(-1)^k}{k+1} = \ln 2$.
We gave an alternate derivation of $\sum_{k=0}^{\infty} \frac{(-1)^k}{k+1} = \ln 2$ in Example 3.4.1.

Remark 7.4.4 Niels Henrik Abel (1802–1829) is a contemporary of Cauchy and Gauss, however he was younger than them and unknown in his lifetime. He is from Norway, and he was a brilliant mathematician who proved the above theorem. But he is more famous for proving that there does not exist an algebraic formula to solve a fifth-degree polynomial in general. Back then, communication among mathematicians was difficult. With financial help from the Norwegian government, and the encouragement of August Leopold Crelle, he went to Germany and later to France hoping to be recognized by Gauss and Cauchy, among others, thinking that might help him get a professorship at a university in Norway. But he was ignored by both of them, and he went home heartbroken, and could not obtain a permanent university job. He soon died of tuberculosis. It is one of the saddest tragedies of a talented mathematician. The government of Norway established the Abel's Prize to honor him in 2002, exactly two hundred years after his birth. It is probably the most prestigious award in the world for a mathematician's lifetime achievement.

August Leopold Crelle is also a mathematician, and famous for establishing the *Journal für die reine und angewandte Mathematik*, commonly known as Crelle's journal. The first issue of the journal was published in 1826, and it contains seven of Abel's papers. It is the oldest journal still being published today in Germany. Abel published almost all of his research papers in Crelle's journal.

7.5   Taylor's Series

Let $\langle a_n \rangle_{n=0}^{\infty}$ be a sequence of real numbers. Let. $R > 0$ be the radius of convergence of the power series $\sum_{n=0}^{\infty} a_n x^n$. Let $f : (-R, R) \to \mathbb{R}$ be a function defined by $f(x) = \sum_{n=0}^{\infty} a_n x^n$ for every $x \in (-R, R)$. By Theorem 7.4.3, we have.

$$f'(x) = \sum_{n=1}^{\infty} n a_n x^{n-1},$$

$$f''(x) = \sum_{n=2}^{\infty} n(n-1) a_n x^{n-2},$$

For each integer $k > 0$, the k-th derivative $f^{(k)}(x)$ of $f(x)$ is given by

$$f^{(k)}(x) = \sum_{n=k}^{\infty} n(n-1) \cdots (n-k+1) a_n x^{n-k}$$

for every $x \in (-R, R)$. Then we have $f(0) = a_0$, $\frac{1}{1!}f^{(1)}(0) = a_1$, $\frac{1}{2!}f^{(2)}(x) = a_2$, $\frac{1}{3!}f^{(3)}(0) = a_3$. In particular, we have $a_n = \frac{1}{n!}f^{(n)}(0)$ for every $n = 0, 1, 2, 3, \ldots$. Hence, we can rewrite $f(x) = \sum_{n=0}^{\infty} a_n x^n$ by $f(x) = \sum_{n=0}^{\infty} \frac{1}{n!}f^{(n)}(0) x^n$, which is <u>Maclaurin series</u> for $f(x)$. This gives us the next definition.

Definition 7.5.1 Suppose $f : X \to \mathbb{R}$ be an infinitely many times differentiabe function. Let $a \in X$. Then the power series $\sum_{n=0}^{\infty} \frac{1}{n!}f^{(n)}(a) \cdot (x-a)^n$ is called the <u>Taylors series for</u> f about a. The power series $\sum_{n=0}^{\infty} \frac{1}{n!}f^{(n)}(0) \cdot x^n$ is called the <u>Maclaurin series for</u> f.

Example 7.5.1

(1) Let us begin with the polynomial $f(x) = (x+1)^5$. By binomial theorem, we have

$$f(x) = (x+1)^5 = x^5 + \binom{5}{1}x^4 + \binom{5}{2}x^3 + \binom{5}{3}x^2 + \binom{5}{1}x + 1 = x^5 + 5x^4 +$$

$10x^3 + 10x^2 + 5x + 1$. However, by differentiating $f(x) = (x+1)^5$, we have

$$\frac{1}{1!}f^{(1)}(x) = 5(x+1)^4, \quad \frac{1}{2!}f^{(2)}(x) = 10(x+1)^3,$$

$$\frac{1}{3!}f^{(3)}(x) = 10(x+1)^2, \quad \frac{1}{4!}f^{(4)}(x) = 5(x+1), \text{ and}$$

$$\frac{1}{5!}f^{(5)}(x) = 1.$$

Hence, by comparing this to the binomial theorem, we have

$$f(x) = (x+1)^5 = 1 + 5x + 10x^2 + 10x^3 + 5x^4 + x^5$$
$$= f(0) + \frac{1}{1!}f^{(1)}(0)x + \frac{1}{2!}f^{(2)}(0)x^2 + \frac{1}{3!}f^{(3)}(0)x^3 + \frac{1}{4!}f^{(4)}(0)x^4 + \frac{1}{5!}f^{(5)}(0)x^5.$$

This could have been the motivation of both the Maclaurin and Taylor's series.

(2) Let $f(x) = \ln x$. Then $f'(x) = x^{-1}$, $f''(x) = (-1)x^{-2}$. In general, if $n \geq 1$ is an integer, then $f^{(n)}(x) = (-1)^{n+1}(n-1)! \cdot x^{-n}$. Since $f(1) = 0$, and $f^{(n)}(1) = (-1)^{n+1}(n-1)!$ For all integer $n \geq 1$, we have the Taylor's series about 1 to be

$$0 + \frac{1}{1!}(x-1) - \frac{1}{2!}(x-1)^2 + \frac{2!}{3!}(x-1)^3 + \cdots + \frac{(-1)^{n+1}(n-1)!}{n!}(x-1)^n + \cdots$$
$$= (x-1) - \frac{1}{2}(x-1)^2 + \frac{1}{3}(x-1)^3 - \cdots + (-1)^{n+1}\frac{1}{n}(x-1)^n + \cdots.$$

The radius of convergence of this series is 1, and the interval of convergence is $(0, 2]$.

Question: Is it true that

$\ln x = (x-1) - \frac{1}{2}(x-1)^2 + \frac{1}{3}(x-1)^3 - \cdots + (-1)^{n+1}\frac{1}{n}(x-1)^n + \cdots$ for all $x \in (0, 2]$?
Alternately, by substituting $(x-1) = t$, this question becomes as follow:
Does $\ln(t+1) = t - \frac{1}{2}t^2 + \frac{1}{3}t^3 - \cdots + (-1)^{n+1}\frac{1}{n}t^n + \cdots$ for all $t \in (-1, 1]$?
Yes, see Example 7.4.4.

Remark 7.5.1 Suppose $f : X \to \mathbb{R}$ is an infinitely many times differentiabe function. Definition 7.5.1 and the above example leads us to the **question**: Suppose $R > 0$ is the radius of convergence of the infinite series $\sum_{n=0}^{\infty} \frac{1}{n!}f^{(n)}(a) \cdot (x-a)^n$.
Is it true that $f(x) = \sum_{n=0}^{\infty} \frac{1}{n!}f^{(n)}(a) \cdot (x-a)^n$ for all $x \in (a-R, a+R)$?
(Here, we are assuming $(a-R, a+R) \subset X$.)
The next theorem and Corollary 7.5.2 provide the answer to this question.

Theorem 7.5.1 (Taylor) Let $R > 0$ or $R = \infty$. Let $a \in \mathbb{R}$. Suppose $f : (a-R, a+R) \to \mathbb{R}$ be an n-times differentiable function for some positive integer n. Let $x \in (a-R, a+R)$. Then there exists a number c between a and x such that

$$f(x) = \sum_{k=0}^{n-1} \frac{1}{k!}f^{(k)}(a) \cdot (x-a)^k + \frac{1}{n!}f^{(n)}(c) \cdot (x-a)^n.$$

Proof Let $x \in (a-R, a+R)$. Suppose M is the unique number such that

$$f(x) = \sum_{k=0}^{n-1} \frac{1}{k!}f^{(k)}(a)(x-a)^k + \frac{M}{n!}(x-a)^n.$$

Let $g(t) = \sum_{k=0}^{n-1} \frac{1}{k!} f^{(k)}(a)(t-a)^k + \frac{M}{n!}(t-a)^n - f(t)$ for every $t \in (a-R, a+R)$. Then for every $i = 1, 2, \ldots, n$, we have

$$g^{(i)}(t) = \sum_{k=i}^{n-1} \frac{i!}{k!} f^{(k)}(a) \cdot (t-a)^{k-i} + \frac{M}{(n-i)!}(t-a)^{n-i} - f^{(i)}(t).$$

Note that $g(a) = 0$ and $g(x) = 0$ by the choice of M.

By Rolle's theorem, there is a number c_1 between a and x such that $g^{(1)}(c_1) = 0$.

But then we have $g^{(1)}(a) = 0$ and $g^{(1)}(c_1) = 0$.

By Rolle's theorem, there is a number c_2 between a and c_1 such that $g^{(2)}(c_2) = 0$. Note that $g^{(2)}(a) = 0$.

By Rolle's theorem, there is a number c_3 between a and c_2 such that $g^{(2)}(c_3) = 0$. Note that $g^{(3)}(a) = 0$.

By continuing this process, we have $c_0 = x, c_1, c_2, \ldots, c_n$ such that $g^{(i)}(a) = 0$, $g^{(i)}(c_i) = 0$, and c_{i+1} between a and c_i such that $g^{(i+1)}(c_{i+1}) = 0$, for every $i = 0, 1, 2, \ldots, n-1$. Therefore, we have $f(x) = \sum_{k=0}^{n-1} \frac{1}{k!} f^{(k)}(a)(x-a)^k + \frac{1}{n!} f^{(n)}(c_n)(x-a)^n$ for some number c_n between a and c_{n-1}. If we let $c = c_n$, then c is between a and x. Therefore, we have proved this theorem.

Remark 7.5.1 Notice that Theorem 7.5.1 is the mean value theorem considered in Theorem 5.2.2 when $n = 1$. The above proof is due to J. Wolf, (1953) *A Proof of Taylor's Formula*, The American Mathematical Monthly, vol.60, page 415 [4].

Corollary 7.5.2 (Taylor) Let $R > 0$ or $R = \infty$. Let $a \in \mathbb{R}$. Suppose $f : (a-R, a+R) \to \mathbb{R}$ be a function differentiable infinitely many times. Suppose The Taylors series $\sum_{n=0}^{\infty} \frac{1}{n!} f^{(n)}(a)(x-a)^n$ has the radius of convergence R. If $\{|f^{(n)}(x)| : x \in (a-R, a+R), n = 1, 2, 3, \ldots\}$ is bounded above by some positive constant C, then we have $f(x) = \sum_{n=0}^{\infty} \frac{1}{n!} f^{(n)}(a)(x-a)^n$ for all $x \in (a-R, a+R)$.

Proof Let $x \in (a-R, a+R)$. Then, for each integer $n > 0$, there exists c_n between a and x such that

$$\left| f(x) - \sum_{k=0}^{n-1} \frac{1}{k!} f^{(k)}(a) \cdot (x-a)^k \right| = \left| \frac{1}{n!} f^{(n)}(c_n) \cdot (x-a)^n \right|.$$

But $\left| f^{(n)}(c_n) \right| \le C$ for every n. Hence,

$$\left| \frac{1}{n!} f^{(n)}(c_n) \cdot (x-a)^n \right| \le \left| \frac{C}{n!} \cdot (x-a)^n \right| \text{ and } \lim_{n \to \infty} \frac{C}{n!} |x-a|^n = 0.$$

This proves that $f(x) = \sum_{n=0}^{\infty} \frac{1}{n!} f^{(n)}(a)(x-a)^n$ for all $x \in (a-R, a+R)$.

Example 7.5.2 Let $R > 0$. Suppose $f : (-R, R) \to \mathbb{R}$ is a function defined by $f(x) = e^x$ for all $x \in (-R, R)$. Then $f^{(n)}(x) = e^x$ for any positive integer n. Moreover, $\left| f^{(n)}(x) \right| < e^R$

for any integer n and any $x \in (-R, R)$. Therefore, $f(x) = \sum_{n=0}^{\infty} \frac{1}{n!} f^{(n)}(0)x^n = \sum_{n=0}^{\infty} \frac{x^n}{n!}$. Since $R > 0$ is arbitrary, we have $e^x = \sum_{n=0}^{\infty} \frac{x^n}{n!}$ for all real number x as we expected.

Example 7.5.3 Suppose $f(x) = \sum_{n=0}^{\infty} a_n x^n$ is a power series with a radius of convergence $R > 0$. Then it is the Taylor's series expansion of $f(x)$ at the origin.

(1) For example, the Taylor's series of $f(x) = \frac{1}{1-x}$ at $x = 0$ is $f(x) = \sum_{n=0}^{\infty} x^n$ using the geometric series. Its radius of convergence is 1. The interval of absolute convergence is $(-1, 1)$. Integration gives us $\sum_{k=0}^{\infty} \frac{x^{k+1}}{k+1} = \ln\frac{1}{1-x}$. See Example 7.4.2.

In particular, the answer to the question in Example 7.5.1 is YES.

(2) We have $f(x) = \frac{1}{x} = \frac{1}{1-(-(x-1))} = \sum_{n=0}^{\infty} (-(x-1))^n = \sum_{n=0}^{\infty} (-1)^n (x-1)^n$.

By integration, we have $\ln x = \sum_{n=0}^{\infty} (-1)^n \frac{(x-1)^{n+1}}{n+1} + C$. Since $\ln 1 = 0$, we have $C = 0$. Hence, the Taylor's series of $\ln x$ at $x = 1$ is $\ln x = \sum_{n=0}^{\infty} (-1)^n \frac{(x-1)^{n+1}}{n+1}$. Its radius of convergence is **1**. The interval of absolute convergence is $(1-1, 1+1) = (0, 2)$. The interval of convergence is $(0, 2]$. By Abel's theorem, $\ln 2 = \sum_{n=0}^{\infty} (-1)^n \frac{1}{n+1}$.

(3) Similarly, we have $f(x) = \frac{1}{x} = \frac{1}{2-(-(x-2))} = \frac{1}{2} \cdot \frac{1}{1-\left(-\frac{(x-2)}{2}\right)}$

$$= \frac{1}{2} \cdot \sum_{n=0}^{\infty} \left(-\frac{(x-2)}{2}\right)^n = \frac{1}{2} \cdot \sum_{n=0}^{\infty} (-1)^n \left(\frac{x-2}{2}\right)^n.$$

By integration, we have $\ln x = \frac{1}{2} \cdot \sum_{n=0}^{\infty} (-1)^n \frac{2}{n+1} \left(\frac{x-2}{2}\right)^{n+1} + C$. When $x = 2$, we have $C = \ln 2$. Hence, we have

$$\ln x = \sum_{n=0}^{\infty} (-1)^n \frac{1}{n+1} \left(\frac{x-2}{2}\right)^{n+1} + \ln 2$$

$$= \sum_{n=0}^{\infty} (-1)^n \frac{1}{n+1} \left(\frac{x-2}{2}\right)^{n+1} + \sum_{n=0}^{\infty} (-1)^n \frac{1}{n+1} \text{ from (2)}$$

for all $x \in (0, 4)$. By Abel's theorem, this identity holds for all $x \in (0, 4]$.
   Its radius of convergence is **2, not 1.**

Problem 7.5.1 Find the Taylor's series of $\ln x$ at $x = b$ when $b > 0$ using geometric series similar to Example 7.5.3. What is the **interval** of convergence of the series?

Example 7.5.4 By replacing x by $-x$ in Example 7.5.3(1), we have $\frac{1}{1+x} = \sum_{n=0}^{\infty} (-x)^n$ with the radius of convergence 1. By replacing x by x^2, we have $\frac{1}{1+x^2} = \sum_{n=0}^{\infty} (-x^2)^n = \sum_{n=0}^{\infty} (-1)^n x^{2n}$. By integrating both sides, we have $\tan^{-1} x = \sum_{n=0}^{\infty} \frac{(-1)^n}{2n+1} x^{2n}$. And it converges absolutely for all $-1 < x < 1$. Also, $\sum_{n=0}^{\infty} \frac{(-1)^n}{2n+1}$ converges. Since $\tan^{-1} 1 = \frac{\pi}{4}$, we have $\frac{\pi}{4} = \sum_{n=0}^{\infty} \frac{(-1)^n}{2n+1}$ by Abel's theorem. The identity $\frac{\pi}{4} = \sum_{n=0}^{\infty} \frac{(-1)^n}{2n+1}$ is due to Leibniz, but Abel's theorem made the proof complete.

Problem 7.5.2

(1) Show that $\cos x = \sum_{n=0}^{\infty} (-1)^n \frac{x^{2n}}{(2n)!}$.
(2) Show that $\sin x = \sum_{n=0}^{\infty} (-1)^n \frac{x^{2n+1}}{(2n+1)!}$.

(Note that the radii of convergence to both of these power series are ∞.)

Example 7.5.5 Let i be the complex number. Then $i^2 = -1$, $i^3 = -i$, $i^4 = 1$. In particular, $i^{2n} = (-1)^n$, $i^{2n+1} = i \cdot (-1)^n$ for every $n = 0, 1, 2, 3, \ldots$. **Euler** noticed that

$$e^{ix} = \sum_{n=0}^{\infty} \frac{(ix)^n}{n!} = \sum_{n=0}^{\infty} \frac{(ix)^{2n}}{(2n)!} + \sum_{n=0}^{\infty} \frac{(ix)^{2n+1}}{(2n+1)!}$$

$$= \sum_{n=0}^{\infty} \frac{(-1)^n x^{2n}}{(2n)!} + i \cdot \sum_{n=0}^{\infty} \frac{(-1)^n x^{2n+1}}{(2n+1)!}$$

$= \cos x + i \cdot \sin x$ from Problem 7.5.2.

Therefore, we have $e^{ix} = \cos x + i \cdot \sin x$. In particular, $e^{i\pi} + 1 = 0$, and $e^{a+ib} = e^a (\cos b + i \cdot \sin b)$ for any $a, b \in \mathbb{R}$.

There are several formulas named after **Euler**, and $e^{ix} = \cos x + i \cdot \sin x$ and $e^{\pi i} + 1 = 0$ may be the most beautiful ones among them.

References

1. H. Katsuura, Continuous nowhere-differentiable functions—an application of contraction mappings. Am. Math. Mon. **98**(5), 411–416 (1991)
2. D.P. Minassian, *An old result on sequences and differentiability—a concise proof*. American Math. Monthly **109**(2), 172 (2002)
3. W. Rudin, Principle of Mathematical Analysis, McGraw-Hill, Inc. (1976)
4. J. Wolf, A proof of Taylor's formula. American Math. Monthly **60**, 415 (1953)